Ein gutes Fundament schaffen !!!!

ALLE · ZEIT · WACH
1842

Günther Ludwig

An Axiomatic Basis for Quantum Mechanics

Volume 1
Derivation of Hilbert Space Structure

With 6 Figures

Springer-Verlag
Berlin Heidelberg New York Tokyo

Günther Ludwig
Fachbereich Physik der Philipps-Universität
Arbeitsgruppe Grundlagen der Physik
Renthof 7, D-3550 Marburg
Federal Republic of Germany

Translated from the German manuscript by Leo F. Boron

ISBN 3-540-13773-4 Springer-Verlag Berlin Heidelberg New York Tokyo
ISBN 0-387-13773-4 Springer-Verlag New York Heidelberg Berlin Tokyo

Library of Congress Cataloging in Publication Data
Ludwig, Günther, 1918- . An axiomatic basis for quantum mechanics. Bibliography: v. 1, p. Includes index. Contents: v. 1. Derivation of Hilbert space structure. 1. Quantum theory. 2. Axioms. I. Title. QC174.12.L82 1985 530.1'2 85-2711 ISBN 0-387-13773-4 (U.S. : v. 1)

Typesetting: Graphischer Betrieb K. Triltsch GmbH, Würzburg
Printing: Mercedes-Druck GmbH, Berlin; Bookbinding: Lüderitz & Bauer-GmbH, Berlin
2153/3020-543210

Preface

This book is the first volume of a two-volume work, which is an improved version of a preprint [47] published in German. We seek to deduce the fundamental concepts of quantum mechanics solely from a description of macroscopic devices. The microscopic systems such as electrons, atoms, etc. must be detected on the basis of the macroscopic behavior of the devices. This detection resembles the detection of the dinosaurs on the basis of fossils.

In this first volume we develop a general description of macroscopic systems by trajectories in state spaces. This general description is a basis for the special description of devices consisting of two parts, where the first part is acting on the second. The microsystems are discovered as systems transmitting the action.

Axioms which describe general empirical structures of the interactions between the two parts of each device, give rise to a derivation of the Hilbert space structure of quantum mechanics. Possibly, these axioms (and consequently the Hilbert space structure) may fail to describe other realms than the structure of atoms and molecules, for instance the "elementary particles".

This book supplements ref. [2]. Both together not only give an extensive foundation of quantum mechanics but also a solution in principle of the measuring problem.

At several places of this book the reader will find references to the book *Grundstrukturen einer physikalischen Theorie* [3]. Previous knowledge of [3] is not necessary for understanding this book (except for chapter XIII of the second volume). Readers familiar with [3], [30], [48] can easily recognize how the general structure of a physical theory is implemented for quantum mechanics.

Several theorems are not proved in this book, since the reader can, word by word, take over their proofs from [2] (the corresponding references are given). Persons not interested in the proofs can read the book as it is.

References in the text are made as follows: For references to other sections of the same chapter, we shall only list the section number of the reference; for example, § 2.3. For references to other chapters, the chapter is also given; for example III § 4.2 refers to section 4.2 of chapter III. The formulas are numbered as follows: (3.2.7) refers to the 7th formula in section 3.2 of the current chapter. References to formulas of other chapters are given, for example, by III (2.1.8). References to the Appendix are given by A II § 2 where A II denotes Appendix II.

I would like to express my gratitude to *Professor Leo F. Boron* for the translation of the manuscript from German to English. I want to thank very much *Professor K. Just* for reading very critically the English and German version of the manuscript, for countless improvements of the text, and for this thorough proof-reading. His extensive collaboration during the summer of 1983 made the formulation of the text clearer at many places.

Marburg, January 1985 G. Ludwig

Contents

Contents of Vol. 2:

Quantum Mechanics and Macrosystems

I The Problem of Formulating an Axiomatics for Quantum Mechanics

In [3], [30] and [48] we formulated the objective of finding, for a physical theory, an axiomatic basis with "physically interpretable axioms" (see [30] and XIII § 2), wherein the physical laws appear distinctly and whereby the problems of interpretation can be solved clearly. In this book, we attack this objective for quantum mechanics.

In XIII we shall improve the analysis from [3] and [30] of the requirements which one may impose on an axiomatic basis. Various versions of such a basis will be discussed. Reviewing the structure analysis from III through IX we shall present such versions for quantum mechanics.

Also for readers not interested in the general discussions of [1] or who do not care *exactly* what an axiomatic basis is, the following discussion is relevant. The phrase "axiomatic basis" need only symbolize a goal to be aspired to, namely to begin a theory solely from the facts already "known" *before* one develops the theory. These facts can be recognized directly or with the help of other, already known theories (the "pretheories").

If, *after* becoming familiar with the foundations of quantum mechanics developed here, such readers are still interested more generally in the structure of physical theories, they will more easily understand [3] and [30]. The subsequent references to [3] are therefore not needed to understand this text. They rather serve two purposes:

(1) The reader who knows the generalities from [3] can recognize the present considerations as examples.
(2) For readers not acquainted with [3], this presentation of quantum mechanics can ease the reading of [3], since then one has an example at hand.

§ 1 Is There an Axiomatic Basis for Quantum Mechanics?

The decisive requirement of an axiomatic basis (see [3], [30], [48] and XIII) is that the interpretation, i.e. the correspondence rules can be deduced solely by means of concepts already interpreted by pretheories. But just of quantum mechanics it is often asserted that it does not admit an interpretation retaining the "classical" concepts from pretheories. Indeed, this retention is said to incur inconsistence, described by phrases such as "semantic inconsistence", "incompatibility between quantum mechanics and classical physics", "paradox of measuring processes in quantum mechanics", or "inseparability of physics and subjective consciousness".

From the literature on the philosophy of quantum mechanics, one recognizes of what importance it would be to establish an axiomatic basis for quantum mechanics, using only those pretheories which in an *objectivating* way describe the devices for experiments with microsystems. Then an "inseparability of physics and subjective consciousness" in quantum mechanics can no longer arise, since it is at least *possible* to erect quantum mechanics itself on "objective" properties of the devices.

A paradox of the measuring process in quantum mechanics cannot arise either, since conversely the experiments with microsystems form the only point from which quantum mechanics receives its interpretation, i.e. its physical meaning.

Semantic inconsistence likewise does not occur since only the semantically known concepts from the pretheories are used, whereas the further concepts derived from quantum mechanics are reducible to those "semantically known" (see § 2). From the problems thus hinted at, only the compatibility between an *extrapolated* quantum mechanics and the classical description of a macroscopic system remains to be discussed in X and XI. But this compatibility does not pose a problem of any conceptual or logical consistence, rather it concerns a relation between physical theories and asks for a more comprehensive (!) theory.

The motive for developing an axiomatic basis for quantum mechanics therefore concerns primarily the philosophy of science and is thus closely related to the question of the best possible "understanding" of quantum mechanics. Secondly, there also arise interesting viewpoints on the possibility of new physical theories, because one recognizes better where physical laws can perhaps be altered. For (non-relativistic) quantum mechanics itself one naturally obtains no "new" effects.

§ 2 Concepts Unsuitable in a Basis for Quantum Mechanics

We shall really not develop an axiomatics for quantum mechanics interpreted "as usual" (in the sense of [3] § 7.3, or [30]). In fact, the basic concepts such as observable, ensemble, state, or yes-no measurement, employed in the "usual" interpretation of quantum mechanics, are themselves not explainable by *known* pretheories. This even holds for the concept of a microsystem (often called a "particle") so that in an axiomatic basis one cannot use those sets which picture such concepts in quantum mechanics. Rather we must as an axiomatics develop a more comprehensive theory for "usual" quantum mechanics. In view of this "comprehension", usual quantum mechanics then appears as a restriction (in the sense of [3] § 8, or VIII § 3).

On the other hand, if one foregoes an axiomatic basis, one can by all means begin an axiomatics with basic concepts that are meant, for example, ontologically, i.e. not immediately provable. Of course the subsequent theory must indicate image sets and image relations with proper interpretation; then it is quite legitimate to start with basic concepts such as microsystem, property, or state. Also in [3] § 10.7 we pointed out such possibilities in general form.

Also the reader who does not refer to [3] can easily see that concepts such as microsystem, state, properties of the microsystem cannot be interpreted without invoking quantum mechanics. Just therefore they are not suitable for our goal of an axiomatic basis for quantum mechanics.

Let us begin with the concept of a microsystem!

Whenever we speak of *single microsystems*, we cannot simply display them as one would display a cup on a table. Neither does the "explanation" of a macrosystem on the basis of its composition from atoms shows us the existence of individual microsystems. It is only in experiments with individual microsystems that these occasion macroscopically registerable effects. There they arise as entities which only by a theory are inferred indirectly from directly established actions.

When one speaks of the *state* of a microsystem, one means either in the ontological sense something that belongs to each individual system (but just as such can be recognized as little as the microsystem itself); or one understands a state as the sum total of how the microsystem is prepared for experiments (which again one can deduce only indirectly from directly established preparation procedures).

If one speaks of the *properties* of a microsystem, one understands either (in the ontological sense) something the microsystem itself possesses (indirectly measurable by complicated devices); or one means an abstraction (to be made more precise) from the direct actions of the microsystem on a measuring device.

But from what can one start an axiomatics for quantum mechanics if we forego concepts such as microsystems and states or properties?

§ 3 Experimental Situations Describable Solely by Pretheories

It is therefore not a question of explaining a macrosystem by many-particle quantum mechanics (see X), but rather of describing experiments with *single* microsystems. Here, "single" does not mean that they are not composite. Composites in the sense of connected systems or of scattering theory (see IX § 2) are also regarded as single systems, whenever the experiments consist of macroscopic processes triggered by single microsystems and frequently repeated for a statistics.

The scattering of an electron on a proton is such an experiment with single microsystems, where each is just an electron-proton pair. A plasma of many electrons and protons, however, is not the object of an experiment with "individual" electron-proton pairs, but rather of a macroscopic experiment in which the plasma as a whole (!) is investigated.

For an axiomatic basis of quantum mechanics we therefore restrict ourselves to experiments with single microsystems, also when there is multiple repetition. But how can we characterize this experimental situation *without* the concept of "microsystem" which is only established by quantum mechanics itself? Just through the interaction of macrosystems, where the microsystems appear later as effect carriers, i.e. as the entities to carry the action from one macrosystem onto another. An axiomatic basis must therefore start from an interacting macrosystem, described without the "interpretation" and "explanation" of the interactions. Therefore, we shall turn in II to the form of such a "description" of processes on macrosystems, where only concepts of pretheories are permitted.

"Description" of processes here does not mean that a theory of their dynamics is given. For example, the motion of mass points along paths $\boldsymbol{r}(t)$ is describable without specifying the Newtonian equations of motion and the forces. The dynamic laws of processes on macrosystems describable by pretheories therefore need not occur in pretheories. On the contrary, we just seek in quantum mechanics such laws of dynamics that certain interactions among macrosystems can be represented mathematically, i.e. by axioms.

The processes to be used in a basis for quantum mechanics are of course not arbitrary processes on arbitrary macrosystems. Rather they are structural in a special way, briefly characterized by a directed action of one macrosystem on another. One might think that this basis is too narrow. Of course one cannot prove *a priori* that is suffices for a complete theory of the microsystems (carrying action from one macrosystem to another). Precisely this (that we can thus find the "total" quantum mechanics) will be shown in II through IX.

The *choice* of the directed action of one macrosystem on another as a basis of quantum mechanics of course does not signify that we as physicists are desinterested in more complex processes. But there is hope that such processes can be described in a more comprehensive theory (developed in principle in X). In XI it will finally be applied to our special point of departure, the directed interaction processes.

Let us consider the reflection of light in a mirror. One macrosystem, the lamp, acts on the other macrosystem, the mirror. We denote the action carrier by the word "light". The action is directed from the lamp to the mirror. If one lets the reflected light act on a third macrosystems, e.g. a photocell, we have a situation that appears frequently in experiments! To this we shall shortly return below. Now let us compare the directed action of the lamp on the mirror with a typical case of an non-directed interaction: a laser. There the "illuminating material" is so coupled (over the "light") with "mirrors" that we cannot even approximately speak of a directed action of the "illuminating material" on the mirror. On the contrary, the coupling is so mutual that the "illuminating material" between the mirrors behaves completely differently from a "free" material. In the directed action of the lamp on the mirror, however, the lamp is not influenced by the mirror. In this context, see the exact formulation of the concept of directed action in III § 3.

Similarly to the first case (the light reflected by the mirror being investigated by a photocell), where the lamp and the mirror are viewed as the single system which in the directed way acts on the photocell (as the second system), one can also let the laser (composed of illuminating material and mirrors) act in the directed way on a photocell. But if one wishes to investigate the light quanta as action carriers, then only the cases of directed action are suitable. The interaction of the illuminating material and the mirrors in the laser can no longer be desribed by the "single" independent light quanta, but only within the "many-particle system" of the laser as a whole (by methods pointed out in X). The elaboration of these methods by the simplest possible means is described in [1] XV, where various applications are shown (e.g. to lasers).

The indicated examples show that directed actions of one macrosystem on another really comprise complex experiments, where the "decomposition" of the

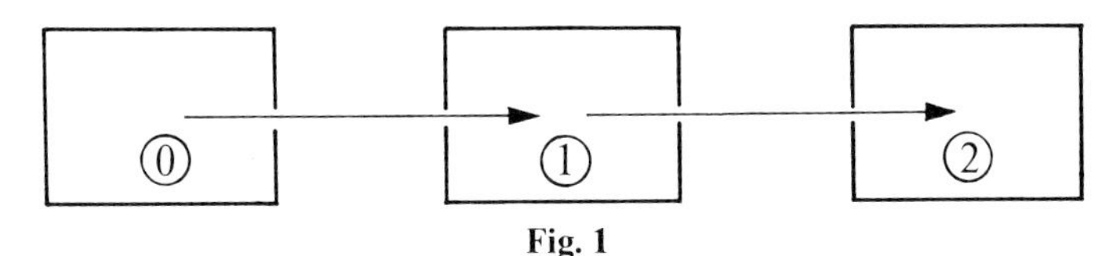

Fig. 1

total macrosystem into parts (that act in a directed way on each other) can often proceed in quite different ways. The entire macrosystem can consist of three parts (see Fig. 1), where the part (0) acts in a directed way on the part (1) and the latter on (2). But this experiment nevertheless belongs to the basis chosen: indeed one can view (0) as the macrosystem which acts in a directed way on the composite system (1) + (2); or one can view (0) + (1) as a system that acts in a directed way on (2). Naturally, here the description of our basis does not include that these two decompositions are only two different decompositions of the *same* entire experiment. But one can easily insert such auxiliary structures later (as in XI and XII).

Of course the basis of the directed processes is itself too broad for a fundamental domain of quantum mechanics. Even such interactions as those between a weapon and a target fall into this category. In order to single out processes relevant for quantum mechanics, one therefore needs certain dynamical laws. Such laws are introduced in VI under the heading "fundamental preparation and registration laws". These laws narrow the region of the theory down to just those interactions which we then denote as transmitted by "microsystems".

This fundamental domain for quantum mechanics is therefore completely describable by known pretheories. It contains nothing that one could a priori designate as something like the properties of microsystems or as a measurement of their observables. Only *subsequently* can one introduce, under certain assumptions concerning the interactions, concepts such as properties, pseudo-properties, observables, states as derived concepts (see in particular V and VII).

§ 4 Mathematical Problems

The mathematics of the basis for quantum mechanics, i.e. the introduction of basic sets, structure terms and axioms of course proceeds under physical points of view. Therefore, in the introduction of structure species the pure mathematician will miss a systematic order of the various more or less rich structures; for, it is no systematics that stands in the foreground.

In IV we shall recognize that a base-normed Banach space $\mathscr{B}$ together with its dual $\mathscr{B}'$ forms a general initial structure. The strengthening of this structure to the $\mathscr{B}$ finally characterized in VIII does not occur by axioms that are perhaps mathematically especially interesting, but rather by physically explicable axioms, interpretable by preparations and registrations. Hence it is physically often meaningful to sharpen derived theorems by axioms whereas one would, from a purely mathematical viewpoint, rather search for sufficient and necessary assumptions for sharpening the theorems. Nevertheless, to the author the axioms in VI appear also of mathematical interest.

A mathematical deficiency in VI through VIII surely consists in that we forego the independence of the axioms. It is therefore possible that certain axioms could be omitted which follow from the others as theorems. The physicist need not prove the independence of axioms (by counterexamples); for, if an axiom later turns out "superfluous", so much the better for physics.

The author has tried, insofar as possible, to derive as many of the theorems as possible from the respective axioms, i.e. to comply with the mathematical desire of asking, which axioms already suffice to derive what theorems. But it should be possible to go still farther (see IV through VIII).

In order to prove the representation theorem for $\mathscr{B}$, $\mathscr{B}'$ in VIII, the author has used the representation theory of modular lattices and of orthocomplementary orthomodular lattices (which during a long historical development arose from projective geometry). This appears as a detour inasmuch as on this path one for a while forgets the embedding of the lattice G in $\mathscr{B}'$. Shouldn't there be a shorter path that immediately uses the entire structure of the dual pair $\mathscr{B}$, $\mathscr{B}'$?

In particular, the question of a more general representation theorem *without* an axiom of the form AV 4 is still unclear.

§ 5 Progress to More Comprehensive Theories

After the basic sets and the structure terms are introduced in III, up to VIII we proceed (by adjoining further axioms) from each theory to a "structure richer" (in the sense of [3] § 8) theory. But in VIII the goal of a theory of electrons, atoms, molecules, etc. is still not reached. Rather, we have in VIII only derived the fundamental description of quantum mechanics by the operators of Hilbert space.

An exhaustive presentation how to proceed by "normal standard extension" (in the sense of [3] § 8) to the quantum mechanics of atoms, molecules, etc. is given in [2]. But so that subsequent to VIII the reader need not study the entire book [2], certain steps of this advance to the closed (to be precise: g. G.-closed; see [3] § 10.3) theory of atoms and molecules are in IX explained in descriptive form, without mathematical formulations. In particular, we shall state how in quantum mechanics one describes composite systems in terms of direct products of Hilbert spaces. This composition is mathematically possible for arbitrarily many microsystems.

But this compositions cannot be physically meaningful for arbitrarily many microsystems because there are not arbitrarily many microsystems in the world. The physical meaning of the "quantum mechanial composition" probably becomes physically questionable when the composite system has the same magnitude as the macrosystems to prepare and register it. Thus a new question arises for a theory to encompass quantum mechanics, to describe not only microsystems as action carriers but also the dynamics of whole macroscopic systems.

But this more comprehensive theory must describe the preparation and registration devices together with their interactions and their dynamical laws, i.e. it must be compatible with the description of macrosystems previously used for quantum mechanics.

This transition to a comprehensive theory of macrosystems no longer proceeds by standard extensions, but rather as an embedding (see again [3] § 8) of the comprehension (into the formal extrapolation of quantum mechanics to "many particles").

In X and XI we shall explain how this embedding could appear and how the mentioned compatibility problem could be solved.

To the extent the author has been able to, an axiomatic basis for quantum mechanics will thus be established. X and XI at least demonstrate the possibility that the developed quantum mechanics (extrapolated to many particles) is compatible with the initial axiomatic basis.

II Pretheories for Quantum

Since we intend to develop an axiomatic basis for q
delve into the pretheories for quantum mechanics
fundamental domain of quantum mechanics (see [1] III §
[3] §§ 3 and 5). Nevertheless, to understand the follow
have read [1] or [3] or [30].

The pretheories of quantum mechanics comprise e
theories such as classical mechanics, electrodynamics, c
The reality domain of these theories ([1] III § 9.6; [3] §
typical feature of all these theories is the *objectivating m*
fact was before quantum mechanics perceived as so [illegible] that many physicists rejected quantum mechanics since it no longer presents an objectivating description of microsystems.

Since we shall in the sequel use abbreviations defined in [3], we shall here note them briefly.

$\mathcal{PT}$ denotes a physical theory, $\mathcal{MT}$ a mathematical theory as a part of $\mathcal{PT}$. The "fundamental domain" of $\mathcal{PT}$ is denoted by $\mathcal{G}$. In this context, "fundamental domain" designates that region of experience to which the theory is applicable *and* which contains *only* those facts that are recognizable directly or through pretheories. Going beyond $\mathcal{G}$, the "reality domain" $\mathcal{W}$ will also encompass those facts which are only recognizable by the theory itself. $\leftrightarrow$ symbolically describes the interpretation of $\mathcal{MT}$ in the given $\mathcal{PT}$ so that often we briefly write $\mathcal{PT} = \mathcal{MT} \leftrightarrow \mathcal{W}$.

§ 1 State Space and Trajectory Space

Let $\mathcal{PT}_m$ be one of the pretheories for quantum mechanics. Of course, here we are not to develop axiomatic bases for the pretheories. It is just the advantage of the methods from [3] that one need not solve all fundamental physical problems in order to present a special theory as correctly as possible in the form $\mathcal{MT} \leftrightarrow \mathcal{W}$. Therefore, here it suffices to *formulate several* structures of the pretheories for quantum mechanics (not *all* their structures), without having to *justify* them within those pretheories. One of these general structures is the objectivating manner of description in a state space.

For each pretheory $\mathcal{PT}_m = \mathcal{MT}_m \leftrightarrow_m \mathcal{W}_m$ we *assume* that the following intrinsic terms ([3] § 7.2) are defined: A set M_m, called the set of the physical systems, a set

Z, called the state space, a time scale Θ, and a mapping $M_m \times \Theta \xrightarrow{f} Z$. By $f(x, t)$ with $x \in M_m$, $t \in \Theta$ one here denotes the "state of the system at the time t". Often one calls M_m the index set for the physical systems.

Concepts such as "system", "state", "time", and "state of the system at time t" (introduced along with the mathematical terms) are simply brief formulations of the correspondence rules from $\mathscr{PT}_m$ (see "Abbildungsprinzipien" in [1] III § 4 and [3] § 5). We will not describe these rules in detail. We can briefly say what can be understood without [3]: We first assume it known how to measure the state of a macrosystem at some time. In § 3 we will go more into the description of the "measurement of the states for various times". Finally, in § 5 we must investigate the relation between the function $M_m \times \Theta \xrightarrow{f} Z$ just introduced and the measuring described in § 3.

It is not customary to introduce the index set M_m explicitly in the pretheories. One speaks only of the state space Z and the "various" trajectories $f\colon \Theta \to Z$. In order to formulate the "special cases" of functions f cleanly, we have introduced the set M_m and instead of "the different $f(t)$" we write $f(x, t)$ where x is the "index" for the *different experiments*. Therefore, note that one must keep $x_1 \neq x_2$ even for two different experiments with perhaps the "same object" (e.g. in the sense of "the same device" from a laboratory). The introduction of the set M_m also promotes the formulation of a statistics (see § 2).

In order to elucidate the general formulation let us briefly give examples.

In point mechanics one can use the Γ-space of the p_i, q_i as the state space. In hydrodynamics one can as points of Z introduce the triples of fields $\{\mu(\boldsymbol{r}), \boldsymbol{u}(\boldsymbol{r}), T(\boldsymbol{r})\}$, where $\mu(\boldsymbol{r})$ is the mass density, $\boldsymbol{u}(\boldsymbol{r})$ the velocity, and $T(\boldsymbol{r})$ the temperature. In the second example one can further use "prescribed" external conditions to characterize the states of Z. But also in point mechanics one can augment the states by perhaps using prescribed external fields as further characterizing the points of a space Z that generalizes Γ. In thermostatics one introduces parameters to describe conditions such as volumes or exterior fields, using them as some of the coordinates of Z. In electrodynamics, the fields: charge density, current density, electric and magnetic field strengths, polarization, magnetization can be regarded as the points of a state space.

Even the most complex macrosystems such as entire devices together with their internal electromagnetic processes can be described in state spaces. It should not be disturbing that several $\mathscr{PT}$ are often used in describing a complex device. In fact, we have emphasized in [3] that physics consists of a coexistence (of course not relationsless) of different $\mathscr{PT}$. We have therefore not spoken above about one but rather of various pretheories $\mathscr{PT}_m$. Only the objectivating manner of description (expressed most clearly in state spaces) is important for an axiomatic basis of quantum mechanics.

The imprecision in the correspondence rules for Z must be described (due to [3] §§ 6 and 9) by a "uniform structure of physical imprecision". The reader who does not refer to [3] can easily visualize the situation by the example of a state space given below. (For the concept of uniform structures, see [5] or [2] AII § 2; for a brief introduction, also see [40].)

The case of a discrete state space Z will not be considered in order to avoid too many special cases. Otherwise we would lose track of the general view (about a discrete Z see [4]). Since in many pretheories it is "not customary" to exhibit the uniform structure of physical imprecision in Z, we must expressively assume that one has enlarged the pretheories by such uniform structures which satisfy the principles from [3] § 9. We cannot show here how to carry out such enlargements in special cases. But one can use methods entirely similar to those by which we shall introduce a uniform structure in the trajectory space Y in terms of those in Z and in Θ. Here, the uniform structures in space and time can serve as the most general point of departure for introducing those in Z ([3] § 9 or [1] III § 5).

The reason why we need a uniform structure in Z (although often one apparently gets away without it) is the need for a statistical description in any pretheory $\mathscr{PT}_m$ for quantum mechanics. We shall recognize this need in the course of developing an axiomatic basis for quantum mechanics. Of course one could for the statistical description get by with a weaker than the uniform structure (e.g. with a "measure space" structure). But it is "more physical" to obtain the measurable sets from a uniform structure (see § 3.3).

Therefore we now assume that a uniform structure is deducible in $\mathscr{MT}_m$, to be used for $\mathscr{PT}_m$ as the uniform structure of the *physical imprecision* in Z.

The meaning of the uniform structure of physical imprecision is that one can use its vicinities to describe the imprecision in the possible measurements of the states $z \in Z$. That two states z_1 and z_2 cannot be distinguished by a specific measurement method can with a suitably chosen vicinity U be expressed by $(z_1, z_2) \in U$. (The reader who wishes quickly to look up details of such imprecisions is referred to [40].)

In the following, Z_p will denote the state space Z with the uniform structure p of physical imprecision (also see [3] § 9). Let $\hat{Z}_p$ be the completion of Z relative to the uniform structure p. According to [3] § 9 we assume that $\hat{Z}_p$ is compact (so that Z_p is precompact), separable and metrizable (if $\hat{Z}_p$ is compact, the metrizability implies that $\hat{Z}_p$ is separable, and conversely; [5] IX § 2, n. 9).

The known pretheories $\mathscr{PT}_m$ are so constructed that there is a further uniform structure in Z (finer than p). In the "usual" formulation of the images $\mathscr{MT}_m$, only this finer structure is "present" (in the Γ-space of point mechanics only that of an $\mathbf{R}^n$). Let Z with this finer structure be Z_g. For Z_p, Z_g we make assumptions that are in generality presented in [3] § 9: Z_g and Z_p have the same topology, Z_g is complete. The elements of $\hat{Z}_p - Z_g$ are often called virtual states.

We have here briefly mentioned the difference between Z_g and Z_p although we shall rarely be concerned with it. But since (as pointed out above) mostly only the structure g is known in the theories $\mathscr{PT}_m$, one could easily confound g and p. In § 3, however, p will be important for the statistical description.

As assumed above, in $\mathscr{MT}_m$ there is a mapping $M_m \times \Theta \xrightarrow{f} Z$ with the physical interpretation that $f(x, t)$ is the state of the system x at the time t. But we have not yet said which values of t the time scale shall comprise. It may be tempting, idealizing (see [3] §§ 6 and 9 for "idealizing") to assume that Θ comprises all $-\infty < t < +\infty$. But if one looks more closely at some pretheories this assumption appears too far-reaching; for, there are pretheories $\mathscr{PT}_m$ which describe irreversible behavior. In such theories it can be meaningless, idealizing to assume that $M_m \times \Theta \xrightarrow{f} Z$ exists for arbitrary $t \to -\infty$. But another idealization, that

$M_m \times \Theta \xrightarrow{f} Z$ exists for arbitrarily large t, is allowed in all pretheories and expresses that in principle no bound of t is "known" (indicating an ignorance; see [3] § 9), beyond which the system x is no longer describable in Z (it may have been "absorbed" in the neighborhood). Of course, a concrete physical system from the fundamental domain of $\mathscr{PT}_m$ can after a certain time cease to "exist", i.e. it can egress from the fundamental domain of $\mathscr{PT}_m$ (e.g. being "destroyed from the outside"). In [3], end of § 10.3, we called the reader's attention to such possibilities.

For these reasons we only assume that Θ comprises an interval $t_0 \leqq t < \infty$. Here, t_0 is a time (in the laboratory scale), *before* which the systems $x \in M_m$ have been created (i.e. prepared). In order to avoid too many terms in this § 1 we think of choosing such a laboratory time scale that we can set $t_0 = 0$. Therefore, we assume that Θ equals the interval $0 \leqq t < \infty$, but reserve the right to displace the "time zero" to points t_0 specified in any other way.

That we consider for Θ "only" the interval $0 \leqq t < \infty$ of course permits that in a particular pretheory the function $M_m \times \Theta \xrightarrow{f} Z$ may be defined for $t < 0$. Furthermore, we must keep in mind that the mathematically "sharp" initial point $t = 0$ will lose its sharpness under a uniform structure of the physical imprecision for trajectories (see below) since one can physically achieve only that the trajectories are observed from approximately $t = 0$ on.

(In principle one can also consider discrete time scales $\Theta = \{n\,\Delta\tau \mid n = 0, 1, 2, \ldots\}$. But here let us not diminish clarity by utmost generality; for discrete scales, see [4]).

The set Z^Θ of all functions $\Theta \to Z$ shall be called the space of *all* trajectories. The $M_m \times \Theta \xrightarrow{f} Z$ defined above then determines a function $M_m \xrightarrow{g} Z^\Theta$, where $g(x)$ represents $\Theta \xrightarrow{f(x,t)} Z$. It is probably impractical to consider the whole space Z^Θ since only the subset $g(M_m) \subset Z^\Theta$ is interesting for physics. One calls $g(M_m)$ the set of physically possible trajectories (see [3] § 10.4 for "physically possible").

In order to express mathematically the imprecision in the measurement of trajectories, we consider the set $\Theta \times Z$. Each trajectory $z(t)$ can be characterized uniquely by

$$G(x(t)) = \{(t, z) \mid t \in \Theta,\ z \in Z \quad \text{and} \quad z = z(t)\} \tag{1.1}$$

(the subset of $\Theta \times Z$ called the graph of the mapping $\Theta \to Z$). One can thus identify Z^Θ with the set of all graphs. This enables us, in a physically meaningful way, to base a uniform structure for the physical imprecision of trajectories on a physical imprecision in

$$X = \Theta \times Z\,. \tag{1.2}$$

We *presumed* that a uniform structure p of physical imprecision is given in Z. A uniform structure g in Θ is defined by the metric $|t - t'|$. Let the product uniform structure in $\Theta_g \times Z_p$ ([5] II § 2, n. 6) be briefly denoted by $g\,p$. It is metrizable because Θ_g and Z_p are. With $\delta(z, z')$ as the metric in Z_p, one can e.g. use $d((t, z), (t', z')) = \max\{|t - t'|, \delta(z, z')\}$ as the metric in X_{gp}. But $g\,p$ is certainly too sharp for a uniform structure of the physical imprecision since Θ_g and therefore X_{gp} is not precompact (required by [3] § 9 or [40] for a uniform structure of physical imprecision).

According to [3] §§ 6 and 9, or [40], we must abolish the idealization $t \to \infty$. How does this idealization go beyond the possibilities of experimental measuring?

For the points in X, i.e. for the pairs (t, z), the experimental difficulty does not consist in that arbitrarily large t can only be determined with greater and greater imprecision. Experience with the theories $\mathscr{P}\mathscr{T}_m$ rather shows that the considered systems cease to exist (as indicated above), *before* the exactness of measuring decreases *very strongly*. We will therefore *not* introduce in Θ a uniform structure p of imprecision (e.g. by the metric $|\arctan t - \arctan t'|$), using $\Theta_p \times Z_p$ to describe the imprecision in X. We rather shall postulate that from some finite time T one can no longer measure. Of course, T must still remain "open", otherwise we didn't have to introduce the idealization $t \to \infty$ at all. Since we still wish to lift the sharpness of the initial point $t=0$ (as pointed out above), it is natural to introduce the following uniform structure p of physical imprecision in X.

Let Θ_0 denote the open interval $0 < t < \infty$ and put $X_0 = \Theta_0 \times Z$. One can describe "finite time intervals" from Θ_0 by compact subsets of Θ_{0g} (with g the uniform structure introduced on Θ). Such a compact subset ϑ of Θ_0 also leaves the sharpness of the origin $t=0$ of Θ "open". It is thus natural to define X_p by the fundamental system of vicinities

$$U_{\vartheta, V} = (X_{\vartheta'} \times X_{\vartheta'}) \cup V. \tag{1.3}$$

While the V are the vicinities of X_{gp} and the ϑ are the compact subsets of Θ_0, we used $X_{\vartheta'} = \vartheta' \times Z$ with $\vartheta' = \Theta \setminus \vartheta$. We evidently have the inclusion

$$U_{\vartheta_1 \cap \vartheta_2, V_1 \cap V_2} \subset U_{\vartheta_1, V_1} \cap U_{\vartheta_2, V_2}.$$

In (1.3) we have expressed that nothing is measured outside ϑ, while inside one measures with the precision characterized by the vicinities V of X_{gp}. Instead of using all the compact subsets ϑ of Θ_0, it suffices to use the special $\vartheta_n = \{t \mid \frac{1}{n} \le t \le n\}$ $(n = 1, 2, \ldots)$ in order to obtain from (1.3) a fundamental system of vicinities for X_p. Likewise, instead of all the V one can use a countable fundamental system of vicinities for X_{gp} since $g\,p$ is metrizable. It follows immediately that X_p is metrizable.

But X_p is also precompact (as required by [3] § 9). That X_p is precompact follows from (1.3) if one prescribes $V = \{(t, z), (t', z') \mid |t - t'| < \varepsilon$ and $(z, z') \in W\}$, with W from a vicinity filter of Z_p. To ϑ then belongs a finite number of t_ν so that for each $t \in \vartheta$ there is a t_ν with $|t - t_\nu| < \varepsilon$ and at least one t_ν (e.g. t_1) is outside ϑ. To W there belongs a finite number of z_μ so that for each $z \in Z$ there is a z_μ with $(z, z_\mu) \in W$. We will show that to each $(t, z) \in X$ there is one of the finitely many pairs (t_ν, z_μ) so that $((t, z), (t_\nu, z_\mu)) \in U_{\vartheta, V}$: If $t \in \vartheta$ one chooses $|t - t_\nu| < \varepsilon$ and $(z, z_\mu) \in W$; if $t \notin \vartheta$ one chooses $t_\nu = t_1 \notin \vartheta$.

Since X_p is precompact and metrizable, X_p is also separable ([5] IX § 2, n. 9). Though p does not separate in X since the points $(0, z)$ are not distinguished, it obviously separates in X_0. Therefore, X_{0p} is a separated, metrizable, precompact, separable space.

With $X_g = \Theta_g \times Z_g$ one sees immediately that X_g is finer than X_{gp} and X_{gp} finer than X_p. The metrizability of X_g follows from that of Θ_g and Z_g in the way shown above for X_{gp}. The topologies of X_{0g} and X_{0p} are identical on X_0. This follows when the identity mappings $X_{0gp} \to X_{0g}$ and $X_{0p} \to X_{0gp}$ are continuous. That $X_{0gp} \to X_{0g}$ is continuous follows immediately from Z_g and Z_p having the same topology. The continuity of $X_{0p} \to X_{0gp}$ follows from the fact that for each compact subset $\vartheta \subset \Theta_0$ the spaces $[\vartheta \times Z]_p$ and $[\vartheta \times Z]_{gp}$ have the same uniform structure, as (1.3) easily shows.

Since we are interested in the trajectories, i.e. in the graphs as subsets of X, let us in $\mathscr{P}_0(X)$ (equal to "$\mathscr{P}(X)$ without the empty set") introduce a uniform structure of physical imprecision based on X_p. It is physically natural to do this in a "canonical" way (see [3] § 8 and also [5], Exercise 5 for II § 1).

Let $\mathcal{N}$ be the vicinity filter of X_p. We introduce the corresponding "canonical" vicinity filter $\tilde{\mathcal{N}}$ of $\mathscr{P}_0(X)$ by the fundamental system of the sets

$$\tilde{U} = \{(A, B) \mid A \in \mathscr{P}_0(X), B \in \mathscr{P}_0(X), A \in U(B), B \in U(A), U \in \mathcal{N}\}, \quad (1.4)$$

where $U(A)$ is the set of those $x \in X$ for which there is an $x' \in A$ with $(x, x') \in U$. In this way $\mathscr{P}_0(X)_p$ is defined (likewise $\mathscr{P}_0(X_0)_p$). As is easily seen, one also can regard $\mathscr{P}_0(X_0)_p$ as a subspace of the uniform space $[\mathscr{P}_0(X)]_p$. But the uniform structure p separates neither in $\mathscr{P}_0(X)$ nor in $\mathscr{P}_0(X_0)$, since A and its closure $\bar{A}$ in X_p (resp. in X_{0p}) obviously have the same neighborhood filter.

If A and B are the graphs of two trajectories, (1.4) with (1.3) yields

$$\tilde{U}_{\vartheta, V} = \{(A, B) \mid A_\vartheta \in V(B), B_\vartheta \in V(A)\} \quad (1.5)$$

as a vicinity of p (in $Z^\Theta \subset \mathscr{P}_0(X)$), where $A_\vartheta = A \cap (\vartheta \times Z)$ and $B_\vartheta = B \cap (\vartheta \times Z)$. In fact, (1.5) follows from (1.4) and (1.3) because $A \cap (\vartheta' \times Z)$ and $B \cap (\vartheta' \times Z)$ are not empty.

Since X_p is precompact, $[\mathscr{P}_0(X)]_p$ is so ([5], Exercise 11 for II § 4). The completion of $[\mathscr{P}_0(X)]_p$ is also metrizable since, with $d(x, x')$ as a metric in X_p, one can in $[\mathscr{P}_0(X)]_p$ use the distance

$$|A, B| = \sup \{d(x, A), d(x', B) \mid x \in B, x' \in A\},$$

where $d(x, A) = \inf \{d(x, x') \mid x' \in A\}$ ([5], Exercise 6 for IX § 2). Therefore $[\mathscr{P}_0(X)]_p$ is also separable, and the corresponding holds for $[\mathscr{P}_0(X_0)]_p$.

$[\mathscr{P}_0(X)]_p$ and $[\mathscr{P}_0(X_0)]_p$ are typical examples of sets in which the uniform structure of physical imprecision does not separate all elements. One should see clearly that just this is "physically reasonable". As shown above, $[\mathscr{P}_0(X)]_p$ and $[\mathscr{P}_0(X_0)]_p$ fulfill all assumptions of [3] § 9. The subset of the singletons (i.e. one-element sets $\{x\}$), considered as a subset of $\mathscr{P}_0(X)$ relative to the uniform structure in $[\mathscr{P}_0(X)]_p$, is isomorphic to X_p ([5], Exercise 6a for II § 2), which is also as it should be on "physical grounds".

Let $\mathscr{F}(X_0)$ be the set of closed subsets of X_0. They do not depend on the uniform structure p or g of X_0 since X_{0p} and X_{0g} have the same topology. Then $[\mathscr{F}(X_0)]_p$, considered as a subspace of $[\mathscr{P}_0(X_0)]_p$, is separated, i.e. p distinguishes the closed subsets of X_0 ([5], Exercise 6b for II § 2).

One can to each $A \in \mathscr{F}(X_0)$ assign its closure $\bar{A}$ in X_g. Since A is closed in X_{0g}, it follows that $A = \bar{A} \cap X_0$. Let $\mathscr{F}_0(X)$ be the set of those closed subsets B of X_g for which there is an $A \in \mathscr{F}(X_0)$ with $B = \bar{A}$. Therefore, $B = \overline{(B \cap X_0)}$ for $B \in \mathscr{F}_0(X)$.

Thus $B \leftrightarrow B \cap X_0$ establishes a one-to-one correspondence between $\mathscr{F}_0(X)$ and $\mathscr{F}(X_0)$ and, with $\mathscr{F}_0(X) \subset [\mathscr{P}_0(X)]_p$ and $\mathscr{F}(X_0) \subset [\mathscr{P}_0(X_0)]_p$ it follows that $[\mathscr{F}_0(X)]_p$ and $[\mathscr{F}(X_0)]_p$ are isomorphic. Therefore, $[\mathscr{F}_0(X)]_p$ is a separated, separable, metrizable, precompact space.

It is customary to denote the set of continuous mappings $\Theta \to \dot{Z}$ by $C(\Theta, Z)$. Since the topologies of Z_g and Z_p are equal, the set $C(\Theta, Z)$ does not depend on whether one uses p or g as uniform structure in Z. The graph (1.1) of a trajectory $z(t) \in C(\Theta, Z)$ is an element of $\mathscr{F}_0(X)$, for, firstly, it is a closed subset of $X_g = \Theta_g \times Z_g$ ([5] I § 8, n. 1) and, secondly, $z(0)$ is the limit of $z(t)$ for $t > 0$ and $t \to 0$.

As the "trajectory space" Y we denote the set of graphs belonging to $C(\Theta, Z)$ and identify $C(\Theta, Z)$ with it. Here, we "mentally" assumed that $M_m \xrightarrow{g} Y$ holds for the mapping $M_m \xrightarrow{g} Z^{\Theta}$ introduced above. But we will not exhibit such an assumption mathematically since we must in § 5 look into the mapping $M_m \xrightarrow{g} Z^{\Theta}$ more closely. An injective mapping $C(\Theta, Z) \to C(\Theta_0, Z)$ is defined by $z(t) \to z(t)|_{\Theta_0}$ so that one can identify $Y_p = C_p(\Theta, Z)$ with a subset of $C_p(\Theta_0, Z)$ since $[\mathscr{F}_0(X)]_p$ and $[\mathscr{F}(X_0)]_p$ are isomorphic.

Since $Y \in \mathscr{F}_0(X)$, the space Y_p is separated, uniform, precompact, separable and metrizable. We view p here as the uniform structure of physical imprecision for trajectories. That we have chosen $Y = C(\Theta, Z)$ as the trajectory space is a statement about the structure of the pretheory $\mathscr{P}\mathscr{T}_m$. One could imagine also to use pretheories where it is practical to admit non-continuous trajectories. But no theory is known which could not describe the trajectories by elements from $\hat{Y}_p$ (the completion of Y_p). Since (as mentioned above) one can identify Y with a subset of $C(\Theta_0, Z) \in \mathscr{F}(X_0)$, and $[\mathscr{F}(X_0)]_p$ is separated, one can identify some of the elements of $\hat{Y}_p$ with elements of $\mathscr{F}(X_0)$, namely with those of $\bar{Y}$ (the closure of Y in $[\mathscr{F}(X_0)]_p$). From this it easily follows that $\bar{Y}$ contains for example such subsets by which one can represent "jumps" of the trajectories (Fig. 2). Therefore, it would be quite thinkable that as the trajectory space Y in a pretheory one chooses any other subset, which is dense in $\hat{Y}_p$ but differs from $C(\Theta, Z)$. Let us not consider such cases in order not to sacrifice clarity for generality. The choice of the trajectory space Y depends intimately on the choice of another uniform structure in Y, which is finer than p but yields the same topology in Y as p does. We will now construct a uniform structure for $C(\Theta_0, Z)$ (and thus for $Y \subset C(\Theta_0, Z)$), relative to which $C(\Theta_0, Z)$ is complete, but which leads to the same topology as $C_p(\Theta_0, Z)$.

Let ϑ be a subset of Θ_0. A mapping $Z^{\Theta_0} \to Z^{\vartheta}$ is defined by $z(t) \to z(t)|_{\vartheta}$ (the mapping of a trajectory $z(t)$ onto its part in ϑ), for which $C(\Theta_0, Z) \to C(\vartheta, Z)$ holds. In Z^{ϑ} we define (by means of Z_g) a uniform structure $g\,u$, called the structure of the uniform convergence on Z^{ϑ}. As a basis of this $g\,u$ let us choose the

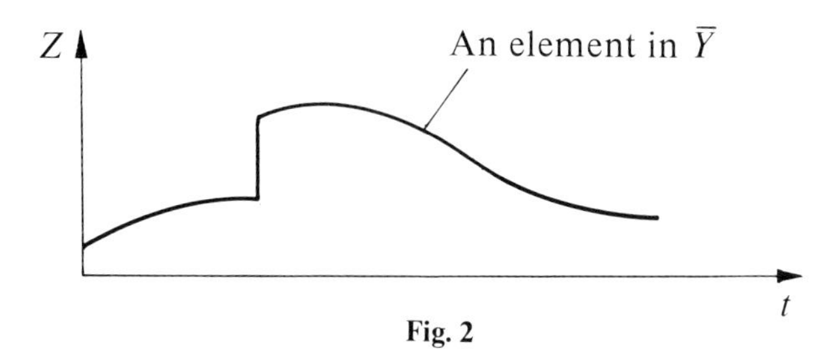

Fig. 2

vicinities

$$\tilde{V} = \{(z_1(t), z_2(t)) \mid (z_1(t), z_2(t)) \in V \text{ for all } t \in \vartheta \text{ with } V \in \mathcal{N}_g\},$$

where $\mathcal{N}_g$ is a vicinity filter of Z_g.

If $\mathcal{H}$ is a subset of $\mathcal{P}(\Theta_0)$, the mappings $Z^{\Theta_0} \to [Z^{\vartheta}]_{gu}$ for all $\vartheta \in \mathcal{H}$ define an initial uniform structure on Z^{Θ_0}, to be denoted $g\,\mathcal{H}$. One sees immediately that for $\vartheta_1 \subset \vartheta_2$ the mapping $[Z^{\vartheta_2}]_{gu} \to [Z^{\vartheta_1}]_{gu}$ is uniformly continuous. If $Z^{\Theta_0} \to [Z^{\vartheta_2}]_{gu}$ is so and $\vartheta_1 \subset \vartheta_2$, then $Z^{\Theta_0} \to [Z^{\vartheta_1}]_{gu}$ is also uniformly continuous. If $\mathcal{H}'$ is a subset of $\mathcal{H}$, so that for each $\vartheta \in \mathcal{H}$ there is a $\vartheta' \in \mathcal{H}'$ with $\vartheta \subset \vartheta'$, the uniform structures $g\,\mathcal{H}$ and $g\,\mathcal{H}'$ are thus equal on Z^{Θ_0}. If $\mathcal{H}$ is the set of compact subsets of Θ_0, one can choose $\mathcal{H}'$ as the set of all intervals $\vartheta_n = \{t \mid \frac{1}{n} \leqq t \leqq n\}$ $(n = 1, 2, \ldots)$. For $\mathcal{H}$ the set of compact subsets of Θ_0 we write $g\,\mathcal{H}$ simply g. Therefore, g is the initial uniform structure for which all the mappings $Z^{\Theta_0} \to [Z^{\vartheta_n}]_{gu}$ $(n = 1, 2, \ldots)$ are uniformly continuous. Since they are countable and Z_g was assumed metrizable, $[Z^{\Theta_0}]_g$ is also metrizable. Therefore, $C(\Theta_0, Z) \subset Z^{\Theta_0}$ implies that $C_g(\Theta_0, Z)$ is also metrizable. But $C_g(\Theta_0, Z)$ is also complete ([5] X § 1, n. 6) since Z_g was assumed complete. We will show that $C_p(\Theta_0, Z)$ and $C_g(\Theta_0, Z)$ have the same topology.

Let us first show that the identity mappings $C_g(\Theta_0, Z) \to C_{pc}(\Theta_0, Z) \to C_p(\Theta_0, Z)$ are uniformly continuous, where p is the uniform structure obtained from Z_p exactly as g is obtained from Z_g. The mapping $C_g(\Theta_0, Z) \to C_{pc}(\Theta_0, Z)$ is uniformly continuous if the mappings $C_g(\Theta_0, Z) \to C_{pu}(\vartheta, Z)$ are so for all compact ϑ since pc is the initial uniform structure generated by $C(\Theta_0, Z) \to C_{pu}(\vartheta, Z)$. Since $C_g(\Theta_0, Z) \to C_{gu}(\vartheta, Z)$ is uniformly continuous, it follows that the mappings $C_g(\Theta_0, Z) \to C_{gu}(\Theta_0, Z) \to C_{pu}(\vartheta, Z)$ are so if $C_{gu}(\vartheta, Z) \to C_{pu}(\vartheta, Z)$ is uniformly continuous, which easily follows from the fact that Z_g is finer than Z_p. Thus we still have to show the uniform continuity of $C_{pc}(\Theta_0, Z) \to C_p(\Theta_0, Z)$.

Let $\tilde{U}_{\vartheta, V}$ be a vicinity of the form (1.5) with

$$V = \{(t, z), (t', z') \mid |t - t'| < \delta, (z, z') \in W\},$$

where W is a symmetric vicinity of Z_p; then $(z(t), z'(t)) \in W$ for all $t \in \vartheta$ implies $((z, t), z'(t)) \in \tilde{U}_{\vartheta, V}$ and thus the uniform continuity of $C_{pc}(\Theta_0, Z) \to C_p(\Theta_0, Z)$.

We have thus shown that the uniform structure g in $C(\Theta_0, Z)$ is finer than p. Therefore, the topologies of $C_p(\Theta_0, Z)$ and $C_g(\Theta_0, Z)$ coincide if the mappings $C_p(\Theta_0, Z) \to C_{pc}(\Theta_0, Z) \to C_g(\Theta_0, Z)$ are continuous. We first show that $C_{pc}(\Theta_0, Z) \to C_g(\Theta_0, Z)$ is continuous.

Since the topology of $C_g(\Theta_0, Z)$ is the initial topology for all mappings $C_g(\Theta_0, Z) \to C_{gu}(\vartheta, Z)$ ([5] II § 2, n. 3), $C_{pc}(\Theta_0, Z) \to C_g(\Theta_0, Z)$ is continuous if the mapping $C_{pc}(\Theta_0, Z) \to C_{gu}(\vartheta, Z)$ is continuous for all compact subsets ϑ of Θ_0. Since $C_{pc}(\Theta_0, Z) \to C_{pu}(\vartheta, Z)$ is continuous, it suffices to show that $C_{pu}(\vartheta, Z) \to C_{gu}(\vartheta, Z)$ is.

One can in a "canonical way" obtain uniform structures $\tilde{g}$ and $\tilde{p}$ for $\mathcal{P}_0(\vartheta \times Z)$ from those of $\vartheta \times Z_g$ and $\vartheta \times Z_p$. According to [5], Exercise 11c for II § 4, the topologies of $[\mathcal{K}(\vartheta \times Z)]_{\tilde{g}}$ and $[\mathcal{K}(\vartheta \times Z)]_{\tilde{p}}$ (with $\mathcal{K}(\vartheta \times Z)$ the set of compact subsets of $\vartheta \times Z$) are equal since $\vartheta \times Z_g$ and $\vartheta \times Z_p$ have the same topology. Since ϑ is compact, the graph of a continuous function $\vartheta \to Z$ is likewise compact, i.e. $C(\vartheta, Z) \in \mathcal{K}(\vartheta \times Z)$. Therefore $C_{\tilde{g}}(\vartheta, Z)$ and $C_{\tilde{p}}(\vartheta, Z)$ have the same topology.

Since $C_{pu}(\vartheta, Z) \to C_{\tilde{p}}(\vartheta, Z)$ is easily recognized as continuous (see above), it therefore suffices to show that $C_{\tilde{g}}(\vartheta, Z) \to C_{gu}(\vartheta, Z)$ is so.

Let W be a vicinity of Z_g and $z_0(t)$ a fixed trajectory. Let W_1 be a vicinity with $W_1^2 \subset W$. Since $z_0(t)$ is uniformly continuous on ϑ (since ϑ is compact), there is an ε such that for $|t-t'| < \varepsilon$ the relation $(z(t), z'(t)) \in W_1$ holds. If we choose a vicinity in $C_{\tilde{g}}(\vartheta, Z)$ by means of ε and W_1, then $z(t)$ lies in the corresponding neighborhood of $z_0(t)$ if there is such a t' with $|t-t'| < \varepsilon$ that $(z(t), z_0(t')) \in W_1$. Since also $(z_0(t'), z_0(t)) \in W_1$, it follows that $(z(t), z_0(t)) \in W$, whereby the continuity of $C_{\tilde{g}}(\vartheta, Z) \to C_{gu}(\vartheta, Z)$ is proved. Therefore $C_{pc}(\Theta_0, Z) \to C_g(\Theta_0, Z)$ is continuous.

It remains to show that $C_p(\Theta_0, Z) \to C_{pc}(\Theta_0, Z)$ is also continuous. Since the topology of $C_{pc}(\Theta_0, Z)$ is the initial topology for the mappings $C(\Theta_0, Z) \to C_{pu}(\vartheta_n, Z)$ with $\vartheta_n = \{t \mid \frac{1}{n} \leqq t \leqq n\}$, it suffices to show that $C_p(\Theta_0, Z) \to C_{pu}(\vartheta_n, Z)$ is continuous for all ϑ_n. Let W be a vicinity of Z_p and $z_0(t)$ a fixed trajectory. Let W_1 be a vicinity with $W_1^2 \subset W$. Since $z_0(t)$ is uniformly continuous on the compact subset ϑ_{n+1}, there is an $\varepsilon > 0$ with $\varepsilon < \dfrac{1}{n} - \dfrac{1}{n+1}$, such that $(z_0(t), z_0(t')) \in W_1$ holds for $t, t' \in \vartheta_{n+1}$ and $|t-t'| < \varepsilon$. With these ϑ_{n+1}, ε and W_1, according to (1.5) with

$$V = \{((t, z), (t', z')) \mid |t-t'| < \varepsilon, (z, z') \in W_1\}$$

we form a neighborhood of $z_0(t)$ by means of $\tilde{U}_{\vartheta_{n+1}, V}$. Then the following holds for $z(t) \in \tilde{U}_{\vartheta_{n+1}, V}(z_0(t))$: For $t \in \vartheta_n \subset \vartheta_{n+1}$ there is a t' with $|t-t'| < \varepsilon$ and $(z(t), z_0(t')) \in W_1$. Because of $t \in \vartheta_n$ and $\varepsilon < \dfrac{1}{n} - \dfrac{1}{n+1}$ we have $t' \subset \vartheta_{n+1}$; hence $(z_0(t'), z_0(t)) \in W_1$ and thus for $t \in \vartheta_n$ the relation $(z(t), z_0(t)) \in W_1^2 \subset W$ holds. Therefore $C_p(\Theta_0, Z) \to C_{pu}(\vartheta_n, Z)$ is continuous.

We have thus finally shown that $C_p(\Theta_0, Z)$ and $C_g(\Theta_0, Z)$ have the same topology. Since $C_p(\Theta_0, Z)$ is separable, $C_g(\Theta_0, Z)$ is also separable, which also follows directly from [5] X § 3, n. 1. Therefore, $C_g(\Theta_0, Z)$ is complete, separable and metrizable. Since one can identify $Y = C(\Theta, Z)$ with a subset of $C(\Theta_0, Z)$, the topologies of Y_p and Y_g also coincide. Although Y_g need not be complete, Y is dense in $C(\Theta_0, Z)$. This easily follows from the fact that for each ϑ_n and a trajectory $z(t) \in C(\Theta_0, Z)$ one can construct the trajectories

$$z_n(t) = \left\{ z(t) \text{ for } t \in \vartheta_n \text{ and } z_n(t) = z\left(\frac{1}{n}\right) \text{ for } t < \frac{1}{n} \text{ and } z_n(t) = z(n) \text{ for } t > n \right\} \in C(\Theta, Z),$$

which converge to $z(t)$ in $C(\Theta_0, Z)$. Since Y is dense in $C(\Theta_0, Z)$ and $C_g(\Theta_0, Z)$ is complete, one can identify the completion of Y_g with $C_g(\Theta_0, Z)$, and hence the completion of Y_p with that of $C_p(\Theta_0, Z)$. In the sequel we shall for the completion $\hat{Y}_p$ of Y_p simply write $\hat{Y}$. We denote the elements of $\hat{Y} \setminus C(\Theta_0, Z)$ as virtual trajectories.

Later, yet the following facts will be important. When ϑ_τ consists of a single point $\tau \in \Theta_0$, it is compact so that $[Z^{\Theta_0}]_g \to [Z^{\vartheta_\tau}]_{gu}$ is uniformly continuous. One easily sees that the mapping $[Z^{\vartheta_\tau}]_{gu} \to Z_g$ for $[z(t)]_{t=\tau} \in Z^{\vartheta_\tau}$ defined by $[z(t)]_{t=\tau}$

$\to z(\tau) \in Z$ is uniformly continuous. Therefore, for each fixed $\tau \in \Theta_0$, the mapping $[Z^{\Theta_0}]_g \xrightarrow{g_\tau} Z_g$ defined by $y = z(t) \to z(\tau) \in Z$ and thus the similarly defined mappings $C_g(\Theta_0, Z) \xrightarrow{g_\tau} Z_g$ and $C_g(\Theta, Z) \xrightarrow{g_\tau} Z_g$ are uniformly continuous. The last, however, does not hold for $\tau = 0$!

Therefore, since the topologies of $C_g(\Theta, Z)$ and $C_p(\Theta, Z)$ coincide and also those of Z_g and Z_p do, $Y_p \xrightarrow{g_\tau} Z_p$ is at least continuous.

§ 2 Preparation and Registration Procedures

In many pretheories $\mathcal{PT}_m$, no statistical description is formulated to begin with; and if there already is such a description (in statistical point mechanics by means of densities ϱ in Γ-space), then it is not of the general form particularly suited for an axiomatic basis of quantum mechanics. Therefore let us extend the mathematical pictures $\mathcal{MT}_m$ of the pretheories by a statistical structure, but retaining the notation $\mathcal{MT}_m$ for the sake of simplicity. In § 3 it will turn out that the statistical description to be introduced encompasses all cases in which there already was a statistical description.

In order to introduce the statistical description we use [3] § 12 and [2] II §§ 2–4. The *general* significance of this method proceeds clearly from [3] § 12. In [2] II §§ 2–4, one finds the most detailed presentation of theorems, but also of the physical interpretation. Here one must note that in [2] II in no way was it used that the set M is an image set (index set) of microsystems. Therefore, one can carry over all the considerations from [2] II, replacing the M from [2] II by the set M_m introduced above in § 1. For that reason, here we can be brief in introducing the statistics and forego proofs of theorems.

The reader can look up an introductory formulation of the statistical description used here in [1] XIII §§ 1 and 2. One can find a brief formulation of this method in [6] §§ 1 and 2.

§ 2.1 Statistical Selection Procedures

When a set $\mathcal{S} \subset \mathcal{P}(M)$ as a theorem or axiom obeys

AS 1.1 $\quad a, b \in \mathcal{S}$ and $a \subset b \Rightarrow b \backslash a \in \mathcal{S}$,

AS 1.2 $\quad a, b \in \mathcal{S} \quad \Rightarrow \quad a \cap b \in \mathcal{S}$

(with $b \backslash a$ the relative complement of a in b), let us name it a structure of species "selection procedures" (shortly SP) over M. For brevity such an $\mathcal{S}$ will simply be called a "set of SP", an element of it a "selection procedure" (SP).

In the following we always denote the set of all $b \in \mathcal{S}$ with $b \subset a$ by $\mathcal{S}(a)$. This $\mathcal{S}(a)$ is a Boolean ring (with a as unit element). For each subset $\mathcal{A}$ of $\mathcal{P}(M)$ there is a smallest set $\mathcal{S}$ of SP with $\mathcal{A} \subset \mathcal{S}$. One calls this $\mathcal{S}$ the set of SP generated by $\mathcal{A}$. (For further details and considerations concerning AS 1.1, 2, see [2] II § 2.)

A set $\mathcal{S}$ of SP shall be called a structure of species "statistical selection procedures" (SSP) or simply a "set of SSP", if there is given a mapping λ of

$$\mathcal{T} = \{(a, b) \mid a, b \in \mathcal{S},\ a \supset b \text{ and } a \neq \emptyset\}$$

into the interval [0, 1] of real numbers for which the following holds:

AS 2.1 $a_1, a_2 \in \mathscr{S};\ a_1 \cap a_2 = \emptyset,\ a_1 \cup a_2 \in \mathscr{S} \Rightarrow \lambda(a_1 \cup a_2, a_1) + \lambda(a_1 \cup a_2, a_2) = 1.$

AS 2.2 $a_1, a_2, a_3 \in \mathscr{S},\ a_1 \supset a_2 \supset a_3,\ a_2 \neq \emptyset \Rightarrow \lambda(a_1, a_3) = \lambda(a_1, a_2)\,\lambda(a_2, a_3).$

AS 2.3 $a_1, a_2 \in \mathscr{S},\ a_1 \supset a_2 \neq \emptyset \Rightarrow \lambda(a_1, a_2) \neq 0.$

AS 2.4.1 If $a_\nu \in \mathscr{S}$ is a decreasing sequence with $\bigcap_{\nu=1}^{\infty} a_\nu = \emptyset$ and $a \in \mathscr{S}$ with $a \supset a_1$, then $\lambda(a, a_\nu) \to 0$.

AS 2.4.2 Each totally ordered subset of $\mathscr{S}$ has an upper bound in $\mathscr{S}$.

One can further adjoin the following axiom without contradicting known applications of probability theory in physics.

AS 2.5 Let $a_i \in \mathscr{S}$, with $a_i \neq \emptyset$ for $i = 1, 2, \ldots, n$, and let $a_1 = a_n$. Then

$$\prod_{i=1}^{n-1} \lambda(a_{i+1}, a_i \cap a_{i+1}) = \prod_{i=1}^{n-1} \lambda(a_i, a_i \cap a_{i+1})$$

holds. The reader may look up further consequences of this axiom in [41]; we do not use it in the sequel.

AS 2.4.2 can somewhat be weakened as described in [3] § 11; there one can also read about the physical interpretation of a set of SSP. Therefore let us only state tersely that physicists compare the real number $\lambda(a, b)$ with the experimental frequencies with which circumstances $x \in M$, selected according to a, do also fulfill the criteria of b.

One calls $\lambda(a, b)$ the *probability* of b relative to a. For our development of an axiomatic basis for quantum mechanics it is important that the concept of probability and its interpretation is already clarified in the pretheories. Quantum mechanics has no need for any *novel* probability concept. On the contrary, all probabilities in quantum mechanics relate in the final analysis to those of the pretheories, i.e. to probabilities in the macroscopic domain of objectivatingly described devices (see III)!

Above, we have pictured the selection procedures $a \in \mathscr{S}$ immediately by subsets $a \subset M$, namely by the "subsets of those systems which are selected according to a". This practice (followed above) is very convenient for many purposes since one gets on "quickly". But this practice happens to be misunderstood; also it can often be necessary to analyse more precisely the selections by a device. For these reasons, we will here briefly formulate at pretheory for the theory of selections, even though we shall seldom use it (see § 2.3 and § 3.1).

Instead of proceeding from a set $\mathscr{S} \in \mathscr{P}(M)$, we start from a *basic* set $\tilde{\mathscr{S}}$ of elements to picture actual procedures. Let $\tilde{\mathscr{S}}$ be equipped with the following structure (in brief formulation): $\tilde{\mathscr{S}}$ is an ordered set; there is a smallest element in $\tilde{\mathscr{S}}$ to be denoted by 0; for each pair of elements $\tilde{a}_1, \tilde{a}_2 \in \tilde{\mathscr{S}}$ there is a lower bound $\tilde{a}_1 \wedge \tilde{a}_2$; each order interval $[0, \tilde{a}]$ is a Boolean ring (with $\tilde{a}$ as the unit).

We then briefly call $\tilde{\mathscr{S}}$ the set of procedures. The physical notion here is this: If $\mathscr{PT}$ involves a set $\tilde{\mathscr{S}}$ with the structure just described, correspondence rules shall be given under which the elements of $\tilde{\mathscr{F}}$ are the pictures of definite physical procedures and the order relation in $\tilde{\mathscr{F}}$ pictures a "finer-coarser" factual relation.

Therefore, $\tilde{\mathscr{F}}$ represents a picture of the procedures *without* already referring to the fact that "systems" $x \in M$ are selected by these procedures. Hence, in order to portray the selection of systems, one needs a mathematical relation between the elements of $\tilde{\mathscr{F}}$ and of M.

Consequently, we proceed from two basic sets $\tilde{\mathscr{F}}$ and M and a two-place relation $\alpha \subset \tilde{\mathscr{F}} \times M$ where $(\tilde{a}, x) \in \alpha$ is the picture of the factual relation that "x is selected by the procedure $\tilde{a}$".

The two-place relation α can also be characterized uniquely by the mapping $\tilde{\mathscr{F}} \xrightarrow{h} \mathscr{P}(M)$ with

$$\tilde{a} \xrightarrow{h} \{x \mid x \in M \quad \text{and} \quad (\tilde{a}, x) \in \alpha\}.$$

Therefore, $h(\tilde{a})$ is the "set of systems selected by $\tilde{a}$".

For the mapping h we demand the following relations (2.1.1) and (2.1.2):

$$h(\tilde{a}_1 \wedge \tilde{a}_2) = h(\tilde{a}_1) \cap h(\tilde{a}_2). \tag{2.1.1}$$

This expresses that $\tilde{a}_1 \wedge \tilde{a}_2$ selects precisely those systems which are selected by $\tilde{a}_1$ as well as $\tilde{a}_2$.

For $\tilde{a}_1 > \tilde{a}_2$ (hence $\tilde{a}_1 \wedge \tilde{a}_2 = \tilde{a}_2$) it follows from (2.1.1) that $h(\tilde{a}_2) = h(\tilde{a}_1) \cap h(\tilde{a}_2)$, i.e. $h(\tilde{a}_1) \subset h(a_2)$ holds. Therefore, h preserves the ordering. Thus it is meaningful, with $\tilde{a}_1 \backslash \tilde{a}_2$ as the complement of $\tilde{a}_2$ in the Boolean ring $[0, \tilde{a}_1]$ to demand:

$$\tilde{a}_1 > \tilde{a}_2 \quad \text{implies} \quad h(\tilde{a}_1 \backslash \tilde{a}_2) = h(\tilde{a}_1) \backslash h(\tilde{a}_2). \tag{2.1.2}$$

This asserts that $\tilde{a}_1 \backslash \tilde{a}_2$ selects precisely those systems which were selected by $\tilde{a}_1$ but do not satisfy the criteria of the finer procedure $\tilde{a}_2$.

If $\tilde{a}_1 \vee \tilde{a}_2$ exists (which is just then the case when there is an $\tilde{a}_3 \in \tilde{\mathscr{F}}$ with $\tilde{a}_1, \tilde{a}_2 < \tilde{a}_3$!), then it follows that $h(\tilde{a}_1 \vee \tilde{a}_2) \supset h(\tilde{a}_i)$ for $i = 1, 2$, hence $h(\tilde{a}_1 \vee \tilde{a}_2) \supset h(\tilde{a}_1) \cup h(\tilde{a}_2)$. On the other hand, from $h(\tilde{a}_1 \vee \tilde{a}_2 \backslash \tilde{a}_1) = h(\tilde{a}_1 \vee \tilde{a}_2) \backslash h(\tilde{a}_1)$ with $\tilde{a}_1 \vee \tilde{a}_2 \backslash \tilde{a}_1 < \tilde{a}_2$ follows $h(\tilde{a}_1 \vee \tilde{a}_2) \backslash h(\tilde{a}_1) \subset h(\tilde{a}_2)$ and hence $h(\tilde{a}_1) \cup [h(\tilde{a}_1 \vee \tilde{a}_2) \backslash h(\tilde{a}_1)] = h(\tilde{a}_1 \vee \tilde{a}_2) \subset h(\tilde{a}_1) \cup h(\tilde{a}_2)$. Therefore, $h(\tilde{a}_1 \vee \tilde{a}_2) = h(\tilde{a}_1) \cup h(\tilde{a}_2)$.

T 2.1.1 The set $\mathscr{S} \overset{\text{def}}{=} h(\tilde{\mathscr{F}})$ is a system of selection procedures.

Proof. If $a_1, a_2 \in \mathscr{S}$, there exist $\tilde{a}_1, \tilde{a}_2 \in \tilde{\mathscr{F}}$ with $a_1 = h(\tilde{a}_1)$, $a_2 = h(\tilde{a}_2)$. From this follows $h(\tilde{a}_1 \wedge \tilde{a}_2) = h(\tilde{a}_1) \cap h(\tilde{a}_2) = \tilde{a}_1 \cap \tilde{a}_2 \in \mathscr{S}$.

From $a_1 \subset a_2$, with $h(\tilde{a}_i) = a_i$ follows $h(\tilde{a}_1 \wedge \tilde{a}_2) = h(\tilde{a}_1) \cap h(\tilde{a}_2) = a_1 \cap a_2 = a_2$ and $h(\tilde{a}_1 \backslash \tilde{a}_1 \wedge \tilde{a}_2) = h(\tilde{a}_1) \backslash h(\tilde{a}_1 \wedge \tilde{a}_2) = a_1 \backslash a_2 \in \mathscr{S}$. □

T 2.1.2 For the set

$$\mathscr{J} = \{\tilde{a} \mid \tilde{a} \in \tilde{\mathscr{F}}, h(\tilde{a}) = \emptyset\}$$

the following assertions hold:

(i) $\tilde{a}_1 \in \mathscr{J}$, $\tilde{a}_2 \in \tilde{\mathscr{F}}$, $\tilde{a}_2 \subset \tilde{a}_1 \Rightarrow \tilde{a}_2 \in \mathscr{J}$;
(ii) $\tilde{a}_1, \tilde{a}_2 \in \mathscr{J}$, $\tilde{a}_1 \vee \tilde{a}_2 \in \tilde{\mathscr{F}} \Rightarrow \tilde{a}_1 \vee \tilde{a}_2 \in \mathscr{J}$.

Proof. Since h preserves the ordering, from $\tilde{a}_1 \subset \tilde{a}_1$ follows that $h(\tilde{a}_2) \subset h(\tilde{a}_1)$ and hence i) holds. ii) follows from $h(\tilde{a}_1 \vee \tilde{a}_2) = h(\tilde{a}_1) \cup h(\tilde{a}_2)$. □

D 2.1.1 An *ideal* in $\tilde{\mathscr{F}}$ is defined as a subset $\mathscr{J} \subset \tilde{\mathscr{F}}$ for which (i), (ii) in T 2.1.2 hold.

T 2.1.3 A partition of $\tilde{\mathscr{F}}$ into classes relative to an ideal $\mathscr{J}$ is determined by the equivalence

$$\tilde{a}_1 \sim \tilde{a}_2 \overset{\text{def}}{\Leftrightarrow} \tilde{a}_1 \backslash \tilde{a}_1 \wedge \tilde{a}_2 \in \mathscr{J} \quad \text{and} \quad \tilde{a}_2 \backslash \tilde{a}_1 \wedge \tilde{a}_2 \in \mathscr{J}$$

with $\tilde{a} \backslash \tilde{b}$ the complement of $\tilde{b}$ in $[0, \tilde{a}]$.

Proof. It follows from $\tilde{a}_1 \sim \tilde{a}_2 \sim \tilde{a}_3$, with the aid of the identity

$$\tilde{a}_1 \backslash (\tilde{a}_1 \wedge \tilde{a}_3) = [(\tilde{a}_1 \backslash \tilde{a}_1 \wedge \tilde{a}_2) \backslash (\tilde{a}_1 \backslash \tilde{a}_1 \wedge \tilde{a}_2) \wedge \tilde{a}_3] \vee [\tilde{a}_1 \wedge (\tilde{a}_2 \backslash \tilde{a}_2 \wedge \tilde{a}_3)],$$

that $\tilde{a}_1 \backslash \tilde{a}_1 \wedge \tilde{a}_3 \in \mathscr{J}$ holds. Likewise follows $\tilde{a}_3 \backslash \tilde{a}_3 \wedge \tilde{a}_1 \in \mathscr{J}$ and hence $\tilde{a}_1 \sim \tilde{a}_3$. □

The set of classes of $\tilde{\mathscr{F}}$ relative to the equivalence in T 2.1.3 will be denoted $\tilde{\mathscr{F}}/\mathscr{J}$. Let k be the canonical mapping $\tilde{\mathscr{F}} \overset{k}{\rightarrow} \tilde{\mathscr{F}}/\mathscr{J}$.

T 2.1.4 With $\mathscr{J}$ from T 2.1.2, the mapping h can be factored according to the diagram

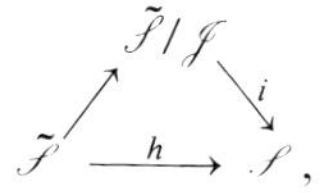

where i is a bijection. If $\mathscr{J}$ contains only the zero element, i.e. $\tilde{\mathscr{F}} \overset{k}{\rightarrow} \mathscr{S}$ is bijective, h is an order isomorphism of $\tilde{\mathscr{F}}$ and $\mathscr{S}$.

Proof. From $h(\tilde{a}_1) = h(\tilde{a}_2)$ follows $h(\tilde{a}_1 \wedge \tilde{a}_2) = h(\tilde{a}_1) = h(\tilde{a}_2)$ and hence $h(\tilde{a}_i \backslash \tilde{a}_1 \wedge \tilde{a}_2) = \emptyset$ for $i = 1, 2$, i.e. $\tilde{a}_1 \sim \tilde{a}_2$. And, if $\tilde{a}_1 \sim \tilde{a}_2$, there follows $h(\tilde{a}_i \backslash \tilde{a}_1 \wedge \tilde{a}_2) = \emptyset$ and hence $h(\tilde{a}_1) = h((\tilde{a}_1 \backslash \tilde{a}_1 \wedge \tilde{a}_2) \vee (\tilde{a}_1 \wedge \tilde{a}_2)) = h(\tilde{a}_1 \wedge \tilde{a}_2)$; likewise, $h(\tilde{a}_2) = h(\tilde{a}_1 \wedge \tilde{a}_2)$, hence $h(\tilde{a}_1) = h(\tilde{a}_2)$. If $\tilde{\mathscr{F}} \overset{h}{\rightarrow} \mathscr{S}$ is bijective, then from $h(\tilde{a}_1) \supset h(\tilde{a}_2)$ and $h(\tilde{a}_1 \wedge \tilde{a}_2) = h(\tilde{a}_1) \cap h(\tilde{a}_2)$ follows $\tilde{a}_1 \wedge \tilde{a}_2 = \tilde{a}_2$, i.e. $\tilde{a}_1 > \tilde{a}_2$. □

The set $\mathscr{J} = \{\tilde{a} \mid \tilde{a} \in \tilde{\mathscr{F}}, h(\tilde{a}) = \emptyset\}$ is formed by those procedures $\tilde{a}$ for which there are no systems $x \in M$ that they could select.

Therefore, if for a $\mathscr{PT}$ one need not formulate more exactly the pretheory with $\tilde{\mathscr{F}}$, then we shall (as at the beginning of this § 2.1) restrict ourselves to considering *only* $\mathscr{S}$.

§ 2.2 Preparation Procedures

After briefly having given the fundamental structure of the statistical description in § 2.1 let us now formulate how to apply this statistics to physical systems. In order to get a statistics for the behavior of the systems $x \in M_m$ one must often repeat trials, i.e. one must prepare a large number of systems $x \in M_m$. The statistics of their behavior depends on the type of preparation. Hence we first introduce a mathematical structure to describe the different preparation possibilities for the systems from M_m (also see [1] XIII § 2; [2] II § 4.1; [3] § 12.1).

As a picture of the preparation, let in $\mathcal{M}\mathcal{T}_m$ a structure $\mathcal{Q}_m \subset \mathcal{P}(M_m)$ be given which obeys:

APS 1 $\mathcal{Q}_m$ is a structure of species statistical selection procedures (SSP).

We denote $\mathcal{Q}_m$ briefly as the "set of preparation procedures". This presents an abbreviated formulation of a correspondence rule from $\mathcal{P}\mathcal{T}_m$. For instance, if a special preparation procedures has been applied in the experiments and if one has denoted it by the letter a, then the relation $a \in \mathcal{Q}_m$ must be recorded in $\leftrightarrow_r$ (1) (see [3] § 5). If a system c has been prepared by the procedure a, then the relation $c \in a$ must be recorded in $\leftrightarrow_r$ (2). Therefore, $x \in a$ (for $x \in M_m$, $a \in \mathcal{Q}_m$) is a picture relation (see [3] § 5).

Without going into details of [3], one can also briefly and in an obvious way formulate the interpretation of $\mathcal{Q}_m$ as follows: An $a \in \mathcal{Q}_m$ is the set of those systems which were prepared (i.e. produced) by a definite procedure a.

The relations characterized here and in the following by $\leftrightarrow_r$ always represent experimental facts, expressed in the language of $\mathcal{M}\mathcal{T}$. Therefore, $c \in a$ is simply the mathematical way of expressing that the concrete system c used in the experiment was prepared (i.e. produced) by the procedure a.

The probability function according to APS 1 defined for $\mathcal{Q}_m$ is in the following denoted by $\lambda_{\mathcal{Q}_m}$.

If a_1, a_2 are two preparation procedures, of which a_2 is finer than a_1 (i.e. $a_1 \supset a_2$), and if $c_1, \ldots, c_{N_1}$ have been produced by the procedure a_1, one can search out those $c_{i_1}, c_{i_2}, \ldots, c_{i_{N_2}}$ among the c_k ($k = 1, \ldots, N_1$) which were produced by the finer procedure. Then in $\leftrightarrow_r$ (2) one must (as experimental result) record $\lambda_{\mathcal{Q}_m}(a_1, a_2) \sim N_2/N_1$ corresponding to the mapping principle for probabilities (see [3] § 11; [2] II § 2; [1] XIII § 1).

§ 2.3 Registration Procedures

More interesting for physics is not the preparation but the registration, i.e. the measurement of the trajectories $g(x)$ (the function $M_m \xrightarrow{g} Y$ introduced in § 1). In order to describe the statistics of such measurements we need in $\mathcal{M}\mathcal{T}_m$ a picture for the *methods* according to which one measures and a picture for the *measured results*. We will describe the results digitally, i.e. with Yes-No indicators of the measuring device, which is "customary" today in the era of computer technology. We shall only in § 3 formulate more precisely how these (frequently very many!) Yes-No statements of a measuring device are interconnected with the trajectories $y \in Y$ "to be measured".

Digital measuring likewise represents a selection procedure since one "selects those systems to which a definite indicator has responded". This is well known to all physicists. Less consciously one is aware that already the *application of a measuring method* (without any sorting for results) represents a selection procedure, namely the (of course, very arbitrary) "selection of those systems to which the measuring method is applied". We must try to describe this experimental situation by selection procedures. In order in § 3 to establish the mentioned connection between "measuring" and "trajectories", let us here begin with the more general representation of procedures as described in the second half of § 2.1.

For this reason we introduce a new basic set $\tilde{\mathscr{R}}_m$ whose elements shall describe the digital indications of the measuring device. We have said "we introduce" since till now it is not customary in a $\mathscr{PT}_m$ to exhibit mathematical terms for the measuring procedure. In § 3 we shall yet go in more detail into the meaning of such an extension of the usual macroscopic theories. But first we must equip $\tilde{\mathscr{R}}_m$, M_m with a structure which mathematically describes that a measuring device is applied.

First we introduce a subset $\tilde{\mathscr{R}}_{0m} \subset \tilde{\mathscr{R}}_m$ which is to describe the measuring methods. While $\varrho_0 \in \tilde{\mathscr{R}}_{0m}$ means that ϱ_0 is the picture of a definite measuring *method*, $\varrho \in \tilde{\mathscr{R}}_m$ means that ϱ is the picture of a certain digital indication. It is very convenient to also view $\varrho_0 \in \tilde{\mathscr{R}}_{0m}$ as an indication from $\tilde{\mathscr{R}}_m$, namely as the "trivial indication" that one has measured with the method ϱ_0. For this reason we introduced the measuring methods as a subset $\tilde{\mathscr{R}}_{0m} \subset \tilde{\mathscr{R}}_m$.

As further structure we assume that $\tilde{\mathscr{R}}_m$ is a set of procedures (in the sense of § 2.1), i.e. an ordered set with the additional requirements from § 2.1. In this context $\varrho_1 < \varrho_2$ (i.e. ϱ_1 finer than ϱ_2) means that ϱ_1 indicates the result "more precisely" than ϱ_2 does, i.e. that (due to the construction of the measuring device) the response of ϱ_1 necessarily implies that of ϱ_2. We here extend the requirements from § 2.1 to the additional one: For each $\varrho \in \tilde{\mathscr{R}}_m$ there is a $\varrho_0 \in \tilde{\mathscr{R}}_{0m}$ with $\varrho < \varrho_0$.

The last demand just states that for each indication there also is a device on which it appears, because $\varrho < \varrho_0$ (i.e. ϱ *finer* than ϱ_0) means that ϱ is finer than the "trivial" indication ϱ_0. Hence, the order interval $[0, \varrho_0]$ in $\tilde{\mathscr{R}}_m$ is just the set of all digital indications on the device ϱ_0.

As in § 2.1, the choice of physical systems by means of procedures from $\tilde{\mathscr{R}}_m$ is described by a mapping $\tilde{\mathscr{R}}_m \xrightarrow{h} \mathscr{P}(M_m)$ on which (2.1.1) and (2.1.2) are imposed. We denote $h(\tilde{\mathscr{R}}_m)$ by $\mathscr{R}_m$ and $h(\tilde{\mathscr{R}}_{0m})$ by $\mathscr{R}_{0m}$,

We must yet describe the physical fact that the norms of the measuring devices fluctuate statistically. All experimental physicists know this. They take pains to keep these fluctuations as small as possible, but take them into consideration in the "error estimates" of their measurements. We describe this situation in two steps: first by the demand that $\tilde{\mathscr{R}}_{0m}$ be a set of procedures in the sense of § 2.1, then by the additional requirement of a statistics which is involved in the relation denoted below by APS 3.

From the theorems proved in § 2.1 follows that $\mathscr{R}_0$ and $\mathscr{R}_{0m}$ are structures of species selection procedure over M_m, that $\mathscr{R}_{0m} \subset \mathscr{R}_m$ holds and also: For each $b \in \mathscr{R}_m$ there is a $b_0 \in \mathscr{R}_{0m}$ with $b \subset b_0$.

In order not to talk about "useless" measuring methods we agree that the mapping $\tilde{\mathscr{R}}_{0m} \xrightarrow{h} \mathscr{R}_{0m}$ be bijective. From T 2.1.4 then follows that $\tilde{\mathscr{R}}_{0m}$ and $\mathscr{R}_{0m}$ are isomorphic. This is the reason why one often identifies $\tilde{\mathscr{R}}_{0m}$ with $\mathscr{R}_{0m}$ and speaks of the $b_0 \in \mathscr{R}_{0m}$ as measuring methods. We will use the following terminology: $\mathscr{R}_{0m}$ is called the set of registration *methods* and $\mathscr{R}_m$ that of registration *procedures.*

We further require that $\mathscr{R}_{0m}$ is not only a structure of species selection procedures (SP), but a structure of species *statistical* selection procedures (SSP).

The corresponding probability function $\lambda_{\mathscr{R}_{0m}}$ shall just describe the above-mentioned fluctuations of the norms of the measuring devices. In fact, because of the isomorphism of $\mathscr{R}_{0m}$ and $\tilde{\mathscr{R}}_{0m}$ one can also view $\lambda_{\mathscr{R}_{0m}}$ as the probability relative to the procedures from $\tilde{\mathscr{R}}_{0m}$.

Therefore, for the structures $\mathscr{R}_m \subset \mathscr{P}(M_m)$ and $\mathscr{R}_{0m}(\mathscr{P}(M_m)$ we find (as *theorems*) the relations

APS 2 $\mathscr{R}_m$ is a structure of species selection procedures.
APS 3 $\mathscr{R}_{0m}$ is a structure of species statistical selection procedures.
APS 4.1 $\mathscr{R}_{0m} \subset \mathscr{R}_m$;
APS 4.2 For each $b \in \mathscr{R}_m$ there is a $b_0 \in \mathscr{R}_{0m}$ such that $b \subset b_0$.

Because of the isomorphism of $\mathscr{R}_{0m}$ and $\tilde{\mathscr{R}}_{0m}$, the designation of $\mathscr{R}_{0m}$ as the "set of registration methods" is only an abbreviated form of the following correspondence rule: For a measuring method b_0 one records $b_0 \in \mathscr{R}_0$ in $\hookrightarrow_r$ (1); for a system $x \in M_m$ which was measured by the method b_0 one records $x \in b_0$ in $\hookrightarrow_r$ (2).

The designation of $\mathscr{R}_m$ as "the set of registration procedures" is likewise only an abbreviated form of the following principles: For a digital indication ϱ one is to record $\varrho \in \tilde{\mathscr{R}}_m$. When for a measurement of the system $x \in M_m$ the digital indication ϱ has occurred, one is to record $x \in b = h(\varrho)$.

As has already been stated, the order interval $[0, \varrho_0]$ is the set of digital indications on the measuring device ϱ_0. As a subset of $\tilde{\mathscr{R}}_m$, this interval $[0, \varrho_0]$ is a Boolean ring well known to computer technologists as the logical switch ring. The image set $h([0, \varrho_0])$ is the set of all $b \in \mathscr{R}_m$ with $b \subset b_0 = h(\varrho_0)$, i.e. the order interval $[\emptyset, b_0]$ in $\mathscr{R}_m$ which we shall write $\mathscr{R}_m(b_0)$ and which is a Boolean set ring. Often one denotes (somewhat léger) the $b \in \mathscr{R}_m(b_0)$ as the digital responses of the registration method b_0 although the mapping $\tilde{\mathscr{R}}_m \xrightarrow{h} \mathscr{R}_m$ need not be injective. In this sense, $x \in b$ with $b \subset b_0$ means that the registration procedure b has "responded" in the measurement of the system x by the method b_0, while $x \in b_0 \backslash b$ correspondingly means that the procedure b has "not responded". One often does not introduce a pretheory in the sense of a basic set $\tilde{\mathscr{R}}_m$, but proceeds straight from the structures $\mathscr{R}_{0m}$, $\mathscr{R}_m$ with the requirements APS 2 through APS 4, as detailed in [2] II § 4.2 and briefly sketched in [3] § 12.1.

§ 2.4 Dependence of Registration on Preparation

The statistics of registering depends decisively on the preparing procedure. We must try to describe this mathematically. For $a \in \mathscr{Q}_m$ and $b_0 \in \mathscr{R}_{0m}$, the set $a \cap b_0$ pictures "the systems prepared by a and measured by the method b_0". Thus, the question immediately arises whether the procedures a and b_0 might stand in each other's way experimentally, hence mathematically, whether perhaps $a \cap b_0 = \emptyset$. The combination problem thus brought up is in various areas of physics by no means trivial (see e.g. [2] II § 4.3; [2] III § 1, and here in III § 5.1). But for the considered pretheories this problem can easily be treated since we have already assumed in § 1 that the systems were prepared "before" the time $t = 0$ and the trajectories measured "after" $t = 0$. We can formulate this presumption mathematically by:

APS 5m From $a \in \mathscr{Q}_m$, $b_0 \in \mathscr{R}_{0m}$, $a \neq \emptyset$, $b_0 \neq \emptyset$ there results $a \cap b_0 \neq \emptyset$.

In the following we write $\mathscr{Q}'_m$ for the set $\mathscr{Q}_m$ without the empty set and $\mathscr{R}'_{0m}$ for $\mathscr{R}_{0m}$ without the empty set. Then APS 5 m can also be formulated as: $\alpha \in \mathscr{Q}'_m$, $b_0 \in \mathscr{R}'_{0m}$ implies $a \cap b_0 \neq \emptyset$. In the notation from [2] II § 4.3 or [3] § 12.1 this assumption simply reads $C = \mathscr{Q}'_m \times \mathscr{R}'_{0m}$. As pointed out in [2] III § 1, for macrosystems $C = \mathscr{Q}'_m \times \mathscr{R}'_{0m}$ is meaningful.

In order to spare unnecessary deliberations (see [2] § 4.4 and [3] § 12.1) we further assume

APS 8 $M_m = \bigcup_{a \in \mathscr{Q}_m} a = \bigcup_{b_0 \in \mathscr{R}_{0m}} b_0$.

This is physically almost meaningless since $x \in a$ resp. $x \in b_0$ does *not* assert that experimentally one has *established* under *which* procedure a the system x was prepared or by which method it was measured. (The relations APS 5.1.3 and APS 5.2 from [2] III § 1 and [2] II § 4.3 resp. [3] § 2.1 follow from APS 8 and APS 5m.)

We first introduce the set

$$\Theta_m = \{c \mid c = a \cap b, \quad \text{where} \quad a \in \mathscr{Q}_m, b \in \mathscr{R}_m\}.$$

The set $a \cap b$ precisely pictures "all systems prepared under a for which the digital indication b has occurred". The frequencies with which the different b occur are of physical interest. As a portrait for this we introduce the structure $\mathscr{S}_m$ of species selection procedure (SP) generated by Θ_m (see § 1) and assume:

APS 6 $\mathscr{S}_m$ is a structure of species *statistical* selection procedures (SSP).

Hence there is a probability function $\lambda_{\mathscr{S}_m}$ so that $\lambda_{\mathscr{S}_m}(c_1, c_2)$ with $c_1, c_2 \in \mathscr{S}_m$ represents the frequency with which systems selected by c_2 are found among the systems selected by c_1.

Corresponding to the notion that b_0 characterizes a measuring method (see [2] II § 4.3 and [3] § 12.1) we assume

APS 7 For a_1, $a_2 \in \mathscr{Q}_m$ with $a_2 \subset a_1$, $a_2 \neq \emptyset$ and b_{10}, $b_{20} \in \mathscr{R}_{0m}$ with $b_{20} \subset b_{10}$, $b_{20} \neq \emptyset$ follows

APS 7.1 $\lambda_{\mathscr{S}_m}(a_1 \cap b_{10}, a_2 \cap b_{10}) = \lambda_{\mathscr{Q}_m}(a_1, a_2)$;

APS 7.2 $\lambda_{\mathscr{S}_m}(a_1 \cap b_{10}, a_1 \cap b_{20}) = \lambda_{\mathscr{R}_{0m}}(b_{10}, b_{20})$.

The function $\lambda_{\mathscr{S}_m}$ completely describes all experimental possibilities for the taking of statistics. But one need not at all know the complete function $\lambda_{\mathscr{S}_m}(c_1, c_2)$ for all $c_2, c_1 \in \mathscr{S}_m$ with $c_2 \subset c_1$, because in [3] II § 4.5 we have proved the important theorem:

The function $\lambda_{\mathscr{S}_m}$ is uniquely determined by the function $\lambda_{\mathscr{Q}_m}$ and the special values

$$\lambda_{\mathscr{S}_m}(a \cap b_0, a \cap b) \tag{2.4.1}$$

for $a \in \mathscr{Q}'_m$, $b_0 \in \mathscr{R}'_{0m}$ and $b \in \mathscr{R}_m$ with $b \subset b_0$ (hence $b \in \mathscr{R}_m(b_0)$).

The function values (2.4.1) are those experimentally measured almost exclusively: One produces a system by a preparation procedure a and measures it under the method b_0 (i.e. systems $x_1, x_2, \ldots, x_N$ from $a \cap b_0$), noting the frequencies of the various digital indications b (i.e. noting the number N_+ of those $x_{i_1}, x_{i_2}, \ldots, x_{i_N}$ for which $x_{i_k} \in b$ holds, and thus getting the frequency N_+/N) and comparing these frequencies with the real numbers $\lambda_{\mathscr{F}_m}(a \cap b_0, a \cap b)$. The experimental physicists will try to find the finest possible preparation procedures, since their experiments tell the more the finer preparation procedures have been chosen. This is the reason why in most experiments the statistics $\lambda_{\mathscr{Q}_m}$ are not talked about except perhaps in "error estimates" of the preparation. But it is theoretically cleaner to introduce $\mathscr{Q}_m$ as a structure of species SSP (statistical selection procedures).

Since only the function values (2.4.1) are physically important, it is natural to introduce the sets

$$\mathscr{F}_m \overset{\text{def}}{=} \{(b_0, b) \mid b_0 \in \mathscr{R}'_{0m}, b \in \mathscr{R}_m(b_0)\}; \tag{2.4.2}$$

$$\mathscr{C}_m \overset{\text{def}}{=} \{(a, f) \mid a \in Q'_m, f \in \mathscr{F}_m\} = \mathscr{Q}'_m \times \mathscr{F}_m. \tag{2.4.3}$$

$\mathscr{F}_m$ is called the set of "effect processes" or "questions".

Then a function μ_m on $\mathscr{C}_m$ is defined by

$$\mu_m(a, f) = \mu_m(a, (b_0, b)) = \lambda_{\mathscr{F}_m}(a \cap b_0, a \cap b), \tag{2.4.4}$$

so that [2] II § 4.5 yields the theorem:

T 2.4.1 The following assertions hold for the function μ_m:

(i) $0 \leqq \mu_m(a, f) \leqq 1$.

(ii) For each $a \in \mathscr{Q}'_m$ there is an $f_0 \in \mathscr{F}_m$ with $\mu_m(a, f_0) = 0$.

(iii) For each $a \in \mathscr{Q}'_m$ there is an $f_1 \in \mathscr{F}_m$ with $\mu_m(a, f_1) = 1$.

(iv) For a decomposition $a = \bigcup_{i=1}^{n} a_i$ (i.e. $a_i \neq \emptyset$ and $a_i \cap a_j = \emptyset$ for $i \neq j$) one gets $\mu_m\left(\bigcup_{i=1}^{n} a_i, f\right) = \sum_i \lambda_i \mu_m(a_i, f)$ for all $f \in \mathscr{F}_m$, where $0 < \lambda_i = \lambda_{\mathscr{Q}_m}(a, a_i) \leqq 1$ and $\sum_{i=1}^{n} \lambda_i = 1$.

(v) For $b_{01}, b_{02} \in \mathscr{R}'_{0m}$, $b_{01} \supset b_{02} \supset b$ and for $f_1 = (b_{01}, b)$, $f_2 = (b_{02}, b)$ follows $\mu_m(a, f_1) = \lambda_{\mathscr{R}_{0m}}(b_{01}, b_{02})\, \mu_m(a, f_2)$ for all $a \in \mathscr{Q}'_m$.

(vi) For $f_i = (b_0, b_i)$ with $b_0 = \bigcup_{i=1}^{n} b_i$ ($b_i \cap b_j = \emptyset$ for $i \neq j$) and all $a \in \mathscr{Q}'_m$ follows $\sum_{i=1}^{n} \mu_m(a, f_i) = 1$.

(vii) For $f = (b_0, b)$ we find $\mu_m(a, f) = 0$ for an $a \in \mathscr{Q}'_m \Leftrightarrow a \cap b = \emptyset$.

(viii) If $b_0 = \bigcup_{i=1}^{n} b_{0i}$ with $b_{0i} \in \mathscr{R}'_{0m}$ ($b_{0i} \cap b_{0j} = \emptyset$ for $i \neq j$) then, for $f = (b_0, b)$ and $f_i = (b_{0i}, b_{0i} \cap b)$ follows $\mu_m(a, f) = \sum_{i=1}^{n} \lambda_{\mathscr{R}_{0m}}(b_0, b_{0i})\, \mu_m(a, f_i)$ and $\sum_{i=1}^{n} \lambda_{\mathscr{R}_{0m}}(b_0, b_{0i}) = 1$.

A certain converse of this theorem also holds (see [2] II § 4.5):

T 2.4.2 If, for given $\lambda_{\mathscr{Q}_m}$, a function μ_m defined on $\mathscr{C}_m$ fulfills (i), (iv), (v), (vi) and (vii), for $\mathscr{S}_m$ there is one, and only one, probability function $\lambda_{\mathscr{S}_m}$ for which

$$\lambda_{\mathscr{S}_m}(a \cap b_0, a \cap b) = \mu_m(a, f)$$

holds with $f = (b_0, b)$. The $\lambda_{\mathscr{S}_m}$ so defined fulfills the conditions APS 7 and the further, just listed relations (ii), (iii), (viii), with $\lambda_{\mathscr{R}_{0m}}$ determined by μ_m because

$$\lambda_{\mathscr{R}_{0m}}(b_{01}, b_{02}) = \mu_m(a(b_{01}, b_{02}))\,.$$

This theorem justifies that it suffices experimentally to test the function μ_m over $\mathscr{C}_m$. Thus the conditions (i), (vi) and (vii) are almost trivially satisfied in an experiment (due to the meaning of μ_m as the picture of a frequency). Condition (iv) expresses the fact that the preparation is not influenced by the registration (one must experiment this way, otherwise one has not carried out a "suitable" experiment!). Condition (v) expresses the fact that a refinement of the measuring method is statistically independent of the preparation (also this *must* hold in experiments).

The description just given shows that some of the axioms are less "prescribed by nature" than containing "orders for correct experimentation". Thus, for example, APS 7 contains an experimentation order. If, say, one has violated APS 7, then one must not use the results to test the theory. When APS 7 is violated, we no longer even talk about "preparations" and "registrations". Therefore, APS 7 is a "natural law" in the sense of [3] and [30] only insofar as it is possible to satisfy its requirements, i.e. insofar as there really "exists" a fundamental domain $\mathscr{G}$ of suitable experiments.

§ 3 Trajectory Preparation and Registration Procedures

The structures of the statistical way of description introduced in § 2 are not yet connected with the objectivating way of description from § 1. In fact the structures from § 2 do not yet distinguish an objectivating from a non-objectivating description. From this we recognize that a structure of the same type $\mathscr{Q}$, $\mathscr{R}_0$, $\mathscr{R}$ as in § 2 can also be used for the description of microsystems (see III § 4). But the microsystems cannot be described in an objectivating manner (see VII §§ 5.3 and 6, and [2] IV § 8).

For the pretheories $\mathscr{P}\mathscr{T}_m$ we must still express mathematically that a registration is a measurement of trajectories (in § 2 pointed out verbally without a mathematical formulation).

§ 3.1 Trajectory Effects

In the "usual" probability theory, which proceeds from a measure space $(\Omega, \mathscr{A}, P)$ (see [3] § 11.1), it would be natural to set $\Omega = Y$, i.e. to regard the trajectories $y \in Y$ as the elementary events and to select a σ-algebra $\mathscr{A}$ of subsets of Y. One would call an element $\tilde{Y} \in \mathscr{A}$ an event, namely the event that the trajectory y lies in $\tilde{Y}$ (that $y \in \tilde{Y}$ holds). Then $P(\tilde{Y})$ would be the probability that the event $\tilde{Y}$

occurs. But in § 2 we consciously have not started from the "customary" form of probability theory (see the remarks in [3] § 11). Thus we shall only later obtain a derived structure of the form $(\Omega, \mathscr{A}, P)$. In this derivation we shall recognize that the elements of $\mathscr{A}$ only correspond to "idealized" registrations, i.e. to only approximately attainable Yes-No registrations.

The foundation of that probability theory on which § 2 has been based is physically more realistic and more appropriate for constructing an axiomatic basis of quantum mechanics. But in order to express that the $\tilde{b} \in \tilde{\mathscr{R}}_m$ should represent the registration of trajectories, we must establish a mathematical relation between $\tilde{\mathscr{R}}_m$ and Y. The physical background for such a relation is the *functioning of the devices by means of which one can measure the trajectories.*

It is physically impossible to measure a trajectory y arbitrarily precisely. Just this ought to be expressed by the uniform structure p of Y. But if it is impossible to measure trajectories precisely, then one also cannot construct any device with digital indications $\varrho \in \tilde{\mathscr{R}}_m$ whose response or non-response decides whether a trajectory y belongs to some subset $\tilde{Y} \subset Y$. The indications can therefore "decide" only with a certain imprecision (described by the uniform structure p) whether $y \in \tilde{Y}$ or $y \notin \tilde{Y}$ holds. How should this imprecision of realistic registrations $\varrho \in \tilde{\mathscr{R}}_m$ be described mathematically?

If one has built a real device $\varrho_0 \in \tilde{\mathscr{R}}_{0m}$ and aks for its response, one recognizes (what every experimenter has long known) that there are trajectories for which a digital indication $\varrho < \varrho_0$ occurs with certainty and other trajectories y for which ϱ certainly does *not* occur, but that there are y in a transition region (describing the "measurement imprecision") such that ϱ sometimes does and sometimes does not occur. Therefore, for the device ϱ_0 there exist probabilities to respond to the trajectories y. The experimenters use these trajectory probabilities to calculate their measurement errors.

It is typical of "classical" measuring that it is *possible* (due to the construction of the measuring device, i.e. by means of pretheories for the $\mathscr{PT}_m$) to describe *what* the device measures; hence briefly: It is typical of classical theories that the observables are directly measurable (for direct and indirect measurability, see [3] §§ 5 and 10 or [1] III §§ 4 and 9).

Contrary to widespread opinion it is also *not* decisive in classical physics that all measurements show a "negligible" interaction with the measured systems. Just the fact that for any classical theory $\mathscr{PT}_m$ there are pretheories to explain how something should be measured, is the reason why in the classical theories it is not "customary" to bring into $\mathscr{PT}_m$ something from the structure of measuring.

The standpoint of many theoreticians that "measurement is a thing for the experimental physicist" can therefore be retained for the classical theories $\mathscr{PT}_m$. But is has turned out that just this standpoint is untenable for quantum mechanics, since the pretheories for quantum mechanics *no longer* allow to describe *what* on the microsystems a device measures (see I § 2). The pretheories for quantum mechanics only permit us, objectivatingly to "describe" the processes within the devices, a fact we shall exploit in III.

For our goal of constructing an axiomatic basis for quantum mechanics it will turn out very advantageous if general structures of measuring are already included in the picture $\mathscr{MT}_m$ of $\mathscr{PT}_m$. In § 2 we have already begun to do so and now we must continue by introducing a probability for the response ϱ of a device ϱ_0 to the various trajectories $y \in Y$, a probability which "customarily" occurs only in the pretheories used by experimental physicists.

We shall later see that by "forgetting" structures one can recover the "customary" form of the theories $\mathscr{P}\mathscr{T}_m$ from the picture $\mathscr{M}\mathscr{T}_m$ we use. In this sense the form of $\mathscr{P}\mathscr{T}_m$ presented here (though more mathematicized!) is more closely connected with experimental physics than those forms in which the measuring methods are not considered.

Consequently, as a structure of measuring methods $\varrho_0 \in \tilde{\mathscr{R}}_{0m}$ we now include in $\mathscr{P}\mathscr{T}_m$ that it is *independently* (!) of the measured systems $x \in M_m$ possible to interpret the occurrences of the individual digital indications $\varrho < \varrho_0$ as measurements of the trajectories (by means of pretheories!). This shall hold even then when there are not systems $x \in M_m$ whatever to let this or that indication occur.

The fact that the measuring devices can be completely described by pretheories shall be expressed by saying that for $\tilde{\mathscr{R}}_m$ there exists a probability structure which yields the response probability of an indication for the various trajectories from Y. Thus we assume a function $Y \times \mathscr{T}_{\tilde{\mathscr{R}}_m} \to [0, 1]$ with

$$\mathscr{T}_{\tilde{\mathscr{R}}_m} = \{(\varrho_1, \varrho_2) \mid \varrho_1, \varrho_2 \in \tilde{\mathscr{R}}_m, \varrho_1 > \varrho_2 \text{ and } \varrho_1 \neq 0\}$$

is defined to furnish the probability that ϱ_2 occurs if the trajectory $y \in Y$ is realized and ϱ_1 has occurred. Let these probability functions (defined by the construction of measuring devices due to the pretheories for $\mathscr{P}\mathscr{T}_m$) be denoted by $\lambda_{\text{Meas}}(y; \varrho_1, \varrho_2)$. For fixed y, this probability shall of course with respect to $\tilde{\mathscr{R}}_m$ satisfy relations similar to the structure relations AS 2 from § 2.1. Corresponding to the meaning of $\tilde{\mathscr{R}}_{0m}$ and $\tilde{\mathscr{R}}_m$, and in analogy to APS 7.2, let us additionally require: For $\varrho_{10}, \varrho_{20} \in \tilde{\mathscr{R}}_{0m}$ with $\varrho_{20} \leqq \varrho_{10}, \varrho_{10} \neq 0$,

$$\lambda_{\text{Meas}}(y; \varrho_{10}, \varrho_{20}) = \lambda_{\mathscr{R}_{0m}}(\varrho_{10}, \varrho_{20}) \tag{3.1.1}$$

holds with $\lambda_{\mathscr{R}_{0m}}$ transferred to $\tilde{\mathscr{R}}_{0m}$ (see § 2.3).

Then it easily follows that λ_{Meas} is completely determined by $\lambda_{\mathscr{R}_{0m}}$ and the special values $\lambda_{\text{Meas}}(y; \varrho_0, \varrho)$ with $\varrho_0 \neq 0$ and $\varrho_0 \in \tilde{\mathscr{R}}_{0m}$, $\varrho \in \tilde{\mathscr{R}}_m(\varrho_0)$, where $\tilde{\mathscr{R}}_m(\varrho_0)$ is the order interval $[0, \varrho_0]$ in $\tilde{\mathscr{R}}_m$.

With the set

$$\Phi = \{(\varrho_0, \varrho) \mid \varrho_0 \in \tilde{\mathscr{R}}_{0m}, \varrho_0 \neq 0, \varrho \in \tilde{\mathscr{R}}_m(\varrho_0)\} \tag{3.1.2}$$

formed similarly to $\mathscr{F}_m$ in (2.4.2), $\lambda_{\text{Meas}}(y; \varrho_0, \varrho)$ represents a function $Y \times \Phi \to [0, 1]$. This function uniquely determines a function $\Phi \xrightarrow{\psi_m} Y^{[0,1]}$, where $Y^{[0,1]}$ is the set of functions $Y \to [0, 1]$. Here the function ψ_m is given by $\psi_m(\varrho_0, \varrho) = k(y)$ with $k(y) = \lambda_{\text{Meas}}(y; \varrho_0, \varrho)$.

The properties of λ_{Meas} imply: If $\varrho_{01}, \varrho_{02} \in \tilde{\mathscr{R}}_{0m}$, $\varrho_{01} > \varrho_{02} \neq 0$, $\varrho_{02} > \varrho$ then

$$\psi_m(\varrho_{01}, \varrho) = \lambda_{\mathscr{R}_{0m}}(\varrho_{01}, \varrho_{02})\, \psi_m(\varrho_{02}, \varrho). \tag{3.1.3}$$

From $\varrho_0 = \bigvee_{i=1}^{n} \varrho_i$ ($\varrho_i \wedge \varrho_j = 0$ for $i \neq j$) follows

$$\sum_{i=1}^{n} \psi_m(\varrho_0, \varrho_i) = \mathbf{1}, \tag{3.1.4}$$

where **1** is that function in $Y^{[0,1]}$ that identically equals 1. For **0** as the function in $Y^{[0,1]}$ that identically equals 0, we have

$$\psi_m(\varrho_0, \varrho) = \mathbf{0} \Leftrightarrow \varrho = 0 . \tag{3.1.5}$$

Conversely, if a function ψ_m obeys (3.1.3) through (3.1.5), then ψ_m uniquely determines a function λ_{Meas} which satisfies the above requirements AS 2 and (3.1.1).

With $\varrho_0 = \varrho_1 \vee \varrho_2$ and $\varrho_1 = \varrho_0, \varrho_2 = 0$ from (3.1.5) and (3.1.3) follows $\psi_m(\varrho_0, \varrho_0) = \mathbf{1}$. With $\varrho_0 = \varrho_1 \vee \varrho_2 \vee \varrho_3$ and $\varrho_0 = \tilde{\varrho} \vee \varrho_3$ (where $\tilde{\varrho} = \varrho_1 \vee \varrho_2$) follows

$$\begin{aligned} \mathbf{1} &= \psi_m(\varrho_0, \varrho_1) + \psi_m(\varrho_0, \varrho_2) + \psi_m(\varrho_0, \varrho_3) \\ &= \psi_m(\varrho_0, \tilde{\varrho}) + \psi_m(\varrho_0, \varrho_3) . \end{aligned}$$

Therefore, for $\varrho_1 \wedge \varrho_2 = 0$ and $\varrho_1, \varrho_2 \in \tilde{\mathscr{R}}_m(\varrho_0)$ we find

$$\psi_m(\varrho_0, \varrho_1 \vee \varrho_2) = \psi_m(\varrho_0, \varrho_1) + \psi_m(\varrho_0, \varrho_2) . \tag{3.1.6}$$

While this can directly be derived from the properties of λ_{Meas}, it states that ψ_m is an additive measure on the Boolean ring $\tilde{\mathscr{R}}_m(\varrho_0)$.

Let us point out the normative demand that a measuring method shall be said to measure *trajectories* only then, when it affects the statistics of these trajectories only in such a way that the changes are not registered by this method itself. If a method violates this demand, it can often be "altered" by erasing those indications $\varrho < \varrho_0$, which are affected by the changed course of the trajectories. In this way one obtains a new method, which measures less but obeys the normative demand. Thus a sand bag can serve to measure the trajectories of bullets, if one *only* registers the entrance hole in the sandbag and not the further course of each bullet in the sand.

Only in § 3.2 shall we show the mathematical formulation of this normative demand on measuring methods, here merely depicted intuitively. Yet the physical meaning of the following considerations becomes more evident, when we already think of this normative demand on the elements of $\tilde{\mathscr{R}}_m$.

The range of values $\psi_m(\Phi) \subset Y^{[0,1]}$ must (due to the fact that the $(\varrho_0, \varrho) \in \Phi$ just describe the measurement of trajectories) be intimately connected with the uniform structure p of the physical imprecision, since this must be identical with the imprecision in the measurement.

But this imprecision due to the registration is, just by the meaning of ψ_m, the initial uniform structure generated by the mappings

$$Y \xrightarrow{k} [0, 1] \quad \text{for all} \quad k \in \psi_m(\Phi) . \tag{3.1.7}$$

Since we always assume (under the presumptions of [3] § 9) that $\tilde{\mathscr{R}}_m$ is a countable set, Φ is also countable and hence that initial structure is metrizable. Since the mappings k occur in [0, 1], also Y (with this initial structure) is precompact. But since p shall describe the measurement imprecisions, we thus must assume that just such measuring procedures $\varrho_0 \in \mathscr{R}_{0m}$ can be constructed that the initial structure determined by (3.1.7) coincides with p. We want to find a necessary and sufficient condition on the set $\psi_m(\Phi)$ to make the initial uniform structure determined by (3.1.7) identical with p.

If follows immediately that the $k \in \psi_m(\Phi)$ must be uniformly continuous functions on Y_p. Since each uniformly continuous function can be extended to the completion $\hat{Y}$ which is compact, the set of uniformly continuous functions on Y can be identified with the set $C(\hat{Y})$ of the continuous functions $\hat{Y} \to \mathbf{R}$. This $C(\hat{Y})$ is an order unit space (see [7] V § 8 or [2] A III § 6). Since $\hat{Y}$ is separable, the Banach space is so ([5] X § 3, no. 3).

When we denote by $L(\hat{Y})$ the order interval $[\mathbf{0}, \mathbf{1}]$ in $C(\hat{Y})$, we find $L(\hat{Y}) = \{k \mid k \in C(\hat{Y}) \text{ and } \mathbf{0} \leqq k \leqq \mathbf{1}\}$. We therefore have $\psi_m(\Phi) \subset L(\hat{Y})$, hence

$$\Phi \xrightarrow{\psi_m} L(\hat{Y})\,. \tag{3.1.8}$$

If this holds, the initial uniform structure determined by (3.1.7) can only be weaker than p. For it to be equal to p, the subset $\psi_m(\Phi)$ of $L(\hat{Y})$ must be "sufficiently large".

Let F_m be the smallest subspace of $C(\hat{Y})$ which contains $\psi_m(\Phi)$ and contains $|k(y)|$ whenever it contains $k(y)$. Then F_m is a vector lattice. Since even the initial structure determined by (3.1.7) must separate the elements of Y, this is a fortiori the case if all the elements of F_m are used for k. By the *Stone-Weierstrass* theorem ([7] V § 8.1), F_m separates the points of Y only if F_m is dense in $C(\hat{Y})$. Thus $F_m \cap L(\hat{Y})$ must be dense in $L(\hat{Y})$ because this is the closure of its open kernel. We briefly write $F_m \cap L(\hat{Y}) = \text{la}\ \psi_m(\Phi)$.

In the following we always presume that la $\psi_m(\Phi)$ is dense in $L(\hat{Y})$. This condition can often be sharpened. For instance, when the statistics of the trajectories of considered systems is rather insensitive to measurements, one can arrive at the later demand X (2.3.12). Stronger than "la $\psi_m(\Phi)$ dense in $L(\hat{Y})$" would be the demand that co $\psi_m(\Phi)$ be dense in $L(\hat{Y})$.

When a presumption analogous to AR 1 from [2] III § 2 is imposed on $\mathscr{R}_{0m}$ and thus on $\tilde{\mathscr{R}}_{0m}$, then "co $\psi_m(\Phi)$ dense in $L(\hat{Y})$" implies that even $\psi_m(\Phi)$ is dense there. Let us emphasize, however, that a sharpening of "la $\psi_m(\Phi)$ dense in $L(\hat{Y})$" is not a trivial demand.

The structure connections among $\tilde{\mathscr{R}}_{0m}$, $\tilde{\mathscr{R}}_m$, $\mathscr{R}_{0m}$, $\mathscr{R}_m$ and Y_p (established in this § 3.1) are meant when we briefly say that the elements of $\mathscr{R}_m$ represent "trajectory registration procedures". The elements of $L(\hat{Y})$ are briefly denoted as "trajectory effects".

It must be emphasized that in the pretheories $\mathscr{P}\mathscr{T}_m$ for quantum mechanics we consider no other registration procedures than the trajectory registrations, but not because one could not think up other procedures on macrosystems. Rather it would go beyond the "pretheories" for quantum mechanics, to consider theories more comprehensive than $\mathscr{P}\mathscr{T}_m$ (about "more comprehensive" see [3] § 8), which would describe still other procedures than trajectory registrations. For this problem see also II § 6.5, § 6.6 and X, XI.

§ 3.2 Trajectory Ensembles

We must now consider the connection between the function μ_m, described as centrally important in § 2.4, and the function ψ_m, introduced in § 3.1. Also the normative demands described behind (3.1.6) enter this connection, which will conclusively be formulated in (3.2.2). Hence this formulation will include the normative demand to consider (as elements of $\tilde{\mathscr{R}}_{0m}$) only measuring methods obeying the following restriction. They shall measure only those *parts* of the trajectories the statistics of which is not affected by the measuring methods themselves!

The following considerations shall make it plausible, how to arrive at the demand (3.2.2). We shall show two different ways (α) and (β) to lead us to (3.2.2).

(α): By $M_m \xrightarrow{g} Y$ (the assignment of trajectories from § 1) a mapping $a \xrightarrow{g} Y$ is given for every $a \in \mathscr{Q}'_m$. The set $g(a)$ consists of the trajectories of the systems prepared by a. Since the normative demands on the measuring of trajectories (as discussed in § 3.1) do not permit the occurrence of an indication b to depend on anything but the statistics of the trajectories $g(x)$ determined by the preparation procedure a, the probability measure $\mu_m(a, (b_0, b))$ can only depend on $\psi_m(\varrho_0, \varrho)$.

With h as the canonical extension of $\tilde{\mathscr{R}}_m \xrightarrow{h} \mathscr{R}_m$ to $\Phi \xrightarrow{h} \mathscr{F}_m$ (i.e. $h(\varrho_0, \varrho) = (b_0, b)$) one can represent this requirement most simply by the diagram

$$\begin{array}{ccc} \Phi & \xrightarrow{\psi_m} & \psi_m(\Phi) \subset L(\hat{Y}) \\ h\downarrow & & \downarrow u_a \\ \mathscr{F}_m & \xrightarrow{\mu_m(a, \ldots)} & [0, 1]\,. \end{array} \tag{3.2.1}$$

It should hold for an appropriate function u_a (for each $a \in \mathscr{Q}'_m$). From $1 = \mu_m(a, (b_0, b_0))$ and $\psi_m(\varrho_0, \varrho_0) = \mathbf{1}$ follows $u_a(\mathbf{1}) = 1$ and likewise $u_a(\mathbf{0}) = 0$.

Since the norm $\| k_1 - k_2 \| = \sup_y |k_1(y) - k_2(y)|$ for $k_1, k_2 \in \psi_m(\Phi)$ represents the maximal deviation of the response probability we should have $|u_a(k_1) - u_a(k_2)| \leq \| k_1 - k_2 \|$; thus u_a should represent a norm continuous mapping on $\psi_m(\Phi)$. When $\psi_m(\Phi)$ is assumed norm dense in $L(\hat{Y})$, we can extend u_a to all of $L(\hat{Y})$.

If for $\mathscr{R}_{0m}$ and hence $\tilde{\mathscr{R}}_{0m}$, one makes an assumption analogous to AR 1 from [2] III § 2, it follows from $\mu_m(a, (h(\varrho_0), h(\varrho))) = u_a \psi_m(\varrho_0, \varrho)$ in a manner analogous to the proof of Th 2.4 in [2] II § 2, that u_a is a rational affine mapping on the rational convex set $\psi_m(\Phi)$. Therefore, u_a can be extended to all of $L(\hat{Y})$ as an affine mapping and thus to all of $C(\hat{Y})$ as a linear norm continuous mapping. Consequently, one can identify u_a with an element of the Banach space $C'(\hat{Y})$ dual to $C(\hat{Y})$.

From $L(\hat{Y}) \xrightarrow{u_a} [0, 1]$ follows $u_a \in K(\hat{Y})$ with $K(\hat{Y})$ the basis of the base normed Banach space $C'(\hat{Y})$ – that is, $K(\hat{Y}) = \{u \mid u \in C'_+(\hat{Y}) \text{ and } \langle u, \mathbf{1} \rangle = 1\}$ with $C'_+(\hat{Y})$ the positive cone of $C'(\hat{Y})$. Hence there is a mapping

$$\mathscr{Q}'_m \xrightarrow{\varphi_m} K(\hat{Y}) \quad \text{with} \quad \mu_m(a, (h(\varrho_0), h(\varrho))) = \langle \varphi_m(a), \psi_m(\varrho_0, \varrho) \rangle\,; \tag{3.2.2}$$

and therefore the elements of $\mathscr{Q}'_m$ are called "trajectory preparation procedures". The elements of $K(\hat{Y})$ are also called "ensembles of trajectories".

(β): If one does not want to presume that $\psi_m(\Phi)$ is dense in $L(\hat{Y})$, one could argue as follows in order to arrive at (3.2.2).

Again one starts from the opinion that a preparation procedure a determines the statistical distribution of the trajectories $y = g(x)$ with $x \in a$. To each $k \in C(\hat{Y})$ this statistical distribution should assign an expectation value, especially to each $k \in L(\hat{Y})$ a value from [0, 1]. To each $a \in \mathscr{Q}'_m$ thus corresponds a mapping

$L(\hat{Y}) \xrightarrow{u_a} [0, 1]$; and vice versa the statistical distribution of the trajectories is described by u_a. For $k_1, k_2 \in L(\hat{Y})$ and $k_1 + k_2 \in L(\hat{Y})$ the concept of an expectation value yields

$$u_a(k_1 + k_2) = u_a(k_1) + u_a(k_2) . \tag{3.2.3}$$

A theorem analogous to Th 4.2.1 from [2] V § 4.2 then implies that u_a can as a linear, norm continuous mapping be extended to all of $C(\hat{Y})$; hence u_a determines a mapping $\mathcal{Q}'_m \xrightarrow{\varphi_m} K(\hat{Y})$ with

$$u_a(k) = \langle \varphi_m(a), k \rangle . \tag{3.2.4}$$

The initially mentioned normative demand that the $\varrho_0 \in \tilde{\mathcal{R}}_{0m}$ only measure the statistics of the prepared trajectories $g(x)$ with $x \in a$, then says that $\mu_m(a, (h(b_0), h(b)))$ must equal the expectation value of $\psi_m(\varrho_0, \varrho)$; hence (3.2.4) leads to (3.2.2). The condition from § 3.1, that la $\psi_m(\Phi)$ be dense in $L(\hat{Y})$, guarantees that the $\psi_m(\varrho_0, \varrho)$ separate the $\varphi_m(a)$ with $a \in \mathcal{Q}'_m$.

The two mappings ψ_m, φ_m form the background because of which in the "customary" presentations of the theories $\mathcal{PT}_m$ one does not speak about the set M_m (stated more precisely, why one transfers all considerations about M_m, $\mathcal{Q}_m$, $\mathcal{R}_{0m}$, $\mathcal{R}_m$, $\tilde{\mathcal{R}}_{0m}$, $\tilde{\mathcal{R}}_m$ to pretheories of $\mathcal{PT}_m$). In the statistical descriptions from $\mathcal{PT}_m$, one then employs only $\hat{Y}$ and the probability measures $u \in K(\hat{Y})$ and even these mostly in a specialized form. From the form of $\mathcal{PT}_m$ that we presented, one can by means of the mappings h, ψ_m, φ_m proceed to a restriction (in the sense of [3] § 8) in which M_m, $\mathcal{Q}_m$, $\mathcal{R}_{0m}$, $\mathcal{R}_m$, $\tilde{\mathcal{R}}_{0m}$, $\tilde{\mathcal{R}}_m$ no longer occur but only $\hat{Y}$, $L(\hat{Y})$ and the norm closure of $\varphi_m(Q'_m)$. We shall study similar restrictions for quantum mechanics in VIII § 6.

As an abbreviation, we write $K_m(\hat{Y}) = \overline{\text{co}}\, \varphi_m(\mathcal{Q}'_m)$ for the norm closure of the convex set generated by $\varphi_m(\mathcal{Q}'_m)$. If one uses a postulate for $\mathcal{Q}_m$ that is analogous to AP 1 from [2] III § 2, then the norm closure of $\varphi_m(\mathcal{Q}'_m)$ equals $K_m(\hat{Y})$.

§ 3.3 The Dynamic Laws and the Objectivating Manner of Description

Also in this section, we shall not strive for the greatest generality, but rather describe the principal structures with regard to their physical meaning; the reader interested in the general structures of macrosystems is again referred to [4].

First let us perceive that the set $K_m(\hat{Y})$ encompasses the dynamical laws, or expressed more precisely: The dynamical laws of $\mathcal{MT}_m$ can be formulated as structure laws over $K_m(\hat{Y})$. But here we need not go into such structure laws in $\mathcal{MT}_m$, since it is just up to an axiomatic basis for quantum mechanics to formulate such structure laws for certain composite macrosystems (see VI). It is not the structure laws of dynamics, but rather the objectivating manner of description of the devices that in quantum mechanics will be taken over from the pretheories $\mathcal{PT}_m$, as to be described in III.

Let $\bar{K}^\sigma_m(\hat{Y})$ denote the closure of $K_m(\hat{Y})$ in the $\sigma(C', C)$-topology. Since $K(\hat{Y})$ is $\sigma(C', C)$-compact, $\bar{K}^\sigma_m(\hat{Y})$ obeys $\bar{K}^\sigma_m(\hat{Y}) \subset K(\hat{Y})$ and is likewise $\sigma(C', C)$-compact. The elements of $\bar{K}^\sigma_m(\hat{Y})$ comprise "idealizations" of trajectory ensembles. Since $\bar{K}^\sigma_m(\hat{Y})$ is compact, due to the Krein-Milman theorem $\bar{K}^\sigma_m(\hat{Y})$ is generated by the set $\partial_e \bar{K}^\sigma_m(\hat{Y})$ of extreme points, i.e. $\bar{K}^\sigma_m(\hat{Y}) = \overline{\text{co}}^\sigma\, \partial_e \bar{K}^\sigma_m(\hat{Y})$. If $\mathcal{E}$ is a $\sigma(C', C)$-dense subset of $\partial_e \bar{K}^\sigma_m(\hat{Y})$, of course also $\bar{K}^\sigma_m(\hat{Y}) = \overline{\text{co}}^\sigma\, \mathcal{E}$ holds. In this sense one can regard all ensembles of $\bar{K}_m^\sigma(\hat{Y})$ and thus also of $K_m(\hat{Y})$ as "mixtures" of ensembles from $\mathscr{E}$. But the ensembles from $\mathscr{E}$ can perhaps be idealized ensembles (see below).

D 3.3.1 An element $u \in \partial_e \bar{K}_m^\sigma(\hat{Y})$ is called an *elementary* trajectory ensemble.

The elements of $\partial_e \bar{K}_m^\sigma(\hat{Y})$ describe the trajectory ensembles which are in idealized form the "finest preparable". The elements of $\partial_e \bar{K}_m^\sigma(\hat{Y})$ therefore describe the dynamics, since they make the statistically "finest possible" assertions about the trajectories of the systems. Therefore, the structure of the set $\partial_e \bar{K}_m^\sigma(\hat{Y})$ determines the dynamics of the systems. In order to analyze this description, we introduce yet the following concepts.

D 3.3.2 We call the set

$$L_0(K_m; \hat{Y}) = \{k \mid k \in L(\hat{Y}) \text{ with } \langle u, k\rangle = 0 \text{ for all } u \in K_m(\hat{Y})\} \qquad (3.3.1)$$

briefly the set of "trajectory null effects".

Since $\langle u, k\rangle$ is $\sigma(C', C)$-continuous in u (for fixed k), thus also $\langle u, k\rangle = 0$ holds for $k \in L_0(K_m; \hat{Y})$ and all $u \in \bar{K}_m^\sigma(\hat{Y})$.

Since we assume due to [3] § 9 that the set $\mathscr{D}'_m$ is countable, $K_m(\hat{Y})$ is norm separable. Therefore, in $K_m(\hat{Y})$ there is a $\bar{u} = \sum_\nu \lambda_\nu u_\nu$ with $\lambda_\nu > 0$, $\sum_\nu \lambda_\nu = 1$ and $\{u_\nu\}$ regarded as a countable set norm dense in $K_m(\hat{Y})$. Hence $L_0(K_m; \hat{Y}) = L_0(\bar{u}; \hat{Y})$ holds for

$$L_0(u; \hat{Y}) = \{k \mid k \in L(\hat{Y}) \text{ with } \langle u, k\rangle = 0\}. \qquad (3.3.2)$$

Let $\mathscr{B}_m$ denote the Banach subspace of $C'(\hat{Y})$ spanned by $K_m(\hat{Y})$, so that $\mathscr{B}_m$ is norm-separable. Denote the set $\mathscr{B}_m \cap K(\hat{Y})$ by $\tilde{K}_m(\hat{Y})$, such that $K_m(\hat{Y}) \subset \tilde{K}_m(\hat{Y})$. The set of linear combinations of elements in $K_m(\hat{Y})$ which are positive and have norm 1 is norm-dense in $\tilde{K}_m(\hat{Y})$.

Of course, due to the above remarks, $L_0(\tilde{K}_m; \hat{Y}) = L_0(K_m; \hat{Y}) = L_0(\bar{u}; \hat{Y})$. Likewise, on $L(\hat{Y})$, the $\sigma(C(\hat{Y}), K_m(\hat{Y}))$-topology is identical with the $\sigma(C(\hat{Y}), \tilde{K}_m(\hat{Y}))$-topology which describes the physical distinguishability of the trajectory effects.

Thus one often uses $\tilde{K}_m(\hat{Y})$ instead of $K_m(\hat{Y})$ in discussing the dynamics. The elementary trajectory ensembles $u \in \partial_e \bar{\tilde{K}}_m^\sigma(\hat{Y})$ can be properly finer than the $u \in \partial_e \bar{K}_m^\sigma(\hat{Y})$. One then says that the dynamics can be described by the elements of $\partial_e \bar{\tilde{K}}_m^\sigma(\hat{Y})$, even when, because of $K_m(\hat{Y}) \subset \tilde{K}_m(\hat{Y})$, not all the elements of $\partial_e \bar{\tilde{K}}_m^\sigma(\hat{Y})$ can be prepared "arbitrarily well", but perhaps only with a finite precision described by $K_m(\hat{Y})$. But one can *imagine* that all elements of $\partial_e \bar{\tilde{K}}_m^\sigma(\hat{Y})$ describe the "actual" dynamics.

Corresponding to the considerations presented in [7] V § 8.1, one can introduce the support of an ensemble u and the support of $K_m(\hat{Y})$: For $k \in L(\hat{Y})$, $O_k = \{y \mid k(y) > 0\}$ is open. The closure of O_k in $\hat{Y}$ is called the support S_k of $k \in L(\hat{Y})$. To each $u \in K(\hat{Y})$ one can assign the set U_u regarded as the largest open subset of $\hat{Y}$ for which $\{k \in L(\hat{Y}),\ S_k \subset U_u \Rightarrow k \in L_0(u; \hat{Y})\}$ holds. The (closed) set S_u complementary to U_u is called the support of u. With U_{K_m} as the largest open set for which $\{k \in L(\hat{Y}), S_k \subset U_{K_m} \Rightarrow k \in L_0(K_m; \hat{Y})\}$ the set $\hat{S}_m$ complementary to U_{K_m} is called the support of $K_m(\hat{Y})$. Consequently, $\hat{S}_m = S_{\bar{u}}$ with the above $\bar{u}$. We have written $\hat{S}_m$ instead of S_m because we shall later introduce the subset $\hat{S}_m \cap Y$ and denote it by S_m.

Let us still note the theorem (see [7] V § 8.1) that $\{k \in L_0(u; \hat{Y}), u \in K(\hat{Y})\}$ is equivalent with $\{k(y) = 0$ for all $y \in S_u\}$, which we shall use below. Therefore, $k \in L_0(K_m; \hat{Y}) = L_0(\bar{K}_m^\sigma; \hat{Y})$ is equivalent with $\{k(y) = 0$ for all $y \in \hat{S}_m\}$. As a closed subset of $\hat{Y}$, $\hat{S}_m$ is compact. After the above considerations, it is natural to adopt the definitions

D 3.3.3 $\hat{S}_m$ is called the set of *admissible* trajectories; $\hat{Y} \backslash \hat{S}_m$ is called the set of *forbidden* trajectories.

It follows from $u \in \bar{K}_m^\sigma(\hat{Y})$ (or $u \in \tilde{\bar{K}}_m^\sigma(\hat{Y})$) that $L_0(u; \hat{Y}) \supset L_0(K_m; \hat{Y})$ and hence that $S_u \subset \hat{S}_m$. Since $\hat{S}_m$ is compact and $\hat{Y}$ is normal, each continuous function $k(y)$ on $\hat{S}_m$ with $0 \leqq k(y) \leqq 1$ can be extended to all of $\hat{Y}$ as a continuous function with the same bounds (see [5] IX § 4, No. 2). Hence, the mapping $k(y) \xrightarrow{s} k(y)|_{\hat{S}_m}$ is a surjective mapping $L(\hat{Y}) \xrightarrow{s} L(\hat{S}_m)$ and thus also a surjective mapping $C(\hat{Y}) \xrightarrow{s} C(\hat{S}_m)$. Since s is surjective, it follows immediately that the adjoint mapping $C'(\hat{S}_m) \xrightarrow{s'} C'(\hat{Y})$ is injective.

If $h(\varrho) = \emptyset$, we obtain

$$0 = \mu_m(a, (h(\varrho_0), h(\varrho))) = \langle u, \psi_m(\varrho_0, \varrho)\rangle$$

for all $u \in \varphi_m(\mathscr{D}'_m)$ and hence also for $u = \bar{u}$. As we shall see below, this implies $s\,\psi_m(\varrho_0, \varrho) = 0$. Thus follows that the diagram

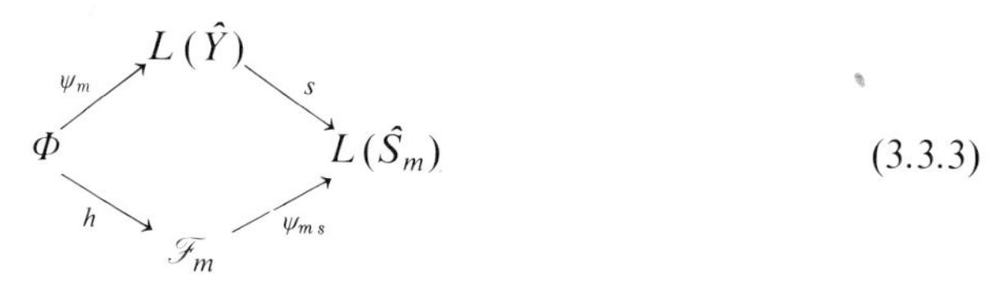

(3.3.3)

uniquely defines a mapping ψ_{ms} from $\mathscr{F}_m$ into $L(\hat{S}_m)$. If one restricts oneself to $\hat{S}_m$ instead of $\hat{Y}$, one can forget $\tilde{\mathscr{R}}_m$, $\tilde{\mathscr{R}}_{0m}$ and need consider only $\mathscr{R}_m$, $\mathscr{R}_{0m}$. Then to each effect procedure from $\mathscr{F}_m$ there uniquely corresponds an element from $L(\hat{S}_m)$!

Furthermore, it is easy to see that

$$s' K(\hat{S}_m) = \{u \mid u \in K(\hat{Y}), S_u \subset \hat{S}_m\}.$$

Being the image of a compact subset, this $s' K(S_m)$ is compact. The subspace $s' C'(\hat{S}_m)$ that is spanned in $C'(\hat{Y})$ by $s' K(\hat{S}_m)$, is closed in the $\sigma(C', C)$-topology and isomorphic with $C'(\hat{S}_m)$. One likewise recognizes the isomorphism in the norm topology.

Because of $K_m(\hat{Y}) \subset \bar{K}_m^\sigma(\hat{Y}) \subset s' K(\hat{S}_m)$ one can, by means of the inverse mapping $(s')^{-1}$, identify the sets $K_m(\hat{Y})$ and $\bar{K}_m^\sigma(\hat{Y})$ with subsets of $K(\hat{S}_m)$ and in this sense write $K_m(\hat{Y}) = K_m(\hat{S}_m)$ and $\bar{K}_m^\sigma(\hat{Y}) = \bar{K}_m^\sigma(\hat{S}_m)$, with $K_m(\hat{S}_m)$, $\bar{K}_m^\sigma(\hat{S}_m)$ as subsets of $K(\hat{S}_m)$. We can likewise regard the Banach space $\mathscr{B}_m$ spanned by $K_m(\hat{Y})$ as a subspace of $C'(\hat{S}_m)$ and identify $\tilde{K}_m(\hat{Y})$ with $\tilde{K}_m(\hat{S}_m) = \mathscr{B}_m \cap K(\hat{S}_m)$.

Therefore, it suffices to consider instead of $\hat{Y}$ its compact subset $\hat{S}_m$, in order to represent the trajectory effects by continuous functions over $\hat{S}_m$ and the trajectory ensembles by elements of $K(\hat{S}_m)$.

Since the support of $K_m(\hat{S}_m) \subset \bar{K}_m^\sigma(\hat{S}_m)$, i.e. of $\bar{u}$, equals the whole $\hat{S}_m$, for $k \in L(\hat{S}_m)$ and $k \neq \mathbf{0}$ follows $\langle \bar{u}, k \rangle \neq 0$.

The elements of $\partial_e K(\hat{S}_m)$ are just all point measures of the form $\langle u_{y_0}, k \rangle = k(y_0)$. Thus one can map $\partial_e K(\hat{S}_m)$ bijectively onto $\hat{S}_m$, whereby the topology of $\hat{S}_m$ coincides with the $\sigma(C', C)$-topology on $\partial_e K(\hat{S}_m)$. In this sense, one can identify $\hat{S}_m$ with $\partial_e K(\hat{S}_m)$ (see [7] V § 8.1).

According to the Riesz representation theorem (see [8] IV § 6.3), $K(\hat{S}_m)$ is the set of all σ-additive measures on the Borel field $\mathscr{B}(\hat{S}_m)$ (see [8] III § 5.10). Therefore, using the characteristic functions $\eta_\sigma(y)$ with $\sigma \in \mathscr{B}(\hat{S}_m)$, we find $\langle u, \eta_\sigma \rangle$ defined for all $\sigma \in \mathscr{B}(\hat{S}_m)$. We write briefly $\langle u, \eta_\sigma \rangle = u(\sigma)$. Using this, we can write $\langle u, k \rangle$ in the usual form of an integral, which is often called the trajectory integral or path integral:

$$\langle u, k \rangle = \int_{\hat{S}_m} k(y)\, du(y). \tag{3.3.4}$$

If, in particular, σ is an open subset of $\hat{S}_m$ (nevertheless, σ need not be open as a subset of $\hat{Y}$!), then $\sigma \in \mathscr{B}(\hat{S}_m)$. Since σ is open (and the compact set $\hat{S}_m$ completely regular!), there is a $k \in L(\hat{S}_m)$ with $k \neq \mathbf{0}$ and $k(y) = 0$ for all $y \notin \sigma$. Then $\langle \bar{u}, k \rangle \leqq \langle \bar{u}, \eta_\sigma \rangle \leqq \bar{u}(\sigma)$ and hence $\langle \bar{u}, k \rangle \neq 0$ implies $u(\sigma) \neq 0$.

If $u_e \in \partial_e \bar{K}_m^\sigma(\hat{S}_m)$ (or $u_e \in \partial_e \bar{\bar{K}}_m^\sigma(S_m)$) due to D 3.3.1 is an elementary trajectory complex, the support S_{u_e} is an element of $\mathscr{B}(\hat{S}_m)$. Precisely as with $\hat{S}_m$, one can with S_{u_e} form the Borel ring $\mathscr{B}(S_{u_e})$ on which u_e is a σ-additive measure, different from zero for all sets open *within* S_{u_e}. The measure u_e indicates the probabilities of the trajectories from S_{u_e}.

If, with the $\bar{u}$ defined above, $\bar{u}(\sigma) = 0$ holds for a $\sigma \in \mathscr{B}(\hat{S}_m)$, then also $u_\nu(\sigma) = 0$ holds for all u_ν from the sum $\bar{u} = \sum_\nu \lambda_\nu u_\nu$. In as much as all elements from $K_m(\hat{S}_m)$ can be approximated arbitrarily closely by elements u_ν in the norm of $C'(\hat{S}_m)$, also $u(\sigma) = 0$ holds for all $u \in K_m(\hat{S}_m)$, since the norm $\| \boldsymbol{u}_1 - \boldsymbol{u}_2 \|$ is also equal to

$$\sup_{\sigma \in \mathscr{B}(\hat{S}_m)} \| u_1(\sigma) - u_2(\sigma) \|$$

(see [8] IV § 6.3). Consequently, the elements of $K_m(\hat{S}_m)$ and $\tilde{K}_m(\hat{S}_m)$ are absolutely continuous measures with respect to $\bar{u}$.

The set $\mathscr{J}(\hat{S}_m)$ of all elements $\sigma \in \mathscr{B}(\hat{S}_m)$ with $\bar{u}(\sigma) = 0$ is a σ-ideal of $\mathscr{B}(\hat{S}_m)$. The factor ring $\Sigma_m = \mathscr{B}(\hat{S}_m)/\mathscr{J}(\hat{S}_m)$ is a complete Boolean ring and $\bar{u}$ is an effective, σ-additive measure on Σ_m. In this context, note that $\mathscr{J}(\hat{S}_m)$ (and thus Σ_m) depends on $\bar{u}$, i.e. on $K_m(\hat{Y}) = K_m(\hat{S}_m)$!

One uses to call Σ_m the *set of objective properties* of the systems from M_m. We shall see in § 5 how this is meant.

§ 3.4 Dynamically Continuous Systems

When in § 1 we "mentally" imagined that $M_m \xrightarrow{g} Y$ holds, the $a \in \mathscr{Q}'_m$ were just to produce systems x with trajectories $g(x)$ from Y and hence with $g(x)$ from

$S_m = \hat{S}_m \cap Y$. Then, for all $u \in \varphi(\mathscr{Q}'_m)$, the u-measure of $\hat{S}_m \backslash S_m$ should vanish. This idea leads us to the definition

D 3.4.1 The systems from M_m are called *dynamically continuous* if there exists a $\sigma \in \mathscr{B}(\hat{S}_m)$ of $\bar{u}$-measure zero (a $\sigma \in \mathscr{J}(\hat{S}_m)$) with $\hat{S}_m \backslash S_m \subset \sigma$ ($\bar{u}$ as in § 3.3).

That σ is of $\bar{u}$-measure zero is equivalent to σ being of u-measure zero for all $u \in \varphi_m(\mathscr{Q}'_m)$. Thus σ is also of u-measure zero for all $u \in K_m(\hat{Y}) = K_m(\hat{S}_m)$ and all $u \in \tilde{K}_m(\hat{S}_m)$.

T 3.4.1 Dynamical continuity implies that S_m is dense in $\hat{S}_m$.

Proof. Since there is a $\sigma \in \mathscr{J}(\hat{S}_m)$ with $\sigma \supset \hat{S}_m \backslash S_m$ no open subset of $\hat{S}_m$ can be contained in $\hat{S}_m \backslash S_m$ since for all open subsets σ of $\hat{S}_m$ the measure $\bar{u}(\sigma)$ is not zero (as shown in § 3.3). If S_m were not dense in $\hat{S}_m$, there would be a neighborhood about a point $y \in \hat{S}_m \backslash S_m$ and hence an open subset which contains no point of S_m. □

If σ is an open subset of S_m, then $\sigma = \tilde{\sigma} \cap S_m$ with $\tilde{\sigma}$ an open subset of $\hat{S}_m$ (see [5] I § 3, n. 1). Therefore $\tilde{\sigma} \to \tilde{\sigma} \cap S_m$ is a surjective mapping of the open subsets of $\hat{S}_m$ onto the open subsets of S_m. Hence $\tilde{\sigma} \to \tilde{\sigma} \cap S_m$ is also a surjective mapping $\mathscr{B}(\hat{S}_m) \to \mathscr{B}(S_m)$ for all $\tilde{\sigma} \in \mathscr{B}(\hat{S}_m)$. Then by $u(\tilde{\sigma} \cap S_m) = u(\tilde{\sigma})$ one can define the measures from $K_m(\hat{S}_m)$ as measures on $\mathscr{B}(S_m)$, for it follows from $\tilde{\sigma}_1 \cap S_m = \tilde{\sigma}_2 \cap S_m$ that the difference between $\tilde{\sigma}_1$ and $\tilde{\sigma}_2$, i.e. $(\tilde{\sigma}_1 \backslash \tilde{\sigma}_1 \cap \tilde{\sigma}_2) \cup (\tilde{\sigma}_2 \backslash \tilde{\sigma}_1 \cap \tilde{\sigma}_2)$, is a subset of $\hat{S}_m \backslash S_m \subset \sigma \in \mathscr{J}(\hat{S}_m)$.

(One can also, what in practice amounts to the same, form the Lebesgue extension $\mathscr{L}(\hat{S}_m)$ of $\mathscr{B}(\hat{S}_m)$. As is known, this is the σ-algebra generated by $\mathscr{B}(\hat{S}_m)$ and by all subsets of the $\sigma \in \mathscr{J}(\hat{S}_m)$. Then the measures $u \in K_m$ can be extended as measures on all of $\mathscr{L}(\hat{S}_m)$, so that S_m and $\hat{S}_m \backslash S_m$ are elements of $\mathscr{L}(\hat{S}_m)$. With $\mathscr{J}_l(S_m)$ as the set of the elements $\sigma \in \mathscr{L}(\hat{S}_m)$ with $\bar{u}(\sigma) = 0$, we then have $\hat{S}_m \backslash S_m \in \mathscr{J}_l(\hat{S}_m)$ so that $u(\tilde{\sigma} \cap S_m) = u(\tilde{\sigma})$ holds for all $\tilde{\sigma} \in \mathscr{L}(\hat{S}_m)$. The Boolean ring $\Sigma_m = \mathscr{B}(\hat{S}_m)/\mathscr{J}(\hat{S}_m)$ is isomorphic with $\mathscr{L}(\hat{S}_m)/\mathscr{J}_l(\hat{S}_m)$, as can easily be seen. Hence Σ_m can also be identified with $\mathscr{L}(S_m)/\mathscr{J}_l(\hat{S}_m)$. Dynamical continuity is equivalent to $\hat{S}_m \backslash S_m \in \mathscr{J}_l(\hat{S}_m)$. All elements of the form $\sigma \cap S_m$, where $\sigma \in \mathscr{L}(\hat{S}_m)$, form a subset of $\mathscr{L}(\hat{S}_m)$ identical with the Lebesgue extension $\mathscr{L}(S_m)$ of $\mathscr{B}(S_m)$. Therefore Σ_m can also be identified with $\mathscr{L}(S_m)/\mathscr{J}_l(S_m)$.)

The mapping $\tilde{\sigma} \to \tilde{\sigma} \cap S_m$ defines a bijective mapping of the elements of $\mathscr{B}(\hat{S}_m)/\mathscr{J}(\hat{S}_m)$ onto $\mathscr{B}(S_m)/\mathscr{J}(S_m)$, with $\mathscr{J}(S_m)$ as the subset of those $\sigma \in \mathscr{B}(S_m)$ for which $\bar{u}(\sigma) = 0$. Thus one can also identify $\Sigma_m = \mathscr{B}(\hat{S}_m)/\mathscr{J}(\hat{S}_m)$ with $\mathscr{B}(S_m)/\mathscr{J}(S_m)$.

The description attained here corresponds to that of "probability" by Kolmogorov's theory. To this end, choose S_m as the space of "elementary events", $\mathscr{B}(S_m)$ (or $\mathscr{L}(S_m)$) as the σ-algebra of the "measurable sets" (the "events"). As the probability measure choose a $u \in K_m(\hat{S}_m)$. Then $(S_m, \mathscr{B}(S_m), u)$ forms an example of the fundamental structure used by Kolmogorov for probability theory.

The mappings $Y_p \xrightarrow{g_\tau} Z_p$ introduced at the end of § 1 are continuous for $\tau > 0$. If ϱ is an open set in Z, therefore Y has the open subset $g_\tau^{-1}(\varrho)$ which we will denote by $\sigma(\varrho;\tau)$. But thus $\sigma(\varrho;\tau)$ is an element of $\mathscr{B}(Y)$ for all $\varrho \in \mathscr{B}(Z)$. Then the elements $S_m \cap \sigma(\varrho;\tau)$ are elements of $\mathscr{B}(S_m)$. We shall write $S_m \cap \sigma(\varrho;\tau)$ simply as $\sigma_m(\varrho;\tau)$. Then, for each $u \in K_m(\hat{S}_m)$ and $\tau > 0$,

$$u_\tau(\varrho) = u(\sigma_m(\varrho, \tau))$$

defines a σ-additive measure u_τ over $\mathscr{B}(Z)$. The support of the set $\{u_\tau | u \in K_m(\hat{S}_m), \tau > 0\}$ of measures is mostly assumed to be Z; otherwise one could simply replace Z by that support.

In the applications, very often *only* the measure u_τ over $\mathscr{B}(Z)$, depending on time, will be considered. This restricted description will of course suffice for many cases in which the dynamical laws have an especially simple structure (also see X § 2.5).

Here in § 3, we will not undertake any typifying of "different" dynamic laws such as "dynamically determined laws" or "stochastic laws". For, we do not intend to develop a detailed theory of macrosystems. We will describe the general structures of macrosystems just to as far as we need them to begin in III the construction of an axiomatic basis for quantum mechanics, and in X to formulate clearly the compatibility problem. *Only then* we shall draw conclusions about the dynamics.

For these purposes, we still need to describe transformations of macrosystems.

§ 4 Transformations of Preparation and Registration Procedures

Transformations of macrosystems play a large role in many respects. Here, we shall not consider all the possibilities of transformations. The first to be investigated are the "time displacements" of the trajectory registration procedure. These transformations will of course play no role in the formation of an axiomatic basis for quantum mechanics, but will be important for the compatibility considerations in X § 2.3.

§ 4.1 Time Translations of the Trajectory Registration Procedures

We begin with a mathematical definition: For each fixed $\tau \geqq 0$ a mapping of $Y = C(\Theta, Z)$ into itself is defined by $T_\tau y = y'$, with $y = z(t)$ and $y' = z'(t) = z(t+\tau)$, because "$z(t)$ is continuous" implies "$z(t+\tau)$ is continuous". One sees immediately that the set of the T_τ with $\tau \geqq 0$ forms a semigroup, since $T_0 = 1$ and $T_{\tau_1} T_{\tau_2} = T_{\tau_1+\tau_2}$.

T 4.1.1 Each T_τ (fixed τ) is a p-uniformly and g-uniformly continuous mapping of Y onto itself, i.e. T_τ is surjective (but *not* injective for $\tau > 0$!).

Proof. As in (1.5), we prescribe a vicinity $\tilde{U}_{\vartheta, V}$, where we regard V as determined by $d((t, z), (t', z')) < \varepsilon$ with the $d(\ldots)$ introduced behind (1.2). We denote this vicinity by $\tilde{U}_{\vartheta,\varepsilon}$. If we now choose an $\varepsilon' > 0$ with $\varepsilon' < \varepsilon$ and $t - \varepsilon' > 0$ for all $t \in \vartheta$, then $\tilde{U}_{\vartheta+\tau,\varepsilon'}$ is a vicinity which by T_τ is mapped into $\tilde{U}_{\vartheta,\varepsilon}$. Therefore, T_τ is p-uniformly continuous.

T_τ is g-uniformly continuous if $C_g(\Theta_0, Z) \xrightarrow{T_\tau} C_g(\Theta_0, Z) \to C_{gu}(\vartheta, Z)$ is uniformly continuous for all ϑ, where $g\,u$ is the uniform structure on Z^ϑ introduced in § 1. But since the composite mapping

$$C_g(\Theta_0, Z) \xrightarrow{T_\tau} C_g(\Theta_0, Z) \to C_{gu}(\vartheta, Z)$$

is just identical with $C_g(\Theta_0, Z) \to C_{gu}(\vartheta + \tau, Z)$, it is uniformly continuous.

In order to show that T_τ is surjective onto Y, take $z(t) \in Y$ and put

$$\tilde{z}(t) = \begin{cases} z(0) & \text{for} \quad t \leqq \tau, \\ z(t-\tau) & \text{for} \quad t \geqq \tau. \end{cases}$$

Then $\tilde{z}(t) \in C(\Theta, Z) = Y$ and $T_\tau \tilde{z}(t) = z(t)$. □

Since T_τ is uniformly continuous on Y_p, it can be extended uniquely to all of $\hat{Y}_p = \hat{Y}$. Therefore, T_τ can also be defined for virtual trajectories. Since $\hat{Y}_p$ is compact. $T_\tau \hat{Y}_p$ is a compact subset of $\hat{Y}_p$. Because $T_\tau Y = Y$, we have $Y \subset T_\tau \hat{Y}_p$. Consequently, since Y is dense in $\hat{Y}_p$, we have $T_\tau \hat{Y}_p = \hat{Y}_p$. Therefore the mapping T_τ is also surjective onto $\hat{Y}$.

For $f \in C(Y)$, the mapping $\hat{Y}_p \xrightarrow{T_\tau} Y_p \xrightarrow{f} \mathbf{R}$ is uniformly continuous; hence a mapping of $C(Y)$ onto itself is defined by

$$V_\tau f(y) = f'(y) = f(T_\tau y). \tag{4.1.1}$$

Because of $T_\tau \hat{Y} = \hat{Y}$, the norm invariance of the mapping V_τ follows immediately.

For $\tau \geqq 0$ the mappings V_τ likewise form a semigroup as do the mappings T_τ.

Since the mapping $\hat{Y}_p \xrightarrow{T_\tau} \hat{Y}_p \xrightarrow{f} \mathbf{R}$ is uniformly continuous, this mapping is already determined uniquely by $Y \xrightarrow{T_\tau} Y \xrightarrow{f} \mathbf{R}$. Thus V_τ from (4.1.1) is also uniquely determined by

$$V_\tau f(z(t)) = f(z(t+\tau)) \tag{4.1.2}$$

for $z(t) \in C(\Theta, Z)$ and for all uniformly continuous mappings $Y_p \xrightarrow{f} \mathbf{R}$.

An $f \in C(\hat{Y})$ is uniquely fixed as a function over Y. We say briefly that f depends only on the course of the trajectory *after* τ, if $y_1 = z_1(t)$, $y_2 = z_2(t)$ and $z_1(t) = z_2(t)$ for $t \geqq \tau$ imply $f(y_1) = f(y_2)$. Correspondingly, we say that f depends only on the course of the trajectory *up to* τ, if $y_1 = z_1(t)$, $y_2 = z_2(t)$ and $z_1(t) = z_2(t)$ for $t \leqq \tau$ imply $f(y_1) = f(y_2)$.

Let the space of all $f \in C(\hat{Y})$ that depend on the course of the trajectory up to τ be denoted by $C(\hat{Y}; \leqq \tau)$. Correspondingly, let $C(\hat{Y}; \geqq \tau)$ denote the space of all f that depend on the trajectory after τ. One sees immediately that $C(\hat{Y}; \leqq \tau)$ and $C(\hat{Y}; \geqq \tau)$ are Banach subspaces (even Banach sublattices) of $C(\hat{Y})$. It easily follows from (4.1.2) that $V_\tau(C(\hat{Y}) \subset C(\hat{Y}; \geqq \tau)$; but we even find

T 4.1.2 V_τ is an isomorphic mapping of the Banach lattice $C(\hat{Y})$ onto $C(\hat{Y}; \geqq \tau)$.

Proof. It is not immediately obvious that $V_\tau C(\hat{Y}) \supset C(\hat{Y}; \geqq \tau)$. Therefore, we must show that for each $f \in C(\hat{Y}; \geqq \tau)$ there is an $f' \in C(\hat{Y})$ with $V_\tau f' = f$, i.e. with $f'(T_\tau y) = f(y)$.

We can factor the mapping T_τ canonically by means of the equivalence relation r: "$y_1 \sim y_2$ for $T_\tau y_1 = T_\tau y_2$" and of the canonical mapping $\hat{Y} \xrightarrow{\varphi} \hat{Y}/r$ as shown in the left part of the diagram

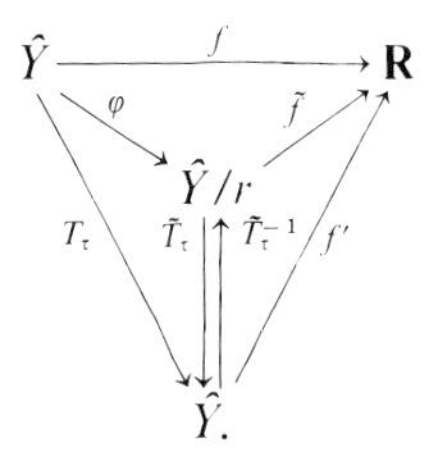

The mappings φ and $\tilde{T}_\tau$ are continuous in the quotient space topology of $\hat{Y}/r$ since this is just the final topology belonging to the mapping φ. Since $\hat{Y}$ is compact, $\hat{Y}/r$ is also compact. Since $\tilde{T}_\tau$ is bijective (T_τ was surjective), $\tilde{T}_\tau$ therefore is a homeomorphic mapping of $\hat{Y}/r$ onto $\hat{Y}$. Thus $\tilde{T}_\tau^{-1}$ is likewise continuous.

If $f \in C(\hat{Y}; \geqq \tau)$, then $f(y_1) = f(y_2)$ for $T_\tau y_1 = T_\tau y_2$, i.e. there is a mapping $\tilde{f}$ with $f(y) = \tilde{f}\varphi(y)$ (see the above diagram). Thus $f' = \tilde{f}\tilde{T}_\tau^{-1}$ is a continuous mapping of $\hat{Y}$ into $\mathbf{R}$, i.e. $f' \in C(\hat{Y})$. Hence the diagram immediately implies $f'(T_\tau y) = f(y)$. □

Using $L(\hat{Y})$ of § 3.1, one calls

$$L(\hat{Y}; \geqq \tau) = L(\hat{Y}) \cap C(\hat{Y}; \geqq \tau) \tag{4.1.3}$$

the set of "trajectory effects after τ" and correspondingly

$$L(\hat{Y}; \leqq \tau) = L(\hat{Y}) \cap C(\hat{Y}; \leqq \tau) \tag{4.1.4}$$

the set of "trajectory effects before τ".

With T 4.2.1 one easily sees

$$V_\tau L(\hat{Y}) = L(\hat{Y}; \geqq \tau). \tag{4.1.5}$$

After this preliminary mathematics let us introduce the easily interpretable time translations of the registration procedures.

For some $\tau \geqq 0$ we consider a mapping $\tilde{R}_\tau$ of $\tilde{\mathscr{R}}_m$ into itself, with $\tilde{\mathscr{R}}_m$ as in § 2.3. This $\tilde{R}_\tau$ maps the set $\tilde{\mathscr{R}}_{0m}$ order isomorphically onto a subset of $\tilde{\mathscr{R}}_{0m}$ and maps the Boolean ring $\tilde{\mathscr{R}}_m(\varrho_0) = \{\varrho \mid \varrho \leqq \varrho_0 \in \tilde{\mathscr{R}}_{0m}\}$ isomorphically onto $\tilde{\mathscr{R}}(\tilde{R}_\tau \varrho_0)$.

The physical interpretation of $\tilde{R}_\tau$ shall be that the registration method $\tilde{R}_\tau \varrho_0$ arises by applying the method ϱ_0 a time τ later (for fixed preparation). We must yet express this "physical meaning" of $\tilde{R}_\tau$ by a mathematical relation between $\tilde{R}_\tau$ and the measurement probability $\lambda_{\text{Meas}}(y; \varrho_1, \varrho_2)$ introduced in § 3.1.

$\tilde{R}_\tau$ is defined canonically by $\tilde{R}_\tau(\varrho_0, \varrho) = (\tilde{R}_\tau \varrho_0, \tilde{R}_\tau \varrho)$ as a mapping of Φ from (3.1.2) onto itself. If ψ_m is the mapping defined in § 3.1, then $f(y) = \psi_m(\varrho_0, \varrho) = \lambda_{\text{Meas}}(y; \varrho_0, \varrho)$. Therefore, if one *shall* represent $\tilde{R}_\tau(\varrho_0, \varrho)$ as the registration displaced just by τ, then $f'(y) = \psi_m(\tilde{R}_\tau(\varrho_0, \varrho))$ must obey

$$f'(y) = f(T_\tau y) \tag{4.1.6}$$

with the above T_τ. This result is due to (4.1.1) equivalent with $f'(y) = V_\tau f(y)$, hence with

$$\psi_m(\tilde{R}_\tau(\varrho_0, \varrho)) = V_\tau \psi_m(\varrho_0, \varrho). \tag{4.1.7}$$

This together with (3.1.8) yields the diagram

$$\begin{array}{ccc} \Phi & \xrightarrow{\psi_m} & L(\hat{Y}) \\ \tilde{R}_\tau \downarrow & & \downarrow V_\tau \\ \Phi & \xrightarrow{\psi_m} & L(\hat{Y}); \geqq \tau). \end{array} \tag{4.1.8}$$

Therefore, we *presume* the diagram (4.1.8) to hold in $\mathscr{PT}_m$. Since we assumed $\tilde{\mathscr{R}}_m$ as countable, we cannot assume that $\tilde{R}_\tau$ is defined for all $\tau \geqq 0$. But we can imagine a countable subset $\{\tau_i\}$ that is dense in $[0, \infty]$ and so chosen that the $\tilde{R}_{\tau_i}$ form a semigroup (for instance, all rational $\tau_i \geqq 0$). Then (4.1.8) will hold for each τ_i.

§ 4.2 Time Translations of the Preparation Procedures

In a way entirely similar to the above for $\tilde{\mathscr{R}}_m$, for $\mathscr{Q}_m$ we can introduce the mappings R'_τ of $\mathscr{Q}_m$ into itself, which are to describe a time translation of the preparation procedure. R'_τ must map each order interval $[\emptyset, a]$ from $\mathscr{Q}_m$ order isomorphically onto $[\emptyset, R'_\tau a]$. Since up to the time $t = 0$ the preparing *must* (!) be terminated, only for $\tau \leqq 0$ can we regard R'_τ as defined on all of $\mathscr{Q}_m$. With the same argument as at the end of § 4.1, let us assume that the $R'_{-\tau_i}$ (defined for the same $\tau_i \geqq 0$ as there) form a semigroup.

Since the mappings $\tilde{R}_\tau$ are injective (as are the R'_τ), $\tilde{R}_\tau^{-1} = \tilde{R}_{-\tau}$ (resp. $R'^{-1}_{-\tau} = R'_\tau$) is defined on $\tilde{R}_\tau \tilde{\mathscr{R}}_m$ (resp. on $R'_{-\tau}\mathscr{Q}_m$) for $\tau \geqq 0$. Therefore, if R'_τ means the displacement of the preparation procedure by the time τ, the probabilities must obey

$$\mu_m(a, h(\varrho_0, \varrho)) = \mu_m(R'_\tau a, h(\tilde{R}_\tau(\varrho_0, \varrho))) \tag{4.2.1}$$

with $h(\varrho_0, \varrho) = (h(\varrho_0), h(\varrho))$, whenever $R'_\tau a$ and $\tilde{R}_\tau \varrho_0$ are defined. In fact the right side of (4.2.1) describes an experiment for which the whole arrangement is displaced only by a time τ; hence we assume (4.2.1) to hold.

Let us assume $\tau \geqq 0$. Then $\tilde{R}_\tau$ is defined on all of $\tilde{\mathscr{R}}_m$. For a, we can put $a = R'_{-\tau}\tilde{a}$ with $\tilde{a}$ an arbitrary element in $\mathscr{Q}_m$. From (4.2.1) with (3.2.4) then follows

$$\langle \varphi_m(R'_{-\tau}\tilde{a}), \psi_m(\varrho_0, \varrho)\rangle = \langle \varphi_m(\tilde{a}), \psi_m \tilde{R}_\tau(\varrho_0, \varrho)\rangle. \tag{4.2.2}$$

Using (4.1.7) we conclude

$$\langle \varphi_m(R'_{-\tau}\tilde{a}), \psi_m(\varrho_0, \varrho)\rangle = \langle \varphi_m(\tilde{a}), V_\tau \psi_m(\varrho_0, \varrho)\rangle = \langle V'_\tau \varphi_m(\tilde{a}), \psi_m(\varrho_0, \varrho)\rangle, \tag{4.2.3}$$

where V'_τ is the operator in $C'(\hat{Y})$ that is dual to V_τ. Since la $\psi_m(\Phi)$ is dense in $L(\hat{Y})$, i.e. since the $\psi_m(\varrho_0, \varrho)$ separate, from (4.2.3) furthermore follows

$$\varphi_m(R'_{-\tau}\tilde{a}) = V'_\tau \varphi_m(\tilde{a}), \tag{4.2.4}$$

as shown in the diagram

$$\begin{array}{ccc} \mathscr{Q}'_m & \xrightarrow{\varphi_m} & K(\hat{Y}) \\ R'_{-\tau} \downarrow & & \downarrow V'_\tau \\ \mathscr{Q}'_m & \xrightarrow{\varphi_m} & K(\hat{Y}). \end{array}$$

Since the left side of (4.2.4) is an element of $\varphi_m(\mathscr{Q}'_m)$, from (4.2.4) follows in particular

$$V'_{\tau_i}\varphi_m(\mathscr{Q}'_m) \in \varphi_m(\mathscr{Q}'_m)$$

for all the τ_i introduced above. Since V'_{τ_i} is norm continuous, we conclude

$$V'_{\tau_i} K_m(\hat{Y}) \subset K_m(\hat{Y}). \tag{4.2.5}$$

Since V'_{τ_i} is even $\sigma(C'(\hat{Y}), C(\hat{Y}))$-continuous,

$$V'_{\tau_i} \bar{K}^\sigma_m(\hat{Y}) \subset \bar{K}^\sigma_m(\hat{Y}) \tag{4.2.6}$$

also holds. If τ_i differs little from τ_j, also the probabilities for $\varphi_m(R'_{-\tau_i}\tilde{a})$ and $\varphi_m(R'_{-\tau_j}\tilde{a})$ must differ but little. For physical reasons we thus must in $\mathscr{P}\mathscr{T}_m$ assume that $V_{\tau'}\varphi_m(\tilde{a})$ with fixed $\tilde{a}$ is norm continuous relative to τ. For fixed $u \in \varphi_m(\mathscr{Q}'_m)$, we obtain

$$\|(V'_\tau - V'_{\tau+\varepsilon})\,u\| \to 0 \quad \text{as} \quad \varepsilon \to 0. \tag{4.2.7}$$

This imposes a condition on $\varphi_m(\mathscr{Q}_m)$; expressed better, it places a condition on $K_m(\hat{Y})$, because (4.2.7) for all $u \in \varphi_m(\mathscr{Q}'_m)$ also implies (4.2.7) for all $u \in K_m(\hat{Y})$. Till now, this is the sole condition on $K_m(\hat{Y})$: For physical reasons, all those $u \in K(\hat{Y})$ must be excluded for which (4.2.7) does not hold.

Because of (4.2.7) one can define $V'_\tau u$ for $u \in \varphi_m(\mathscr{Q}'_m)$ and *all* $\tau \geqq 0$ and obtain

$$V'_\tau \varphi_m(\mathscr{Q}'_m) \subset K_m(\hat{Y}) \tag{4.2.8}$$

for all $\tau \geqq 0$. Thence it follows as above that

$$V'_\tau K_m(\hat{Y}) \subset K_m(\hat{Y}) \tag{4.2.9}$$

and

$$V'_\tau \bar{K}^\sigma_m(\hat{Y}) \subset \bar{K}^\sigma_m(\hat{Y}) \tag{4.2.10}$$

hold for all $\tau \geqq 0$.

From $h(\varrho) = \emptyset$ with a $\varrho \in \tilde{\mathscr{R}}_m$ follows

$$\mu_m(a, h(\varrho_0, \varrho)) = \langle \varphi_m(a), \psi_m(\varrho_0, \varrho) \rangle = 0$$

for all $a \in \mathscr{Q}'_m$. From (4.2.1) further follows

$$\mu_m(\tilde{a}, h(\tilde{R}_\tau(\varrho_0, \varrho))) = 0$$

for all $\tilde{a} \in \mathscr{Q}'_m$. Therefore, $\tilde{a} \cap h(\tilde{R}_\tau \varrho) = \emptyset$ holds for all $\tilde{a} \in \mathscr{Q}'_m$, which by APS 8 implies $h(\tilde{R}_\tau \varrho) = \emptyset$.

Hence $h(\varrho_1) = h(\varrho_2)$ implies $h(\tilde{R}_\tau \varrho_1) = h(\tilde{R}_\tau \varrho_2)$. Therefore a mapping R_τ is uniquely defined by the diagram

$$\begin{array}{ccc} \tilde{\mathscr{R}}_m & \xrightarrow{h} & \mathscr{R}_m \\ \tilde{R}_\tau \downarrow & & \downarrow R_\tau \\ \tilde{\mathscr{R}}_m & \xrightarrow{h} & \mathscr{R}_m . \end{array} \tag{4.2.11}$$

This explains why one often forgets $\tilde{\mathscr{R}}_m$ and speaks in general only about $\mathscr{R}_m$; for, also the time displacement operator R_τ is uniquely defined on $\mathscr{R}_m$. One can immediately carry the diagram (4.2.11) over to Φ, $\mathscr{F}_m$ and compose it with (4.1.8) to

$$\begin{array}{ccccc} \mathscr{F}_m & \xleftarrow{h} & \Phi & \xrightarrow{\psi_m} & L(\hat{Y}) \\ \downarrow R_\tau & & \downarrow \tilde{R}_\tau & & \downarrow V_\tau \\ \mathscr{F}_m & \xleftarrow{h} & \Phi & \xrightarrow{\psi_m} & L(\hat{Y}; \geqq \tau). \end{array} \tag{4.2.12}$$

We will now complete this diagram by (3.3.3). To this end, we must think about the behavior of the mappings s and V_τ. The latter is first defined only in $L(\hat{Y})$. Let f_1 and f_2 be two elements in $L(\hat{Y})$ and assume $sf_1 = sf_2$. Then f_1 and f_2 agree on the subset $\hat{S}_m$ of $\hat{Y}$, so that

$$f_+ - f_- = f_1 - f_2,$$

where f_+ is the positive and f_- the negative part of $f_1 - f_2$. This is to say $f_+, f_- \in L(\hat{Y})$ while f_+ and f_- are equal to zero on $\hat{S}_m$, so that $sf_+ = sf_- = 0$. Therefore, if we show that $f \in L(\hat{Y})$ and $sf = 0$ imply $sV_\tau f = 0$, then $sf_1 = sf_2$ quite generally implies $sV_\tau f_1 = sV_\tau f_2$.

$sf = 0$ is equivalent to $\mu_m(u, f) = 0$ for all $u \in K_m(\hat{Y})$. Because of (4.2.9), thus $\mu_m(u, V_\tau f) = \mu_m(V'_\tau u, f) = 0$ follows for all $u \in K_m(\hat{Y})$ so that $sV_\tau f = 0$. Then, with the abbreviation $V_\tau^{(s)} = sV_\tau$ we also find

$$V_{\tau_1}^{(s)} V_{\tau_2}^{(s)} = s V_{\tau_1}^{(s)} s V_{\tau_2}^{(s)} = s V_{\tau_1} V_{\tau_2}^{(s)} = s V_{\tau_1+\tau_2} = s V_{\tau_1+\tau_2}^{(s)}.$$

Hence, the $V_\tau^{(s)}$ form a semigroup; furthermore we get the diagram

$$\begin{array}{ccc} L(\hat{Y}) & \xrightarrow{s} & L(\hat{S}_m) \\ \downarrow V_\tau & & \downarrow V_\tau^{(s)} \\ L(\hat{Y}; \geqq \tau) & \xrightarrow{s} & L(\hat{S}_m; \geqq \tau). \end{array} \tag{4.2.13}$$

Together with (3.3.3) and (4.2.12), this implies

$$\begin{array}{ccc} \mathcal{F}_m & \xrightarrow{\psi_{ms}} & L(\hat{S}_m) \\ \downarrow R_\tau & & \downarrow V_\tau^{(s)} \\ \mathcal{F}_m \supset R_\tau \mathcal{F}_m & \xrightarrow{\psi_{ms}} & L(\hat{S}_m; \geqq \tau). \end{array} \tag{4.2.14}$$

§ 4.3 Further Transformations of Preparation and Registration Procedures

Here we will make remarks about other transformations possibly defined in $\mathcal{PT}_m$, but without aspiring an exact formulation.

For this purpose one must realize that for preparation and registration devices it is pertinent, how they are oriented to a laboratory coordinate system. We have already assumed in the preceding §§ 4.1 and 4.2 an orientation relative to the laboratory time scale; otherwise the mappings $\tilde{R}_\tau$ and R'_τ would have been meaningless. We have also represented the action of these mappings mathematically. The action of $\tilde{R}_\tau$ could be given by the diagram (4.1.8), since the trajectories depend on the laboratory time scale.

As the preparation and registration procedures have an orientation to the time scale, they also need an orientation in the spatial laboratory coordinate system. Therefore, procedures must also be regarded as different if they only differ in their orientations to the laboratory. Without formulating this more precisely, one recognizes that one can for spatial translations and rotations introduce mappings similar to $\tilde{R}_\tau$, R'_τ. The same can be done for imparting a uniform velocity (relative to the laboratory system), i.e. for every element of the Galileo resp. Poincaré group. Exactly as with $\tilde{R}_\tau$ and R'_τ one can consider such transformations separately for the preparation and registration procedures. Then the transformations of the registration procedures are again connected with the trajectories, since we must due to the meaning of the state space Z assume that the laboratory system enters the description of Z. We have given an example in § 1, where the points of Z were given by $\{\mu(\boldsymbol{r}), \boldsymbol{u}(\boldsymbol{r}), T(\boldsymbol{r})\}$ with $\boldsymbol{r}$ as position vector in the laboratory system.

Naturally, only the transformations of the registration procedures for fixed preparation procedures and of the preparation procedures for fixed registration procedures are physically interesting for the structure of the dynamics of the systems from M_m. If both procedures are transformed in the *same* way, then physically nothing is changed. This fact can as at (4.2.1) be exploited to connect the transformations of the preparation and registration procedures with each other.

But also the joint, identical transformation of preparation and registration procedures can acquire an important physical meaning, if for a composite system one transforms (as to described in III) the procedures of a *sub*system, keeping fixed the procedures of the other subsystem (see III § 7).

§ 5 The Macrosystems as Physical Objects

We introduced in § 1 the objectivating manner of describing the systems from M_m by the function $M_m \xrightarrow{g} Y$, but later, when considering the trajectory registration procedures, referred only "intuitively" to the physical meaning of $M_m \xrightarrow{g} Y$. In fact we have yet not shown any mathematical connection between the mapping $M_m \xrightarrow{g} Y$ introduced in § 1 and the registration procedures. Therefore we must again ask, to what extent the function $M_m \xrightarrow{g} Y$ may be conceived as a physically real structure, if only the fact of the preparation and registration of trajectories is used as a description of the actual experiments. These investigations are conceptually very important since otherwise one could call the function $M_m \xrightarrow{g} Y$ a "pure fantasy" which one can "imagine" but which represents no "physically real" connection (that can be justified in the sense of [3] § 10).

We formulated briefly in § 1 that we "first of all" assume it known how to measure the state of a system at some time. In § 3 we described this "measuring" by mathematical structures. Now, we must ask to what extent the structures introduced in § 3 determine a mapping $M_m \xrightarrow{g} Y$ as "physically real". In this connection, we must of course assume that due to pretheories the elements of the sets M_m, $\tilde{\mathscr{R}}_m$, $\tilde{\mathscr{R}}_{0m}$, $\mathscr{Q}_m$, the connection (3.1.8), and the mapping $\tilde{\mathscr{R}}_m \xrightarrow{h} \mathscr{P}(M_m)$ are recognized as "physically real". To what extent can one then say that, by means of the thus described registration processes, one can "in principle measure" the trajectory $g(x)$ of a system x and thus really recognize it as physically real?

For these deliberations, it is very important to apply the methods from [3] § 10, since one does not get away with only an intuitive notion of "real". If one has measured (approximately) the trajectory for a system x occurring in an experiment (i.e. employed a $\varrho \in \tilde{\mathscr{R}}_m$ for which $\psi_m(\varrho_0, \varrho)$ differs from zero only in a small neighborhood of a trajectory y) and has obtained $x \in h(\varrho)$, then one can rightly say that the trajectory y physically pertains to this x. But what shall it mean that a trajectory pertains to an $x \in M_m$, even if it is not or only partly measured?

According to [3] § 10, it depends on the form of the theory how one has to judge whether $M_m \xrightarrow{g} Y$ describes a physically real structure. If one proceeds from a theory as sketched in § 1, then M_m and Y will in any case be viewed as sets of "real facts" ascertained by the pretheories (in the sense of [3] § 10). Then $M_m \xrightarrow{g} Y$ represents a relation between an $x \in M_m$ and some $y \in Y$. In the construction of a theory according to § 1 one must likewise assume (!) that by direct measurements (in the sense of [3] § 10), i.e. due to pretheories, this relation between x and y is verifiable experimentally, i.e. can be read from a real text (see [3] § 5). Then $M_m \xrightarrow{g} Y$ is to be viewed as a physically real relation, but as an idealized relation which for a comparison with experiment must first be smeared by means of imprecision sets (described in [3] § 6). The imprecision sets are thereby determined by the uniform structure of Y_p (see again [3] § 6). But this path sketched in § 1 has not been followed here, since it furthers the erection of an axiomatic basis for quantum mechanics to bring in the "pretheories" of registering trajectories, as exhibited in § 3. There we also recognized the connection between the uniform structure p for Y, as introduced in § 1, and the registration procedure. The imprecision sets for Y were thus determined just by the digital registration procedures. Therefore, if one chooses the path described in § 2 and § 3, one must rightly ask whether a relation $M_m \xrightarrow{g} Y$ is "determined" by the structures laid down in § 2 and § 3 (for "determined" see [3] § 10).

On the basis of the theory from § 2 and § 3, the following sets (inner terms, [3] §§ 7 and 10) can be derived:

$$\mathscr{T}_x = \{y \mid y \in Y;\ \text{for all } \varrho \in \tilde{\mathscr{R}}_m \text{ with } x \in h(\varrho) \text{ holds with } k = \psi_m(\varrho_0, \varrho): k(y) \neq 0\}; \tag{5.1}$$

$$\mathscr{Q}_{mx} = \{a \mid a \in \mathscr{Q}_m \text{ and } x \in a\}. \tag{5.2}$$

With S_a the support of $\varphi_m(a)$, we then define

$$\mathscr{V}_x = \bigcap_{a \in \mathscr{Q}_x} S_a; \tag{5.3}$$

$$\mathscr{W}_x = \mathscr{T}_x \cap \mathscr{V}_x. \tag{5.4}$$

If $\mathscr{W}_x$ is a singleton for all x, then a function $M_m \xrightarrow{g} Y$ is defined by $x \to y \in \mathscr{W}_x$. Then one can justifiably assert that each system x "in reality" has a trajectory.

Unfortunately, "$\mathscr{W}_x$ is a singleton for all x" cannot be derived as a theorem from the axioms demanded in §§ 2 and 3. But also "$\mathscr{W}_x$ is not a singleton for all x" is not a theorem. Such undeciable situations occur frequently in physical theories. For that reason, in [3] § 10 we introduced for hypotheses, together with "theoretically existent", also the almost equivalent qualifer "certain". That $\mathscr{W}_x$ is a singleton, is (in the sense of [3] § 10) certain for all x, whereas "$\mathscr{W}_x$ is not a singleton" is not certain for any x.

Very intuitively this can due to [3] § 10 be expressed as follows: *Every* experiment of preparing and registering can be imagined as complemented in such a way that $\mathscr{W}_x$ becomes a singleton. But then we can as a additional axiom simply require: $\mathscr{W}_x$ is a singleton for all x. The theory obtained in this way leads to no contradiction with experience.

One often demands that $\mathcal{T}_x$ is already a singleton. This has a very intuitive physical interpretation, namely, that one can in principle improve the exactness of the trajectory measurement for each system x; this is a typical "classical" assumption (see [3] §§ 6 and 9).

The physical interpretation of such assumptions (which lead to the existence of a function $M_m \xrightarrow{g} Y$) must of course be mollified by the imprecise mappings in Y (see § 3.1 and [3] § 6).

The existence of the function g now lets us introduce a set of "objective" properties $\mathscr{E}$ of M_m. Due to (5.3) we have $g(M_m) \subset \hat{S}_m$. As the set of objective properties we define

$$\mathscr{E} = \{E \mid E = g^{-1}\sigma, \sigma \in \mathscr{B}(\hat{S}_m)\} \subset \mathscr{P}(M_m). \tag{5.5}$$

All preparations and trajectory registrations can be interpreted by means of these objective properties. That in this sense there exists a "sufficient" (see V § 10) set of objective properties for the macrosystems rests on the fact that the registration of macrosystems can be described as a measurement of trajectories by pretheories. The distinction to the considerations to follow in V § 10 therefore just arises because for the action carriers to be considered in V § 10 there even do *not* exist pretheories which would allow the *interpretation* of the registrations as the measurement of "anything on the microsystems".

Since the probability measures vanish for $\sigma \in \mathscr{J}(\hat{S}_m)$, for such σ the set $g^{-1}\sigma$ must be empty. Conversely the set $g^{-1}\sigma$ cannot be empty for any open set $\sigma \in \mathscr{B}(\hat{S}_m)$, since for such a σ there exist nonzero probability measures. Hence, one often calls the Boolean ring $\Sigma_m = \mathscr{B}(\hat{S}_m)/\mathscr{J}(\hat{S}_m)$ the set of objective properties. For the compatibility problem, Σ_m will be central in X § 2.3.

The set M_m, equipped with the structures from § 2 and § 3 (augmented by the axiom that $\mathscr{W}_x$ is a singleton), by the structures (5.5) becomes a set of "physical objects" (see V § 10; VII § 5.3 and [2] II § 1; [2] III § 4.1). But despite this, M_m is in general not a set of "classical" systems (in the sense of VII § 5.3) for, though due to § 3.3 and X § 2.3 one can of course identify $K_m(\hat{S}_m)$ with a subset of $K(\Sigma_m)$, in general $K_m(\hat{S}_m) = K(\Sigma_m)$ need *not* hold. Hence one also uses another definition of objective properties.

For a dynamically continuous system, to each $\varrho \in \mathscr{B}(Z)$ and $\tau \in \Theta$ there is assigned (due to § 3.4) a $\sigma_m(\varrho; \tau) \in \mathscr{B}(\hat{S}_m)$. Thereby, for each $\tau \in \Theta$ there is defined a mapping $\mathscr{B}(Z) \to \mathscr{B}(\hat{S}_m)$ and hence also $\mathscr{B}(Z)/\mathscr{J}(Z) \to \Sigma_m = \mathscr{B}(\hat{S}_m)/\mathscr{J}(\hat{S}_m)$. Here, $\mathscr{J}(Z)$ is the set of all $\varrho \in \mathscr{B}(Z)$ such that $u_\tau(\varrho) = u(\sigma_m(\varrho; \tau)) = 0$ for all τ and all $u \in K_m(\hat{S}_m)$. Then, with $\Sigma_Z = \mathscr{B}(Z)/\mathscr{J}(Z)$ and $K(\Sigma_Z)$ the set of all σ-additive measures over Σ_Z, for each τ there is defined a mapping $K_m(\hat{S}_m) \underset{\tau}{\to} K(\Sigma_Z)$ by $u \to u_\tau$. As explained in § 3.4, one can presume the support of all measures $\{u_\tau \mid u \in K_m(\hat{S}_m), \tau > 0\}$ to be Z. There exist many systems for which $\sigma_m(\varrho; \tau)$ is defined as an element of $\mathscr{B}(\hat{S}_m)$ even for $\tau = 0$ (systems called continuous also for $\tau = 0$). Then $u_0 = u(\sigma_m(\varrho; 0))$ is also defined. The mapping $K_m(\hat{S}_m) \underset{0}{\to} K(\Sigma_Z)$ thus defined can then in all known cases be presumed surjective, which simply means that one can at the time $\tau = 0$ prepare all ensembles of states from Z. How the u_τ then appear depends on the $K_m(\hat{S}_m)$, i.e. on the dynamics. But in general it is not true that $K_m(\hat{S}_m) \underset{0}{\to} K(\Sigma_Z)$ is injective and hence bijective.

Therefore, in general to a prepared $u_0 \in K(\Sigma_Z)$ there does not correspond a unique $u \in K_m(\hat{S}_m)$ and thus not a unique measure u_τ for $\tau > 0$. But there are many systems for which one can presume $K_m(\hat{S}_m) \underset{0}{\rightarrow} K(\Sigma_Z)$ to be injective. In fact, general theoretical questions of statistical mechanics have mainly been limited to such systems. This is the background for the very customary definition of Σ_Z as the set of objective properties. One calls $g^{-1}\sigma_m(\varrho;\tau)$ the set of systems which at the time τ have the property $\varrho \in \mathscr{B}(Z)$, often not distinguishing between the $\varrho \in \mathscr{B}(Z)$ and the elements of Σ_Z (an entirely physical custom).

If $K_m(\hat{S}_m) \underset{0}{\rightarrow} K(\Sigma_Z)$ is bijective, one can identify $K_m(\hat{S}_m)$ with $K(\Sigma_Z)$ and (by means of the mapping $K(\Sigma_Z) \underset{0}{\rightarrow} K_m(\hat{S}_m)$ inverse to $K_m(\hat{S}_m) \underset{0}{\rightarrow} K(\Sigma_Z)$) introduce a "dynamical" mapping $u_0 \xrightarrow{W_\tau} u_\tau$ by

$$K(\Sigma_Z) \underset{0}{\rightarrow} K_m(\hat{S}_m) \underset{\tau}{\rightarrow} K(\Sigma_Z). \tag{5.6}$$

One can then consider M_m as a set of classical systems (in the sense of VII § 5.3 with $K(\Sigma_Z)$ the set of ensembles and Σ_Z that of objective properties) whose time variation is described by (5.6). We shall denote such systems briefly as systems "classically describable in the state space Z".

For considerations in X, let us yet note that, by means of V'_τ from § 4.2, one can write the mapping (5.6) as

$$K(\Sigma_Z) \underset{0}{\rightarrow} K_m(\hat{S}_m) \xrightarrow{V'_\tau} K_m(\hat{S}_m) \underset{0}{\rightarrow} K(\Sigma_Z). \tag{5.7}$$

III Base Sets and Fundamental Structure Terms for a Theory of Microsystems

As already pointed out in I, we intend to construct an axiomatic basis for quantum mechanics, beginning only with sets and structures *interpretable* by known pretheories. Therefore, we cannot start with a set M interpreted as a set of microsystems as in [2] (see [3] § 5 for abbreviated formulations such as "set of microsystems"). We rather "question" the existence of a microsystem, i.e. we will theoretically retrace the discovery of microsystems. In the theory to be constructed we do this by obtaining the microsystems as physical realities only through a physically real set (*derived* from the basic sets) with a physically real structure (in the sense of [3] § 10.5).

§ 1 Composite Macrosystems

Therefore, to begin with, we "forget" the microsystems. Each experiment with microsystems (described in I) is really "only" a complex macroscopic device in which various macroscopic processes occur, i.e. processes described by pretheories. We will elaborate a "typical" structure of these frequently complex experiments and describe it mathematically.

This typical structure involves devices composed of two macrosystems, the interaction of which is investigated experimentally. In this characterization, it is *not* excluded that the devices may be composed of more than two systems. But the subsystems can be partitioned into *two* groups, each regarded as a subsystem so that the entire system appears again composed of *two* subsystems (the two groups). That it is possible in this way to consider a device in *different* ways as composed of two subsystems is an interesting structure often used experimentally (also see I § 3). We have investigated such structures in [2] XVII and shall return to this question in XII § 1. In order to construct an axiomatic basis, we at first need not introduce any mathematical description for the *different* possibilities available for a device to be decomposed into subsystems. Thus, in the first step we will mathematically map only the structure of the composition of two subsystems.

In order to describe the composition of *two* systems, we start with *two* base sets of the theory $\mathscr{MT}_\Sigma$*) to be constructed: M_1 and M_2. Corresponding to [3] § 9, we require as axiom (not yet giving it a number):

---

*) The index Σ on $\mathscr{MT}$ now emphasizes that we construct the mathematical theory as a structure of species Σ. A structure of species Σ is characterized by base sets, relations (structures) and axioms (see XIII § 2.1, [3] and [48]).

The set M_1 and M_2 are countable.

In the relevant experiments, a system from M_1 will always be "composed" with a system from M_2. Therefore, we must in $\mathscr{M}\mathscr{T}_\Sigma$ introduce a relation to portray this "composition". But before we do this, we adopt the following convention for the mapping principles of M_1 and M_2, which will turn out very practical (already used in II § 1 for the sets M_m): For two different experiments (e.g. $x \in M_1$, $x' \in M_1$) we set $x \neq x'$ also when the two experiments with the subsystems x and x' are carried out with the "same" subdevice in the sense that the same technical device is used. A subaspect of this reality structure (that x and x' can be "the same" device) is grasped in the preparation process (see II § 2.2) namely by regarding $x \in a$ and $x' \in a$ to hold for the *same* preparation a.

In characterizing *different* experiments by different elements of M_1 and M_2, we also simplify the mathematical relation for a *composition*. It is natural to describe "x_1 and x_2 are composed" by a subset $M \subset M_1 \times M_2$. For the structure M we then require the axiomatic relations

AZ 1 $(x_1, x_2) \in M$ and $(x_1', x_2) \in M$ implies $x_1 = x_1'$;
$(x_1, x_2) \in M$ and $(x_1, x_2') \in M$ implies $x_2 = x_2'$.

This axiom expresses the following fact: If x_1 is composed with x_2, then x_2 cannot be composed with "another" system x_1' since otherwise (x_1, x_2) would not even describe "the system composed of x_1 and x_2".

Since it is uninteresting to consider the systems $x_1 \in M_1$ resp. $x_2 \in M_2$ not composed with other systems, to AZ 1 we add the axiom

AZ 2 For each $x_1 \in M_1$ there is an $x_2 \in M_2$ with $(x_1, x_2) \in M$;
for each $x_2 \in M_2$ there is an $x_1 \in M_1$ with $(x_1, x_2) \in M$.

AZ 1, 2 are equivalent to saying that M is the graph of a bijective mapping $M_1 \to M_2$.

The mapping principle "$(x_1, x_2) \in M$ is the image relation ([3] § 5) for the real relation that the systems x_1 and x_2 are composed" is often summarized in the brief statement: M is the set of composed systems (see [3] § 5 for such abbreviations).

Since M_1, M_2 are countable, also M can at most be countable. We require that M be neither empty nor of finite cardinality.

The axioms AZ 1, 2 determine the application of the concept "composed of two parts". Therefore, these axioms are to be regarded as "laws of nature" only in the sense, that there is a fundamental domain (of real situations) to which the concept specified by AZ 1, 2 is applicable. It is immaterial whether we "encounter" such real situations or "produce" them technically. Just this production, i.e. the experimentation, is of fundamental importance for all physical theories (see about "physically possible" processes in [3] §§ 10 and 11).

In order to emphasize the objectivating description of the systems from M_1, M_2, M by the pretheories from II, we enrich the structure introduced in M by trajectory spaces. This enrichment is of course *not necessary* for an axiomatic basis of quantum mechanics since one obtains the essential structures of quantum mechanics just by "forgetting" the detailed structures of the devices.

The systems from M_1, M_2, M shall be systems describable by the pretheories from II. Let us for instance consider M_1. Certain elements of M_1 (i.e. a subset of M_1) can be described by trajectories in an appropriate state space from *one* pretheory. Other elements from M_1 might be describable in a state space from another pretheory. If one has several state spaces Z_λ (λ an index), one can also introduce $Z = \bigcup_\lambda Z_\lambda$ as a state space. These considerations make it natural to demand that to M_1 there corresponds one state space Z_1 so that the systems from M_1 can be described by trajectories in Z_1. Thus, for the physical interpretation of the points in Z_1 several pretheories may be required.

As described in II, one can corresponding to Z_1, Z_2 introduce trajectory spaces Y_1, Y_2 in which uniform structures g and p are defined, p being the uniform structure of physical imprecision.

As further base sets for quantum mechanics, we introduce Y_1, Y_2 with the physical interpretation (given by pretheories) that they are the trajectory spaces of the systems from M_1 resp. M_2. As axioms we first introduce only those to define the uniform structures of physical imprecision in Y_1 and Y_2. We shall not write out these axioms explicitly, but only indicate them by

AT 1 Axioms for the uniform structures p in Y_1, Y_2.

The introduction of the base sets Y_1, Y_2 is typical for an axiomatic basis of a theory $\mathscr{PT}$, if complex pretheories are already required for its formulation. In the pretheories, Y_1, Y_2 are derived with a complex structure. In $\mathscr{PT}$ they are base terms so that there one need not formulate the complex structures with which these terms are equipped in the pretheories. Rather, these complex structures are innate in the mapping principles of $\mathscr{PT}$ which, just by means of the pretheories, enable us to interpret the elements of Y_1, and Y_2.

As mentioned above, one can formulate an axiomatic basis for quantum mechanics so that the terms Y_1, Y_2 do not appear at all, i.e. one can leave the description of the trajectory registration entirely to the pretheories. But in order to better discuss in XI the measurement process, and to emphasize the objectivating method of describing the experiments, we shall also formulate an axiomatic basis which includes the base sets Y_1, Y_2. Therefore, we shall give two parallel formulations of an axiomatic basis for quantum mechanics. We shall by AT denote all the axioms for that extension by Y_1, Y_2.

In §§ 2 through 5 we first begin with the restricted form (without Y_1, Y_2). In § 6, this is extended to the form with Y_1, and Y_2. This extended form of the axiomatic basis will not be used in IV through IX.

We hope that the presentation of two forms of the axiomatic bases here in III will not confuse the reader. The physical meaning of the mathematical relations will be clear without a detailed idea of the "axioms". In chapter IX, we shall conclude with an overview over the erection of the axiomatic basis, intended for readers interested to know that it is realy possible to erect quantum mechanics without "theoretic auxiliary concepts" (in the sense of [30]). In XIII we shall show that of course there exists an axiomatic basis for quantum mechancis in which only the image relations between elements of the base sets are used. Therefore, XIII is intended only for those readers who, proceeding beyond an understanding of quantum mechanics, have an interest in the philosophy of science.

§ 2 Preparation and Registration Procedures for Composite Macrosystems

The fact that the systems from M are "composite", shall relative to preparations and registrations be expressed as follows. We start with the structures of preparing and registering the *subsystems* from M_1 and M_2, and only then derive corresponding structures for M (as sets of pictures for real situations, see [3] § 10.5). We can forego discussing the physical meaning of the structure terms for M_1 and M_2 as well as of the axioms in view of the explanations in II § 2; [1] XIII; [2] II; [3] § 12.

Therefore, we introduce two structure terms $\mathscr{Q}_1 \subset \mathscr{P}(M_1)$, $\mathscr{Q}_2 \subset \mathscr{P}(M_2)$ with the axiom

APSZ 1 $\mathscr{Q}_1$ and $\mathscr{Q}_2$ are statistical selection procedures (see II § 2.2 and also [2] II § 4.1 [3] § 12).

According to [3] § 9 we require that $\mathscr{Q}_1$ and $\mathscr{Q}_2$ be countable sets; their probability functions be $\lambda_{\mathscr{Q}_1}, \lambda_{\mathscr{Q}_2}$. The sets $\mathscr{Q}_1$ and $\mathscr{Q}_2$ consist of the preparation procedures of the systems from M_1 resp. from M_2. This is – to emphasize it again – an abbreviation for a mapping principle. The interpretation of the elements of $\mathscr{Q}_1$ and $\mathscr{Q}_2$ is given by pretheories, just by identification with elements of the sets $\mathscr{Q}_m$ (from perhaps different pretheories; see [3] § 10.5).

But here let us recall an aspect of the physical interpretation of $\mathscr{Q}_1$ and $\mathscr{Q}_2$ that we briefly sketched in II § 4.3: In the meaning of the preparation procedures from $\mathscr{Q}_1$ resp. $\mathscr{Q}_2$ there enters the orientation of the systems from M_1 resp. M_2 to the "laboratory system". Although we shall only in § 7 express this fact mathematically by further structures, we have again pointed out this aspect of the elements of $\mathscr{Q}_1$ and $\mathscr{Q}_2$ so that the reader will not get false notions about their physical interpretation.

Important is the connection among the structures $\mathscr{Q}_1$, $\mathscr{Q}_2$ and M. The physical significance of these three structures will suggest further axioms for this connection. First we introduce, for the later formulation of such axioms, definitions in which we use words to remind us of the physical interpretation of $\mathscr{Q}_1, \mathscr{Q}_2, M$.

For brevity, in many relations we write

$$\mathscr{Q}_i' = \{a \mid a \in \mathscr{Q}_i, a \neq \emptyset\} \quad \text{for} \quad i = 1, 2.$$

If for $a_i \in \mathscr{Q}_i'$ the set $a_1 \times a_2 \cap M$ is empty, by the interpretation of $\mathscr{Q}_1, \mathscr{Q}_2, M$ this means that it is not "physically possible" (see [3] § 10.4) to compose systems prepared by a_1 and a_2. This problem of which systems can be composed in order to yield a "meaningful" experiment, of course has great physical significance. By demanding a "meaningful" experiment we have emphasized that this "combination" *not only* requires that the devices can be joined in the laboratory; it rather involves a physical choice of experiments. Only step by step can we by axioms map the experimental situations (explained intuitively in I) into the mathematical image.

D 2.1 Two preparation procedures $a_i \in \mathcal{Q}_i'$ are said to be *mutually exclusive* if $a_1 \times a_2 \cap M = \emptyset$.

If we pick any two (not mutually exclusive) preparation procedures $a_i \in \mathcal{Q}_i'$ it may happen that there are finer procedures $a_i' \subset a_i$ with $a_i' \in \mathcal{Q}_i'$ and $a_1' \times a_2' \cap M = \emptyset$. This would mean that indeed the preparation procedures a_1, a_2 do not exclude the physical possibility ([3] § 10.4) of composing some of the systems from a_1 and a_2 into a meaningful experiment but that there are finer procedures a_1', a_2' that exclude each other. In this case it is not true that "arbitrary" systems from a_1 can be composed with "arbitrary" systems from a_2. But then one cannot expect that $a_1 \times a_2 \cap M$ is a statistical selection procedure over M, since systems from a_1 and a_2 cannot be composed "statistically independently". These considerations suggest, not for every pair $a_i \in \mathcal{Q}_i$ to regard the set $a_1 \times a_2 \cap M$ as a "preparation procedure" of composite systems from M. Hence let us first adopt the definition

D 2.2 We say that $a_i \in \mathcal{Q}_i'$ with $i = 1, 2$ may be combined if $a_i' \subset a_i$, $a_i' \in \mathcal{Q}_i'$ always implies $a_1' \times a_2' \cap M \neq \emptyset$.

For abbreviation we introduce the set

$$\Gamma_{12} = \{a_1 \times a_2 \cap M \mid a_1 \in \mathcal{Q}_1', a_2 \in \mathcal{Q}_2', a_1 \text{ and } a_2 \text{ may be combined}\}. \tag{2.1}$$

T 2.1 From "$a_1 \in \mathcal{Q}_1', a_2 \in \mathcal{Q}_2'$ may be combined" there follows

(i) $a_1 \times a_2 \cap M \neq \emptyset$;
(ii) $\{a_i' \in \mathcal{Q}_i', a_i' \subset a_i, i = 1, 2\} \Rightarrow a_1', a_2'$ may be combined;
(iii) $\{a_1^{(k)} \in \mathcal{Q}_1', a_2^{(l)} \in \mathcal{Q}_2', k = 1, \ldots, n; l = 1, \ldots, m$ and there exist an $a_1 \in \mathcal{Q}_1$ with $a_1^{(k)} \subset a_1$ and an $a_2 \in \mathcal{Q}_2$ with $a_2^{(l)} \subset a_2$ and each $a_1^{(k)}$ may be combined with each $a_2^{(l)}\} \Rightarrow \bigcup_{k=1}^{n} a_1^{(k)} \in \mathcal{Q}_1', \bigcup_{l=1}^{m} a_2^{(l)} \in \mathcal{Q}_2'$ and $\bigcup_{k=1}^{n} a_1^{(k)}$ may be combined with $\bigcup_{l=1}^{m} a_2^{(l)}$.

Proof. (i) and (ii) follow directly from D 2.2.
(iii): Since $\mathcal{Q}_1$ is a structure of selection procedures and $a_1^{(k)} \subset a_1 \in \mathcal{Q}_1$ holds, $\bigcup_{k=1}^{n} a^{(k)} \in \mathcal{Q}_1$ also holds; likewise, $\bigcup_{l=1}^{m} a_2^{(l)} \in \mathcal{Q}_2$. If $a_1' \subset \bigcup_{k=1}^{n} a_1^{(k)}$ and $a_2' \subset \bigcup_{l=1}^{m} a^{(l)}$, then from $\emptyset \neq a_1' = \bigcup_{k=1}^{n} (a_1^{(k)} \cap a_1')$ follows that there is a k with $a_1^{(k)} \cap a_1' \neq \emptyset$ and likewise some l with $a_1^{(l)} \cap a_2' \neq \emptyset$. Since $a_1^{(k)}$ and $a_2^{(l)}$ may be combined, we have $(a_1^{(k)} \cap a_1') \times (a_2^{(l)} \cap a_2') \cap M \neq \emptyset$ and because of $a_1' \times a_2' \cap M \supset (a_1^{(k)} \cap a_1') \times (a_2^{(l)} \cap a_2') \cap M$ also $a_1' \times a_2' \cap M \neq \emptyset$, i.e. $\bigcup_{k=1}^{n} a_1^{(k)}$ and $\bigcup_{l=1}^{m} a_2^{(l)}$ may be combined. □

We shall frequently use (i) and (ii) from T 2.1 without indicating it!

T 2.2 If $\mathcal{V}$ is a structure of selection procedures over M_1, then $\tilde{\mathcal{V}} = \{a \times M_2 \cap M \mid a \in \mathcal{V}\}$ is a structure of selection procedures over M, and the mapping $a \to a \times M_2 \cap M$

represents an isomorphism $\mathscr{V} \to \tilde{\mathscr{V}}$. The same holds for a structure $\mathscr{W}$ over M_2 with $\tilde{\mathscr{W}} = \{M_1 \times a \cap M \mid a \in \mathscr{W}\}$.

Proof. It suffices to present the proof for $\mathscr{V}, \tilde{\mathscr{V}}$. From AZ 2 follows for $a \in \mathscr{V}'$ ($\mathscr{V}'$ equals $\mathscr{V}$ without the empty set) that $a \times M_2 \cap M \neq \emptyset$.

From $a \times M_2 \cap M = a' \times M_2 \cap M$ follows $(a \cap a') \times M_2 \cap M = a \times M_2 \cap M$ and thus $(a \backslash a \cap a') \times M_2 \cap M = \emptyset$. This due to $a \backslash a \cap a' \in \mathscr{V}$ implies $a \backslash a \cap a' = \emptyset$, i.e. $a = a \cap a'$. Likewise follows $a' = a \cap a'$ and thus $a = a'$. Therefore, the mapping $\mathscr{V} \to \tilde{\mathscr{V}}$ is bijective. The isomorphism of $\mathscr{V}$ and $\tilde{\mathscr{V}}$ (regarded as structures of selection procedures) follows in an elementary way. □

By application of T 2.2 to $\mathscr{D}_1$ and $\mathscr{D}_2$ one obtains

T 2.3 The sets $\tilde{\mathscr{D}}_1 = \{a_1 \times M_2 \cap M \mid a \in \mathscr{D}_1\}$ and $\tilde{\mathscr{D}}_2 = \{M_1 \times a_2 \cap M \mid a_2 \in \mathscr{D}_2\}$ are structures of species SP, and the mappings $a_1 \to a_1 \times M_2 \cap M$ and $a_2 \to M_1 \times a_2 \cap M$ represent isomorphisms $\mathscr{D}_1 \to \tilde{\mathscr{D}}_1$ and $\mathscr{D}_2 \to \tilde{\mathscr{D}}_2$. □

With $a_1 \in \mathscr{D}_1$, $a_2 \in \mathscr{D}_2$, from $a_1 \times a_2 \cap M = (a_1 \times M_2 \cap M) \cap (M_1 \times a_2 \cap M)$ follows that a_1, a_2 may be combined if and only if the corresponding $\tilde{a}_1 = a_1 \times M_2 \cap M \in \tilde{\mathscr{D}}_1$ and $\tilde{a}_2 = M_1 \times a_2 \cap M \in \tilde{\mathscr{D}}_2$ obey

$$\tilde{a}_i' \in \tilde{\mathscr{D}}_i', \; \tilde{a}_i' \subset \tilde{a}_i \in \tilde{\mathscr{D}}_i', \quad i = 1, 2 \;\Rightarrow\; \tilde{a}_1' \cap \tilde{a}_2' \neq \emptyset \,. \tag{2.2}$$

Therefore let us in an entirely natural way call two elements $\tilde{a}_i \in \tilde{\mathscr{D}}_i'$ combinable if (2.2) holds. To say that a_1, a_2 may be combined is thus equivalent to saying that the corresponding $\tilde{a}_1, \tilde{a}_2$ may. In particular, (2.1) therefore becomes

$$\Gamma_{12} = \{\tilde{a}_1 \cap \tilde{a}_2 \mid \tilde{a}_1 \in \tilde{\mathscr{D}}_1, \; \tilde{a}_2 \in \tilde{\mathscr{D}}_2, \; \tilde{a}_1 \text{ and } \tilde{a}_2 \text{ may be combined}\} \,. \tag{2.3}$$

According to [3] § 10.5, the structure $\mathscr{D}_{12}$ of selection procedures over M generated by Γ_{12} (see [2] II Satz 2.2) is a set of pictures of real situations – briefly, $\mathscr{D}_{12}$ is a set of real situations.

T 2.4 For Q_{12}, a probability $\lambda_{\mathscr{D}_{12}}$ is uniquely defined by the two requirements

$$\lambda_{\mathscr{D}_{12}}(a_1 \times a_2 \cap M, \; a_1' \times a_2 \cap M) = \lambda_{\mathscr{D}_1}(a_1, a_1') \,,$$
$$\lambda_{\mathscr{D}_{12}}(a_1 \times a_2 \cap M, \; a_1 \times a_2' \cap M) = \lambda_{\mathscr{D}_2}(a_2, a_2') \,.$$

It follows

$$\lambda_{\mathscr{D}_{12}}(a_1 \times a_2 \cap M, \; a_1' \times a_2' \cap M) = \lambda_{\mathscr{D}_1}(a_1, a_1') \, \lambda_{\mathscr{D}_2}(a_2, a_2') \,.$$

Proof. Using the elements $\tilde{a}$ that due to T2.3 correspond to the a by the isomorphisms $\mathscr{D}_1 \to \tilde{\mathscr{D}}_1, \mathscr{D}_2 \to \tilde{\mathscr{D}}_2$, one can write

$$\lambda_{\mathscr{D}_{12}}(\tilde{a}_1 \cap \tilde{a}_2, \; \tilde{a}_1' \cap \tilde{a}_2) = \lambda_{\mathscr{D}_1}(a_1, a_1') \,,$$
$$\lambda_{\mathscr{D}_{12}}(\tilde{a}_1 \cap \tilde{a}_2, \; \tilde{a}_1 \cap \tilde{a}_2') = \lambda_{\mathscr{D}_2}(a_2, a_2') \,.$$

This immediately implies

$$\begin{aligned}\lambda_{\mathscr{D}_{12}}(a_1 \times a_2 \cap M, \; a_1' \times a_2' \cap M) &= \lambda_{\mathscr{D}_{12}}(\tilde{a}_1 \cap \tilde{a}_2, \; \tilde{a}_1' \cap \tilde{a}_2') \\ &= \lambda_{\mathscr{D}_{12}}(\tilde{a}_1 \cap \tilde{a}_2, \; \tilde{a}_1' \cap \tilde{a}_2) \, \lambda_{\mathscr{D}_{12}}(\tilde{a}_1' \cap \tilde{a}_2, \; \tilde{a}_1' \cap \tilde{a}_2') \\ &= \lambda_{\mathscr{D}_1}(a_1, a_1') \, \lambda_{\mathscr{D}_2}(a_2, a_2') \,. \end{aligned} \tag{2.4}$$

In order to prove the uniqueness and existence of $\lambda_{\mathscr{D}_{12}}$, one can think of $\lambda_{\mathscr{D}_1}$ and $\lambda_{\mathscr{D}_2}$ as being transferred (by the isomorphisms $\mathscr{D}_1 \to \tilde{\mathscr{D}}_1$, $\mathscr{D}_2 \to \tilde{\mathscr{D}}_2$) to $\tilde{\mathscr{D}}_1$, $\tilde{\mathscr{D}}_2$ as the functions $\lambda_{\tilde{Q}_1}$, $\lambda_{\tilde{Q}_2}$.

Then the requirements of the theorem become

$$\lambda_{\mathscr{D}_{12}}(\tilde{a}_1 \cap \tilde{a}_2, \tilde{a}_1' \cap \tilde{a}_2) = \lambda_{\tilde{\mathscr{D}}_1}(\tilde{a}_1, \tilde{a}_1'),$$
$$\lambda_{\mathscr{D}_{12}}(\tilde{a}_1 \cap \tilde{a}_2, \tilde{a}_1 \cap \tilde{a}_2') = \lambda_{\tilde{\mathscr{D}}_2}(\tilde{a}_2, \tilde{a}_2');$$

hence one can immediately adopt the methods from the proof of theorems Th 4.5.1, Th 4.5.2 and Th 4.5.4 in [2] II:

Pick $a, a' \in \mathscr{D}_{12}$ with $a' \subset a$. Due to Th 4.5.1 from [2] II, it follows that there are an element $\tilde{a}_1 \cap \tilde{a}_2 \in \Gamma_{12}$ (Γ_{12} from (2.3)) with $a \subset \tilde{a}_1 \cap \tilde{a}_2$ and elements $\tilde{a}_1^{(i)}$, $\tilde{a}_1^{(v)} \in \tilde{\mathscr{D}}_1'$; $\tilde{a}_2^{(k)}, \tilde{a}_2^{(\mu)} \in \tilde{\mathscr{D}}_2'$ with $\tilde{a}_1^{(i)} \subset \tilde{a}_1$, $\tilde{a}_1^{(v)} \subset \tilde{a}_1$, $\tilde{a}_2^{(k)} \subset \tilde{a}_2$, $\tilde{a}_2^{(\mu)} \subset \tilde{a}_2$ and $\tilde{a}_1^{(i)} \cap a_1^{(k)} = \emptyset$ for $i \neq k$, $\tilde{a}_1^{(v)} \cap \tilde{a}_1^{(\varrho)} = \emptyset$ for $v \neq \varrho$, $\tilde{a}_2^{(k)} \cap \tilde{a}_2^{(l)} = \emptyset$ for $k \neq l$, $\tilde{a}_2^{(\mu)} \cap \tilde{a}_2^{(\sigma)} = \emptyset$ for $\mu \neq \sigma$ and

$$a' = \bigcup_{(v,\mu) \in B} \tilde{a}^{(v)} \cap \tilde{a}_2^{(\mu)}, \qquad a = \bigcup_{(i,k) \in A} \tilde{a}_1^{(i)} \cap \tilde{a}_1^{(k)}, \tag{2.5}$$

where B is a subset of the pairs (v, μ) and A is a subset of the pairs (i, k). From

$$\lambda_{\mathscr{D}_{12}}(\tilde{a}_1 \cap \tilde{a}_2, a') = \lambda_{\mathscr{D}_{12}}(\tilde{a}_1 \cap \tilde{a}_2, a)\, \lambda_{\mathscr{D}_{12}}(a, a')$$

we then get

$$\lambda_{\mathscr{D}_{12}}(a, a') = \frac{\lambda_{\mathscr{D}_{12}}(\tilde{a}_1 \cap \tilde{a}_2, a')}{\lambda_{\mathscr{D}_{12}}(\tilde{a}_1 \cap \tilde{a}_2, a)}. \tag{2.6}$$

With (2.5) follows

$$\lambda_{\mathscr{D}_{12}}(\tilde{a}_1 \cap \tilde{a}_2, a') = \sum_{(v,\mu) \in B} \lambda_{\mathscr{D}_{12}}(\tilde{a}_1 \cap \tilde{a}_2, \tilde{a}_1^{(v)} \cap \tilde{a}_2^{(\mu)}),$$
$$\lambda_{\mathscr{D}_{12}}(\tilde{a}_1 \cap \tilde{a}_2, a) = \sum_{(i,k) \in A} \lambda_{\mathscr{D}_{12}}(\tilde{a}_1 \cap \tilde{a}_2, \tilde{a}_1^{(i)} \cap \tilde{a}_2^{(k)}).$$

By (2.4) one finally obtains

$$\begin{aligned} \lambda_{\mathscr{D}_{12}}(\tilde{a}_1 \cap \tilde{a}_2, a') &= \sum_{(v,\mu) \in B} \lambda_{\mathscr{D}_1}(a_1, a_1^{(v)})\, \lambda_{Q_2}(a_2, a_2^{(\mu)}). \\ \lambda_{\mathscr{D}_{12}}(\tilde{a}_1 \cap \tilde{a}_2, a) &= \sum_{(i,k) \in A} \lambda_{\mathscr{D}_1}(a_1, a_1^{(i)})\, \lambda_{Q_2}(a_2, a_2^{(k)}). \end{aligned} \tag{2.7}$$

This together with (2.6) shows that $\lambda_{\mathscr{D}_{12}}$ is determined uniquely by $\lambda_{\mathscr{D}_1}$ and $\lambda_{\mathscr{D}_2}$.

In order to show that one can define a function $\lambda_{\mathscr{D}_{12}}$ for prescribed $\lambda_{\mathscr{D}_1}$, $\lambda_{\mathscr{D}_2}$, one uses (2.6) with (2.7) as defining equations. As in the proof of Th 4.5.4 in [2] II, one shows that these definitions are meaningful, i.e. do not depend on the choice of $\tilde{a}_1$, $\tilde{a}_2$ and of the $\tilde{a}_1^{(i)}$, $\tilde{a}_2^{(k)}$, $\tilde{a}_1^{(v)}$, $\tilde{a}_2^{(\mu)}$. For the $\lambda_{Q_{12}}$ thus defined one verifies AS 2.1 through AS 2.3 from II § 2.1. Since the methods of proof are given in detail for the theorem Th 4.5.4 from [2] II, these hints may suffice. □

Q_{12} together with the probability $\lambda_{\mathscr{D}_{12}}$ defined according to T 2.4 will be called the set of preparation procedures of the composite systems from M. This $\mathscr{D}_{12}$ as well as $\lambda_{\mathscr{D}_{12}}$ are pictures of physically real situations (see [3] § 10.5), whose physical interpretation arises from the construction of Q_{12} and the general interpretation of a probability.

Since we shall only consider systems that are somehow prepared, we further require

APSZ 8.1
$$\bigcup_{a \in \mathscr{Q}_{12}} a = M .$$

This axiom does not render any profound physical structure; if it were not valid, one could simply go over from M to the set $\tilde{M} = \bigcup_{a \in \mathscr{Q}_{12}} a$. Therefore, APSZ 8.1 only characterizes the fundamental domain of the theory we envisage.

Since (as stated in the proof of T 2.4) every $a \in \mathscr{Q}_{12}$ is a subset of an $\tilde{a} \in \Gamma_{12}$, from APSZ 8.1 also follows

$$M = \bigcup_{a \in \Gamma_{12}} a . \tag{2.8}$$

For each $x_1 \in M_1$, due to AZ 2 there is an $x_2 \in M_2$ with $(x_1, x_2) \in M$. Hence by (2.8) we have an $a_1 \in \mathscr{Q}_1'$ and an $a_2 \in \mathscr{Q}_2'$ with $(x_1, x_2) \in a_1 \times a_2 \cap M$. Consequently there is an a_1 with $x_1 \in a_1$, so that APSZ 8.1 yields

$$M_1 = \bigcup_{a \in \mathscr{Q}_1} a \tag{2.9}$$

and likewise

$$M_2 = \bigcup_{a \in \mathscr{Q}_2} a . \tag{2.10}$$

Similarly to AZ 1, 2, the axiom APSZ 8.1 determines what situations we want to consider in the fundamental domain of the theory. Hence the same holds as elucidated about APSZ 6, 7 on page 57. Also the introduction of $\mathscr{Q}_{12}$ with $\lambda_{\mathscr{Q}_{12}}$ (in T 2.4) only determines more precisely how the composite systems should be prepared. Thus $\mathscr{Q}_{12}$ with $\lambda_{\mathscr{Q}_{12}}$ is a derived structure, i.e. the physical possibility to prepare as prescribed by $\mathscr{Q}_{12}$ is already given by the axioms APSZ 1, APSZ 8.1, as due to APSZ 8.1 the set Γ_{12} cannot be empty (see (2.8)).

Having over M introduced a structure $\mathscr{Q}_{12}$ of the preparation procedure species, we must yet describe the registration (interpretable by pretheories) of the systems from M_1 and M_2. To this end, we introduce four structure terms $\mathscr{R}_{10}$, $\mathscr{R}_1$, $\mathscr{R}_{20}$, $\mathscr{R}_2$ with $\mathscr{R}_{10} \subset \mathscr{P}(M_1)$, $\mathscr{R}_1 \subset \mathscr{P}(M_1)$, $\mathscr{R}_{20} \subset \mathscr{P}(M_2)$, $\mathscr{R}_2 \subset \mathscr{P}(M_2)$. Due to the general axioms of a registration procedure (see II § 2.3 or [2] II § 4.2 or [3] § 12) we first require

APSZ 2 $\mathscr{R}_1$ and $\mathscr{R}_2$ are structures of species SP.
APSZ 3 $\mathscr{R}_{10}$ and $\mathscr{R}_{20}$ are structures of species SSP.
APSZ 4.1 $\mathscr{R}_{10} \subset \mathscr{R}_1$, $\mathscr{R}_{20} \subset \mathscr{R}_2$;
APSZ 4.2 For each $b \in \mathscr{R}_i$ there is a $b_0 \in \mathscr{R}_{i0}$ with $b \subset b_0$, $i = 1, 2$.

Let the probability functions for $\mathscr{R}_{10}$ and $\mathscr{R}_{20}$ be denoted by $\lambda_{\mathscr{R}_{10}}$, $\lambda_{\mathscr{R}_{20}}$. Due to [3] § 9 we require that $\mathscr{R}_1$, $\mathscr{R}_2$ are countable sets; hence $\mathscr{R}_{10}$, $\mathscr{R}_{20}$ are countable.

Since the zero point of the laboratory time scale has been chosen so that the preparation is completed before $t = 0$ and the registration begins after $t = 0$ (a convention adopted in II), no restrictions should occur for combining a preparation and a registration. Moreover, we will for simplicity assume that no restrictions occur for combining the registration of the systems from M_1 and M_2, since "trajectory registration processes" are concerned. We express this by

APSZ 5.1 From $a_1 \in \mathscr{Q}_1'$, $a_2 \in \mathscr{Q}_2'$, $b_{10} \in \mathscr{R}_{10}'$, $b_{20} \in \mathscr{R}_{20}'$ and a_1, a_2 being combinable there follows $(a_1 \cap b_{10}) \times (a_2 \cap b_{20}) \cap M \neq \emptyset$.

We might replace this by a weaker axiom. It would allow preparation procedures which terminate at some times $t > 0$ and registration procedures which begin already at some $t < 0$. But then we would need the following condition for combining preparation and registration procedures. They may only be combined when the preparation procedure terminates before the registration procedure begins. Also it would be necessary to introduce several trajectory spaces corresponding to the times at which the trajectories begin. We avoid all these complications by the stronger axiom APSZ 5.1.

Since this axiom APSZ 5.1 is normative, one cannot ask for facts "in nature" which contradict it. One can only ask, whether the normative postulate APSZ 5.1 excludes experiments essential for the erection of quantum mechanics. This seems not to be the case.

APSZ 5.1 implies the relation: a_1, a_2 being combinable implies $a_1 \cap b_{10} \neq \emptyset$. From AZ 2 and (2.8) follows that for arbitrary $a_1 \in \mathscr{Q}_1'$ there is an $(x_1, x_2) \in M$ with $x_1 \in a_1$ and furthermore a combinable pair a_1'', a_2 with $(x_1, x_2) \in a_1'' \times a_2$. Therefore $a_1' = a_1 \cap a_1''$ may also be combined with a_2. Hence we find $a_1' \cap b_{10} \neq \emptyset$, and thus also $a_1 \cap b_{10} \neq \emptyset$. Therefore, $a_1 \in \mathscr{Q}_1'$, $b_{10} \in \mathscr{R}_{10}'$ implies $a_1 \cap b_{10} \neq \emptyset$, i.e. we have a relation analogous to APS 5 m from II § 2.4. Likewise follows that $a_2 \in \mathscr{Q}_2'$, $b_{20} \in \mathscr{R}_{20}'$ implies $a_2 \cap b_{20} \neq \emptyset$.

By virtue of (2.8), from APSZ 5.1 also follows that $b_{10} \in \mathscr{R}_{10}'$, $b_{20} \in \mathscr{R}_{20}'$ implies $b_{10} \times b_{20} \cap M \neq \emptyset$.

For the definition D 2.2 carried over from $\mathscr{Q}_1, \mathscr{Q}_2$ to $\mathscr{R}_{10}, \mathscr{R}_{20}$, this asserts that all the elements of $\mathscr{R}_{10}'$ may be combined with all the elements of $\mathscr{R}_{20}'$. Therefore, T 2.4 immediately carries over to

T 2.5 On the structure $\mathscr{R}_{120}$ (of species SP) generated by the set

$$\{b_{10} \times b_{20} \cap M \mid b_{10} \in \mathscr{R}_{10},\ b_{20} \in \mathscr{R}_{20}\}$$

one can uniquely define a probability $\lambda_{\mathscr{R}_{120}}$ by the requirements

$$\lambda_{\mathscr{R}_{120}}(b_{10} \times b_{20} \cap M,\ b_{10}' \times b_{20} \times M) = \lambda_{\mathscr{R}_{10}}(b_{10}, b_{10}')\,,$$

$$\lambda_{\mathscr{R}_{120}}(b_{10} \times b_{20} \cap M,\ b_{10} \times b_{20}' \cap M) = \lambda_{\mathscr{R}_{20}}(b_{20}, b_{20}')\,;$$

then one finds

$$\lambda_{\mathscr{R}_{120}}(b_{10} \times b_{20} \cap M,\ b_{10}' \times b_{20}' \cap M) = \lambda_{\mathscr{R}_{10}}(b_{10}, b_{10}')\,\lambda_{\mathscr{R}_{20}}(b_{20}, b_{20}')\,.$$

From T 2.2 follows immediately

T 2.6 The sets $\tilde{\mathscr{R}}_1 = \{b_1 \times M_2 \cap M \mid b \in \mathscr{R}_1\}$ and $\tilde{\mathscr{R}}_2 = \{M_1 \times b_2 \cap M \mid b_2 \in \mathscr{R}_2\}$ are structures of species SP and the mappings $b_1 \to b_1 \times M_2 \cap M$ and $b_2 \to M_1 \times b_2 \cap M$ represent isomorphisms $\mathscr{R}_1 \to \tilde{\mathscr{R}}$, $\mathscr{R}_2 \to \tilde{\mathscr{R}}_2$.

By $\mathscr{R}_{12}$ let us denote the selection procedure structure generated over M by the set

$$\{b_1 \times b_2 \cap M \mid b_1 \in \mathscr{R}_1, b_2 \in \mathscr{R}_2\} = \{\tilde{b}_1 \cap \tilde{b}_2 \mid \tilde{b}_1 \in \tilde{\mathscr{R}}_1, b_2 \in \tilde{\mathscr{R}}_2\}.$$

$\mathscr{R}_{120}$ from T 2.5 and $\mathscr{R}_{12}$ therefore satisfy, because of T 2.5, the relations APS 2, APS 3 and APS 4.1 (as formulated in II § 2.3 or [2] II § 4.2, or [3] § 12.1). We will show that APS 4.2 also holds.

According to T 2.6, $\mathscr{R}_{12}$ is the SP structure generated by $\{\tilde{b}_1 \cap \tilde{b}_2 \mid \tilde{b}_1 \in \tilde{\mathscr{R}}_1, \tilde{b}_2 \in \tilde{\mathscr{R}}_2\}$, where $\tilde{\mathscr{R}}_1$ and $\tilde{\mathscr{R}}_2$ are SP structures. By Th 4.5.1 in [2] II, there thus exists for each $b \in \mathscr{R}_{12}$ a pair $\tilde{b}_1 \in \tilde{\mathscr{R}}_1$, $\tilde{b}_2 \in \tilde{\mathscr{R}}_2$, where $b \subset \tilde{b}_1 \cap \tilde{b}_2$. Because of $\tilde{b}_1 = b_1 \times M_2 \cap M$, $\tilde{b}_2 = M_1 \times b_2 \cap M$, where $b_1 \in \mathscr{R}_1$, $b_2 \in \mathscr{R}_2$, by APSZ 4.2 there exist $b_{10} \supset b_1$, $b_{20} \supset b_2$, i.e. a $\tilde{b}_{10} = b_{10} \times M_2 \cap M$ and a $\tilde{b}_{20} = M_1 \times b_{20} \cap M$ such that $b \subset \tilde{b}_{10} \cap \tilde{b}_{20} \in \mathscr{R}_{120}$, whereby APS 4.2 is proved.

Calling $\mathscr{R}_{120}$, $\mathscr{R}_{12}$ the registration structure of the composite systems from M, we need not emphasize again that $\mathscr{R}_{120}$, $\mathscr{R}_{12}$ and $\lambda_{\mathscr{R}_{120}}$ portray real situations. Just as little need we speak of the physical interpretation.

Corresponding to APSZ 8.1 we also demand

APSZ 8.2
$$\bigcup_{b \in \mathscr{R}_{12}} b = M.$$

Since $\mathscr{R}_{120}$, $\mathscr{R}_{12}$ satisfy APS 4.2, it follows that

$$M = \bigcup_{b_0 \in \mathscr{R}_{120}} b_0 \tag{2.11}$$

and similar to (2.8):

$$M = \bigcup_{\substack{b_{10} \in \mathscr{R}_{10} \\ b_{20} \in \mathscr{R}_{20}}} (b_{10} \times b_{20} \cap M). \tag{2.12}$$

From this again follows

$$M_1 = \bigcup_{b_{10} \in \mathscr{R}_{10}} b_{10} \tag{2.13}$$

and

$$M_2 = \bigcup_{b_{20} \in \mathscr{R}_{20}} b_{20}. \tag{2.14}$$

Again corresponding to D 2.3, let us define

D 2.4 An $a \in \mathscr{Q}'_{12}$ and a $b_0 \in \mathscr{R}_{120}$ are called combinable if $\emptyset \neq a' \subset a$, $\emptyset \neq b'_0 \subset b_0$ always imply $a' \cap b'_0 \neq \emptyset$.

We introduce the set

$$C_{12} = \{(a, b_0) \mid a \in \mathscr{Q}'_{12}, b_0 \in \mathscr{R}'_{120}, a \text{ and } b_0 \text{ may be combined}\}. \tag{2.15}$$

T 2.7 $C_{12} = \mathscr{Q}'_{12} \times \mathscr{R}'_{120}$.

Proof. We need only show that $a \in \mathscr{Q}'_{12}$, $b_0 \in \mathscr{R}'_{120}$ implies $a \cap b_0 \neq \emptyset$. Due to T 4.5.1 in [2] II, each $a \in \mathscr{Q}'_{12}$ is a union of elements of the form $\tilde{a}_1 \cap \tilde{a}_2 \in \Gamma_{12}$, where $\tilde{a}_1 = a_1 \times M_2 \cap M$, $\tilde{a}_2 = M_1 \times a_2 \cap M$. Similarly follows that each element of $\mathscr{R}'_{120}$ is a union of elements $\tilde{b}_{01} \cap \tilde{b}_{02}$, where $\tilde{b}_{01} = b_{10} \times M_2 \cap M$, $\tilde{b}_{02} = M_1 \times b_{20} \cap M$ and $b_{10} \in \mathscr{R}'_{10}$, $b_{20} \in \mathscr{R}'_{20}$.

The theorem is thus proved when $\tilde{a}_1 \cap \tilde{a}_2 \cap \tilde{b}_{01} \cap \tilde{b}_{02} \neq \emptyset$ in case $\tilde{a}_1 \cap \tilde{a}_2 = a_1 \times a_2 \cap M \in \Gamma_{12}$. But this is just the assertion of APSZ 5.1. □

For brevity, we write

$$\Theta_{12} = \{a \cap b \mid a \in \mathscr{Q}'_{12},\ b \in \mathscr{R}_{12}\}, \tag{2.16}$$

and by $\mathscr{S}_{12}$ denote the SP structure generated over M by Θ_{12}.

As axiom to describe the coupled systems we impose on $\mathscr{R}_{120}$, $\mathscr{R}_{12}$, $\mathscr{S}_{12}$ the general axioms from [2] II or [3] § 12 (for M_m already mentioned in II § 2):

APSZ 6 $\mathscr{S}_{12}$ is an SSP.

For the probability function λ_{12} corresponding to $\mathscr{S}_{12}$, let us require

APSZ 7 For $a_1, a_2 \in \mathscr{Q}_{12}$ with $a_2 \subset a_1$, and for $b_{01}, b_{02} \in \mathscr{R}_{120}$ with $b_{02} \subset b_{01}$ and $(a_1, b_{01}) \in C_{12}$, we presume

APSZ 7.1 $\lambda_{12}(a_1 \cap b_{01}, a_2 \cap b_{01}) = \lambda_{\mathscr{Q}_{12}}(a_1, a_2)$;
APSZ 7.2 $\lambda_{12}(a_1 \cap b_{01}, a_1 \cap b_{02}) = \lambda_{\mathscr{R}_{120}}(b_{01}, b_{02})$.

Because of T 2.7, the structures trivially satisfy all the axioms APS 5 in [3] § 12.1 resp. in [2] III § 1. This together with the axioms APSZ 6 through APSZ 8 implies that M with the structures $\mathscr{Q}_{12}$, $\mathscr{R}_{120}$, $\mathscr{R}_{12}$ forms a set of "physical systems" (in the sense of [3] § 12.1 resp. [2] II § 4.4).

Also note again that the axioms APSZ 6 and APSZ 7 not so much describe structures readable from nature. Rather they characterize the fundamental domain of the theory, i.e. the domain of the "correctly" performed experiments. APSZ 6 requires to consider experiments to which the probability concept is applicable. APSZ 7 says that in a "correct" experiment the preparation procedures and the registration methods must not disturb each other. Therefore, APSZ 6, 7 are "natural laws" in the sense that it is possible to experiment as these axioms demand.

T 2.8 The function λ_{12} is completely determined by $\lambda_{\mathscr{Q}_1}$, $\lambda_{\mathscr{Q}_2}$ and by the special values

$$\lambda_{12}(a_1 \times a_2 \cap M \cap b_{10} \times b_{20},\ a_1 \times a_2 \cap M \cap b_1 \times b_2)$$

for $a_1 \times a_2 \cap M \in \Gamma_{12}$ (with Γ_{12} as in (2.1)) and $b_{10} \in \mathscr{R}_{10}$, $b_{20} \in \mathscr{R}_{20}$, $b_1 \in \mathscr{R}_1$, $b_2 \in \mathscr{R}_2$, $b_1 \subset b_{01}$, $b_2 \subset b_{02}$.

Proof. From Th 4.5.2 in [2] II follows that λ_{12} is completely determined by $\lambda_{\mathscr{Q}_{12}}$ and by the special values $\lambda_{12}(a \cap b_0, a \cap b)$, where $a \in \mathscr{Q}'_{12}$, $b_0 \in \mathscr{R}_{120}$, $b \in \mathscr{R}_{12}$, $b \subset b_0$. According to T 2.4, $\lambda_{\mathscr{Q}_{12}}$ is determined by $\lambda_{\mathscr{Q}_1}$ and $\lambda_{\mathscr{Q}_2}$.

Due to Th 4.5.1 of [2] II, a is of the form $a = \bigcup_{(i,k) \in A} \tilde{a}_1^{(i)} \cap \tilde{a}_2^{(k)}$, with $\tilde{a}_1^{(i)} \in \tilde{\mathscr{Q}}_1$, $\tilde{a}_2^{(k)} \in \tilde{\mathscr{Q}}_2$ and with $\tilde{a}_1^{(i)} \subset \tilde{a}_1 \in \tilde{\mathscr{Q}}_1$, $\tilde{a}_2^{(k)} \subset \tilde{a}_2 \in \tilde{\mathscr{Q}}_2$ and $\tilde{a}_1^{(i)} \cap \tilde{a}_1^{(j)} = \emptyset$ for $i \neq j$, $\tilde{a}_2^{(k)} \cap a_2^{(l)} = \emptyset$ for $k \neq l$. Thus follows

$$\begin{aligned}\lambda_{12}(\tilde{a}_1 \cap \tilde{a}_2 \cap b_0, a \cap b) &= \lambda_{12}(\tilde{a}_1 \cap \tilde{a}_2 \cap b_0, a \cap b_0)\, \lambda_{12}(a \cap b_0, a \cap b) \\ &= \lambda_{\mathscr{Q}_{12}}(\tilde{a}_1 \cap \tilde{a}_2, a)\, \lambda_{12}(a \cap b_0, a \cap b),\end{aligned}$$

and furthermore

$$\lambda_{12}(\tilde{a}_1 \cap \tilde{a}_2 \cap b_0, a \cap b) = \sum_{(i,k) \in A} \lambda_{12}(\tilde{a}_1 \cap \tilde{a}_2 \cap b_0, \tilde{a}_1^{(i)} \cap \tilde{a}_2^{(k)} \cap b).$$

Therefore, λ_{12} is determined by the values $\lambda_{12}(a \cap b_0, a \cap b)$ with the special $a = a_1 \times a_2 \cap M \in \Gamma_{12}$.

As for $a \in \mathscr{D}_{12}$, for $b_0 \in \mathscr{R}_{120}$ and $b \in \mathscr{R}_{12}$ we can from Th 4.5.1 of [2] II conclude

$$b_0 = \bigcup_{(i,k) \in A} \tilde{b}_{10}^{(i)} \cap \tilde{b}_{20}^{(k)}, \ \tilde{b}_{10}^{(i)} \in \tilde{\mathscr{R}}_{10}, \ \tilde{b}_{20}^{(k)} \in \tilde{\mathscr{R}}_{20}, \ \tilde{b}_{10}^{(i)} \subset \tilde{b}_{10} \in \tilde{\mathscr{R}}_{10}, \ \tilde{b}_{20}^{(k)} \subset \tilde{b}_{20} \in \mathscr{R}_{20}, \text{ etc.};$$

$$b = \bigcup_{(\nu,\mu) \in B} \tilde{b}_1^{(\nu)} \cap \tilde{b}_2^{(\mu)}, \ \tilde{b}_1^{(\nu)} \in \tilde{\mathscr{R}}_1, \text{ etc.}$$

Because of $b \subset b_0$ we obtain

$$b = b \cap b_0 = \bigcup_{\substack{(i,k) \in A \\ (\nu,\mu) \in B}} (\tilde{b}_{10}^{(i)} \cap b_1^{(\nu)}) \cap (\tilde{b}_{20}^{(k)} \cap \tilde{b}_2^{(\mu)});$$

thus we can also assume $\tilde{b}_1^{(\nu)} \subset \tilde{b}_{10}$, $\tilde{b}_2^{(\mu)} \subset \tilde{b}_{20}$. Then follows

$$\lambda_{12}(a \cap \tilde{b}_{10} \cap \tilde{b}_{20}, a \cap b) = \lambda_{12}(a \cap \tilde{b}_{10} \cap \tilde{b}_{20}, a \cap b_0)\, \lambda_{12}(a \cap b_0, a \cap b),$$

hence furthermore

$$\lambda_{12}(a \cap \tilde{b}_{10} \cap \tilde{b}_{20}, a \cap b_0) = \sum_{(i,k) \in A} \lambda_{12}(a \cap \tilde{b}_{10} \cap \tilde{b}_{20}, a \cap \tilde{b}_{10}^{(i)} \cap \tilde{b}_{20}^{(k)})$$

and

$$\lambda_{12}(a \cap \tilde{b}_{10} \cap \tilde{b}_{20}, a \cap b) = \sum_{(\nu,\mu) \in B} \lambda_{12}(a \cap \tilde{b}_{10} \cap \tilde{b}_{20}, a \cap b_1^{(\nu)} \cap b_2^{(\mu)}). \quad \square$$

Theorem T 2.8 suggests to consider the function

$$\mu_{12}(a, (b_0, b)) = \lambda_{12}(a \cap b_0, a \cap b), \tag{2.17}$$

(defined analogously to (2.4.4) of II), where $a \in Q'_{12}$, $b_0 \in \mathscr{R}_{120}$, $b \in \mathscr{R}_{12}(b_0)$ and $(a, b_0) \in C_{12}$. In particular, for $a = a_1 \times a_2 \cap M$, $b_0 = b_{10} \times b_{20} \cap M$, $b = b_1 \times b_2 \cap M$ it becomes

$$\mu_{12}(a_1 \times a_2 \cap M, (b_{10} \times b_{20} \cap M, b_1 \times b_2 \cap M))$$
$$= \lambda_{12}(a_1 \times a_2 \cap M \cap b_{10} \times b_{20}, a_1 \times a_2 \cap M \cap b_1 \times b_2). \tag{2.18}$$

According to T 2.8, $\mu_{12}(a_1 \times a_2 \cap M, (b_{10} \times b_{20} \cap M, b_1 \times b_2 \cap M))$ together with $\lambda_{\mathscr{D}_1}, \lambda_{\mathscr{D}_2}$ therefore determines all of λ_{12} and thus the interaction of the systems.

Let it yet be noted that the discription of the registration procedure (to be introduced in § 6 as the "trajectory registration procedure") will in § 6.3 imply that

$$\lambda_{12}(a_1 \times a_2 \cap M \cap b_{10} \times b_{20}, a_1 \times a_2 \cap M \cap b_{10} \times b_2)$$

does not depend on b_{10} and

$$\lambda_{12}(a_1 \times a_2 \cap M \cap b_{10} \times b_{20}, a_1 \times a_2 \cap M \cap b_1 \times b_{20})$$

does not depend on b_{20} (see the explanations after (6.3.11)). Therefore we will *not* introduce the corresponding axioms here; but see the partial implications of APSZ 9 and APSZ 5.3 in § 3.

§ 3 Directed Interactions

We now decisively narrow the fundamental domain $\mathscr{G}$ ([3] §§ 1–6) of our theory for the purpose of discovering the microsystems. Thus we consider only those composite systems (x_1, x_2), where x_1 acts on x_2 but *not* conversely. We briefly say that the interaction is directed from x_1 to x_2. But we must first formulate by a mathematical relation what is meant by "x_2 does not act on x_1" (expressed differently, "the interaction is *directed* from x_1 to x_2"). The mathematical relation to define the directedness in the action of x_1 on x_2 shall at once be imposed as the axiom

APSZ 9 The function

$$\lambda_{12}(a_1 \times a_2 \cap M \cap b_{10} \times b_{20}, a_1 \times a_2 \cap M \cap b_1 \times b_{20})$$
$$= \lambda_{12}((a_1 \cap b_{10}) \times (a_2 \cap b_{20}) \cap M, (a_1 \cap b_1) \times (a_2 \cap b_{20}) \cap M), \tag{3.1}$$

defined for $a_1 \times a_2 \cap M \in \Gamma_{12}$, $b_{10} \in \mathscr{R}_{10}$, $b_1 \in \mathscr{R}_1$, $b_{20} \in \mathscr{R}_{20}$, does not depend on a_2 and b_{20}. (That (3.1) does not depend on b_{20} holds in greater generality; see the remarks at the end of § 2, then the remarks after (6.3.11) and (6.4.1).)

That the function (3.1) does not depend on a_2 and b_{20} expresses the fact that on the systems from M_1 the effect procedures (b_{10}, b_1) yield frequencies *independent* of the selection $a_2 \cap b_{20}$ of the systems from M_2 that are coupled in. Thus the trajectories of the subsystems x_1 in the composite systems (x_1, x_2) do not depend on which systems are "close by" (whenever x_1, x_2 are combinable) and by which method the x_2 are registered.

The introduction of axiom APSZ 9 shows directly that in no way is it a law found in nature, but rather a "choice" of the fundamental domain. In the philosophy of science it is often emphasized that such "laws" must be distinguished from the laws of nature. Protophysics (see the introduction in [42]) has coined the word "norms" for such "selection laws" in order to express that they are "laws imposed" by the people treating the problem. In [3] and [30], we have for "all" laws introduced the designation "natural laws". But axiom APSZ 9 only insofar describes "realities" as it declares some selection to be satisfiable, i.e. that it is possible to select a nonempty fundamental domain obeying APSZ 9.

In [3] and [30] we have not distinguished the various sorts of natural laws in the axiomatic basis since a further classification appears in general very difficult. For, there often occur in one and the same axiom selection principles as well as invariant structures of reality and also idealizations. Up to now, only for each particular case it appears possible to point out the various aspects of an axiom. Only in rare cases *just one* of those three aspects enter e.g. in APSZ 9. In AVid from VI § 3, we shall meet an axiom that introduces *only* an idealization, i.e. that has nothing to do with "norms" or "invariant real structures".

These remarks about the difficulties in classifying natural laws (the concept of natural law meant as broadly as in [3], [30]) shall of course not hinder us in contemplating such classifications. Rather we may try to formulate each axiom of an $\mathscr{MT}_\Sigma$ so that it either contains only invariant structures of reality (and perhaps idealizations) or only the norms to determine the fundamental domain (and perhaps idealizations) or only idealizations. In XIII we shall in view of this question review the axiomatic basis presented here.

Physicists prefer to investigate *directed* interactions when they want to find out some regularity. The general interaction situation, where APSZ 9 is violated, can contain complicated recouplings of the systems $x_1 \in M_1$ and $x_2 \in M_2$, which are not so simple to analyze, but just for technical purposes can be very interesting.

T 3.1 If $a_1 \in \mathscr{Q}'_1$, $b_1 \in \mathscr{R}_1$ and $a_1 \cap b_1 \neq \emptyset$, then there is an $a'_1 \in \mathscr{Q}'_1$ with $a'_1 \subset a_1$ and an $a'_2 \in \mathscr{Q}'_2$ combinable with a'_1, so that $(a'_1 \cap b_1) \times a'_2 \cap M \neq \emptyset$.

Proof. By AZ 2 follows from $a_1 \cap b_1 \neq \emptyset$ that there is an $(x_1, x_2) \in M$ with $x_1 \in a_1 \cap b_1$. From (2.8) follows that there are two combinable a''_1, a'_2 with $(x_1, x_2) \in a''_1 \times a'_2$. For $a'_1 = a_1 \cap a''_1$ thus follows that a'_1, a'_2 may be combined and that $(x_1, x_2) \in (a'_1 \cap b_1) \times a'_2 \cap M$ holds. □

T 3.2 If an $a_1 \in \mathscr{Q}'_1$ may be combined with $a_2 \in \mathscr{Q}'_2$ and makes $a_1 \cap b_1 \neq \emptyset$, then $(a_1 \cap b_1) \times (a_2 \cap b_{20}) \cap M \neq \emptyset$ for all $b_{20} \in \mathscr{R}'_{20}$.

Proof. For $b_{10} \in \mathscr{R}'_{10}$ and $b_{10} \supset b_1$ and $b_{20} \in \mathscr{R}_{20}$, from APSZ 5.1 follows $(a_1 \cap b_{10}) \times (a_2 \cap b_{20}) \cap M \neq \emptyset$. Therefore

$$\lambda_{12}((a_1 \cap b_{10}) \times (a_2 \cap b_{20}) \cap M, (a_1 \cap b_1) \times (a_2 \cap b_{20}) \cap M)$$

is defined (see T 2.7 and (2.16)). For $a'_1 \subset a_1$ follows

$$\lambda_{12}((a_1 \cap b_{10}) \times (a_2 \cap b_{20}) \cap M, (a_1 \cap b_1) \times (a_2 \cap b_{20}) \cap M)$$
$$\geqq \lambda_{12}((a_1 \cap b_{10}) \times (a_2 \cap b_{20}) \cap M, (a'_1 \cap b_1) \times (a_2 \cap b_{20}) \cap M).$$

If a_1, a_2 are combinable, so are a'_1, a_2, and thus APSZ 5.1 gives $(a' \cap b_{10}) \times (a_2 \cap b_{20}) \cap M \neq \emptyset$. Since a'_1 may also be combined with a'_2 (a'_2 as in T 3.1), also $(a_1' \cap b_{10}) \times (a'_2 \cap b_{20}) \cap M \neq \emptyset$ follows. Using APSZ 9, we therefore obtain

$$\begin{aligned}
&\lambda_{12}((a_1 \cap b_{10}) \times (a_2 \cap b_{20}) \cap M, (a'_1 \cap b_1) \times (a_2 \cap b_{20}) \cap M) \\
&\quad = \lambda_{12}((a_1 \cap b_{10}) \times (a_2 \cap b_{20}) \cap M, (a'_1 \cap b_{10}) \times (a_2 \cap b_{20}) \cap M) \\
&\qquad \cdot \lambda_{12}((a'_1 \cap b_{10}) \times (a_2 \cap b_{20}) \cap M, (a'_1 \cap b_1) \times (a_2 \cap b_{20}) \cap M) \\
&\quad = \lambda_{\mathscr{Q}_1}(a_1, a'_1)\, \lambda_{12}((a'_1 \cap b_{10}) \times (a'_2 \cap b_{20}) \cap M, (a'_1 \cap b_1) \times (a'_2 \cap b_{20}) \cap M).
\end{aligned}$$

Because $a'_1 \neq \emptyset$, we have $\lambda_{\mathscr{Q}_1}(a_1, a'_1) \neq 0$. Due to T 3.1, $(a'_1 \cap b_1) \times a'_2 \cap M \neq \emptyset$. If $(a'_1 \cap b_1) \times (a'_2 \cap b_{20}) \cap M$ were equal to $\emptyset$ for all $b_{20} \in \mathscr{R}'_{20}$, then (2.14) would also yield $(a'_1 \cap b_1) \times a'_2 \cap M = \emptyset$; hence there is a b_{20} with

$$\lambda_{12}((a'_1 \cap b_{10}) \times (a'_2 \cap b_{20}) \cap M, (a'_1 \cap b_1) \times (a'_2 \cap b_{20}) \cap M) \neq \emptyset.$$

But since, according to APSZ 9, all this is independent of b_{20}, it also holds for all $b_{20} \in \mathscr{R}'_{20}$. Thus we also conclude

$$\lambda_{12}((a_1 \cap b_{10}) \times (a_2 \cap b_{20}) \cap M, (a_1 \cap b_1) \times (a_2 \cap b_{20}) \cap M) \neq 0$$

and hence

$$(a_1 \cap b_1) \times (a_2 \cap b_{20}) \cap M \neq \emptyset \text{ for all } b_{20} \in \mathscr{R}_{20}. \quad \square$$

For a directed interaction, the following SP structure over M_1 plays a large role: Let $\mathscr{S}_1$ denote the SP structure generated over M_1 by the set

$$\{a_1 \cap b_1 \mid a_1 \in Q_1, b_1 \in \mathscr{R}_1\}. \tag{3.2}$$

This structure describes the possibilities of selecting systems from M_1. Because of APSZ 9 we can expect that by λ_{12} the structure $\mathscr{S}_1$ becomes an SSP structure, since it is "immaterial" which systems from M_2 are combined with the systems from M_1.

D 3.1 We say that a $c_1 \in \mathscr{S}_1$ with $c_1 \neq \emptyset$ may be combined with an $a_2 \in \mathscr{Q}_2'$ if $\emptyset \neq c_1' \subset c_1, c_1' \in \mathscr{S}_1$ and $a_2' \cap a_2, a_2' \in \mathscr{Q}_2'$ imply $c_1' \times a_2' \cap M \neq \emptyset$.

T 3.3 To say that c_1 may be combined with a_2 is equivalent to saying that there exist an $a_1 \in \mathscr{Q}_1'$ and a $b_1 \in \mathscr{R}_1$ with $c_1 \subset a_1 \cap b_1$ so that a_1 may be combined with a_2. If, in particular, $c_1 = a_1 \cap b_1$, then c_1 may be combined with a_2 if and only if a_1 may be combined with a_2.

Proof. There always exist an $a_1 \in \mathscr{Q}_1'$ and a $b_1 \in \mathscr{R}_1'$ with $c_1 \subset a_1 \cap b_1$.
Let $c_1' \subset c_1 \subset a_1 \cap b_1$. With Th 4.5.1 from [2] II, we easily find

$$c_1 = \bigcup_{(i,k) \in A} a_1^{(i)} \cap b_1^{(k)} \quad \text{and} \quad c_1' = \bigcup_{(i,k) \in B} a_1^{(i)} b_1^{(k)}, \tag{3.3}$$

where $a_1^{(i)} \subset a_1$, $b_1^{(k)} \subset b_1$ and B is a subset of A. For $a_2' \subset a_2$ follows

$$c_1' \times a_2' \cap M = \bigcup_{(i,k) \in B} [(a_1^{(i)} \cap b_1^{(k)}) \times a_2' \cap M].$$

If a_1 may be combined with a_2, because $a_1^{(i)} \subset a_1$ and $a_2' \subset a_2$ also $a_1^{(i)}$ may be combined with a_2'. Thus T 3.2 yields $(a_1^{(i)} \cap b_1^{(k)}) \times a_2' \cap M \neq \emptyset$ and hence $c_1' \times a_2' \cap M \neq \emptyset$. Therefore, c_1 may be combined with a_2.

Conversely, if c_1 may be combined with a_2 and $a_2' \subset a_2$, then in (3.3) one can in particular put

$$a_1 = \bigcup_i a_1^{(i)} \quad \text{and} \quad b_1 = \bigcup_k b_1^{(k)}, \tag{3.4}$$

with the unions taken over all i for which there is a k with $(i, k) \in A$ and over all k for which there is an i with $(i, k) \in A$. Moreover, one can in particular set $c_1' = a_1^{(i_0)} \cap b_1^{(k_0)}$ for some pair $(i_0, k_0) \in A$. Therefore follows $c_1' \times a_2' \cap M = (a_1^{(i_0)} \cap b_1^{(k_0)}) \times a_2' \cap M \neq \emptyset$ and hence $a_1^{(i_0)} \times a_2' \cap M \neq \emptyset$. Instead of $a_1^{(i_0)}$, a_2' one could also have chosen the subsets $a_1'' \subset a_1^{(i_0)}$, $a_2'' \subset a_2'$ with $a_1'' \in \mathscr{Q}_1'$, $a_2'' \in \mathscr{Q}_2'$. Consequently all $a_1^{(i)}$ may be combined with a_2. By T 2.1 and (3.4) follows that a_1 may be combined with a_2. $\square$

T 3.4 $\mathscr{S}_{12}$ is also the SP structure over M that is generated by

$$A_1 = \{c_1 \times (a_2 \cap b_2) \cap M \mid c_1 \in \mathscr{S}_1, a_2 \in Q_2, b_2 \in \mathscr{R}_2, c_1 \text{ may be combined with } a_2\}.$$

Proof. $\mathscr{S}_{12}$ is equal to the SP structure generated by

$$A_2 = \{c_1 \cap b_1) \times (a_2 \cap b_2) \cap M \mid a_1 \in \mathscr{Q}_1', a_2 \in \mathscr{Q}_2',$$
$$a_1 \text{ may be combined with } a_2, b_2 \in \mathscr{R}_1, b_2 \in \mathscr{R}_2\};$$

this follows easily from (2.16) and (2.1). Since (due to T 3.3) $c_1 = a_1 \cap b_1$ may be combined with a_2 when a_1 and a_2 are combinable, we have $A_2 \subset A_1$.

When c_1 is combinable with a_2 and (3.3) holds, all $a_1^{(i)} \cap b_1^{(k)}$ may be combined with a_2. Due to T 3.3, all $a_1^{(i)}$ may then be combined with a_2 so that $(a_1^{(i)} \cap b_1^{(k)}) \times (a_2 \cap b_2) \cap M \in A_2$. By T 3.3 there also is an $a_1 \cap b_1$ with $c_1 \subset a_1 \cap b_1$ and a_1 combinable with a_2 such that $(a_1 \cap b_1) \times (a_2 \cap b_2) \cap M \in A_2$. Hence $c_1 \times (a_2 \cap b_2) \cap M$ must be an element of the SP structure generated by A_2, i.e. an element of $\mathscr{S}_{12}$. Therefore $A_1 \subset \mathscr{S}_{12}$. □

T 3.5 If $\emptyset \neq c_1 \in \mathscr{S}_1$ and a_2 may be combined with c_1, we have $c_1 \times (a_2 \cap b_{20}) \cap M \neq \emptyset$ for all $b_{20} \in \mathscr{R}'_{20}$.

The proof follows from the representation (3.3) of c_1 together with T 3.3 and T 3.2.

Thus T 3.5 presents a generalization of T 3.2.

Similarly to the axioms APSZ 8 (which exclude "uninteresting" cases from the fundamental domain rather describing a physical structure), let us postulate

APSZ 5.2 For each $a_1 \in \mathscr{Q}'_1$ there is an $a_2 \in \mathscr{Q}'_2$ which may be combined with a_1.

This extension of APSZ 5.1 will simplify the following considerations.

Physically, there is no reason (unless one considers the entire universe as a preparation device) why it should not be possible, given a preparation procedure a_1 for the systems from M_1, to find another preparation procedure a_2 for the systems from M_2 so that one can arbitrarily combine systems from a_1 and a_2.

T 3.6 For $c_1, c'_1 \in \mathscr{S}_1$, $c_1 \neq \emptyset$, a function $\lambda_{\mathscr{S}_1}$ is given by

$$\lambda_{\mathscr{S}_1}(c_1, c'_1) \stackrel{\text{def}}{=} \lambda_{12}(c_1 \times (a_2 \cap b_{20}) \cap M, c'_1 \times (a_2 \cap b_{20}) \cap M), \tag{3.5}$$

where $b_{20} \in \mathscr{R}'_{20}$, $a_2 \in \mathscr{Q}'_2$ while a_2 may be combined with c_1. This $\lambda_{\mathscr{S}_1}$ makes $\mathscr{S}_1$ into an SSP structure. With $a_1, a'_1 \in \mathscr{Q}_1$ and $b_{10}, b'_{10} \in \mathscr{R}_{10}$ it satisfies

$$\lambda_{\mathscr{S}_1}(a_1 \cap b_{10}, a'_1 \cap b_{10}) = \lambda_{\mathscr{Q}_1}(a_1, a'_1);$$
$$\lambda_{\mathscr{S}_1}(a_1 \cap b_{10}, a_1 \cap b'_{10}) = \lambda_{\mathscr{R}_{10}}(b_{10}, b'_{10}).$$

Proof. Because $c'_1 \subset c_1$ while c_1 may be combined with a_2, also c'_1 may be combined with a_2. Hence T 3.4 yields

$$c_1 \times (a_2 \cap b_{20}) \cap M \in \mathscr{S}_{12} \quad \text{and} \quad c'_1 \times (a_2 \cap b_{20}) \cap M \in \mathscr{S}_{12},$$

so that the right side of (3.5) is defined. Because of $c_1 \neq \emptyset$, by T 3.5 we also have $c_1 \times (a_2 \cap b_{20}) \cap M \neq \emptyset$.

With a_1, b_1 as in T 3.3 follows $\emptyset \neq (a_1 \cap b_1) \times (a_2 \cap b_{20}) \cap M \in \mathscr{S}_{12}$ and

$$\lambda_{12}(c_1 \times (a_2 \cap b_{20}) \cap M, c'_1 \times (a_2 \cap b_{20}) \cap M)$$
$$= \frac{\lambda_{12}((a_1 \cap b_1) \times (a_2 \cap b_{20}) \cap M, c'_1 \times (a_2 \cap b_{20}) \cap M)}{\lambda_{12}((a_1 \cap b_1) \times (a_2 \cap b_{20}) \cap M, c_1 \times (a_2 \cap b_{20}) \cap M)}.$$

With c_1, c_1' as in (3.3) follows

$$\lambda_{12}((a_1 \cap b_1) \times (a_2 \cap b_{20}) \cap M, c_1 \times (a_2 \cap b_{20}) \cap M)$$
$$= \sum_{(i,k) \in A} \lambda_{12}((a_1 \cap b_1) \times (a_2 \cap b_{20}) \cap M, (a_1^{(i)} \cap b_1^{(k)}) \times (a_2 \cap b_{20}) \cap M)$$

and a similar equation for c_1' instead of c_1. Due to these two equations and APSZ 9, the right side of (3.5) does not depend on a_2, b_{20}; defining $\lambda_{\mathscr{S}_1}$ by (3.5) therefore makes sense.

By (3.5) a definition is given for all $c_1 \in \mathscr{S}_1$ and $c_1' \in \mathscr{S}_1$ with $c_1' \subset c_1$, if for each c_1 there is an a_2 that may be combined with c_1. This follows from T 3.3 together with APSZ 5.2 since for each c_1 a pair a_1, b_1 with $c_1 \subset a_1 \cap b_1$ exists according to Th 4.5.1 in [2] II.

It remains to be shown that $\lambda_{\mathscr{S}_1}$ fulfills AS 2.1 through AS 2.3 from II § 2.1.

By (3.5) follows from $\lambda_{\mathscr{S}_1}(c_1, c_1') = 0$ that $c_1' \times (a_2 \cap b_{20}) \cap M = \emptyset$ holds and thus by T 3.5 also $c_1' = \emptyset$, which corresponds to AS 2.3.

AS 2.1 follows immediately from $c_1' \subset c_1$, $c_1'' \subset c_1$ and $c_1' \cap c_1'' = \emptyset, c_1' \cup c_1'' = c_1$ and (3.5). Equally easily AS 2.2 results.

The following statements follow step by step from (3.5):

$$\lambda_{\mathscr{S}_1}(a_1 \cap b_{10}, a_1' \cap b_{10}) = \lambda_{12}((a_1 \times a_2) \cap (b_{10} \times b_{20}) \cap M, (a_1' \times a_2) \cap (b_{10} \times b_{20}) \cap M)$$
$$= \lambda_{\mathscr{Q}_{12}}(a_1 \times a_2 \cap M, a_1' \times a_2 \cap M) = \lambda_{\mathscr{Q}_1}(a_1, a_1');$$

$$\lambda_{\mathscr{S}_1}(a_1 \cap b_{10}, a_1 \cap b_{10}') = \lambda_{12}((a_1 \times a_2) \cap (b_{10} \times b_{20}) \cap M, (a_1 \times a_2) \cap (b_{10}' \times b_{20}) \cap M)$$
$$= \lambda_{\mathscr{R}_{120}}(b_{10} \times b_{20} \cap M, b_{10}' \times b_{20} \cap M) = \lambda_{\mathscr{R}_{10}}(b_{10}, b_{10}'). \quad \square$$

T 3.6 means that $\mathscr{Q}_1$, $\mathscr{R}_{10}$, $\mathscr{R}_1$ together with $\mathscr{S}_1$ and $\lambda_{\mathscr{Q}_1}$, $\lambda_{\mathscr{R}_{10}}$, $\lambda_{\mathscr{S}_1}$ form a preparation-registration structure on M_1, i.e. a structure that fulfills APS 1 through APS 8 (see [3] § 12.1 or [2] II, III). Thus all the axioms APS 5 are trivially fulfilled, because each $a_1 \in Q_1'$ may be combined with each $b_{10} \in \mathscr{R}_{10}'$ as $a_1 \cap b_{10} \neq \emptyset$ follows from $a_1 \in \mathscr{Q}_1$, $b_{10} \in \mathscr{R}_{10}'$ (see the remarks after APSZ 5.1).

In general, due to T 2.8 only the special probabilities

$$\mu_{12}(a_1 \times a_2 \cap M, (b_{10} \times b_{20} \cap M, b_1 \times b_2 \cap M)) =$$
$$= \lambda_{12}(a_1 \times a_2 \cap M \cap b_{10} \times b_{20}, a_1 \times a_2 \cap M \cap b_1 \times b_2)$$

are physically interesting. If APSZ 9 holds, there follows

$$\begin{aligned}
&\lambda_{12}(a_1 \times a_2 \cap M \cap b_{10} \times b_{20}, a_1 \times a_2 \cap M \cap b_1 \times b_2)\\
&\quad = \lambda_{12}((a_1 \cap b_{10}) \times (a_2 \cap b_{20}) \cap M, (a_1 \cap b_1) \times (a_2 \cap b_{20}) \cap M)\\
&\quad \cdot \lambda_{12}((a_1 \cap b_1) \times (a_2 \cap b_{20}) \cap M, (a_1 \cap b_1) \times (a_2 \cap b_2) \cap M)\\
&\quad = \lambda_{\mathscr{S}_1}(a_1 \cap b_{10}, a_1 \cap b_1)\, \lambda_{12}((a_1 \cap b_1) \times (a_2 \cap b_{20}) \cap M,\\
&\qquad (a_1 \cap b_1) \times (a_2 \cap b_2) \cap M).
\end{aligned} \tag{3.6}$$

The first factor is the probability of the (trajectory) registrations on the subsystems $x_1 \in M$, a probability not dependent on the systems $x_2 \in M_2$ that are coupled on. The second factor is the conditional probability for the registrations on the subsystems $x_2 \in M_2$, namely under the "condition" that the systems x_2 are influenced by systems $x_1 \in M_1$ selected according to $a_1 \cap b_1$. This conditional probability

$$\lambda_{12}((a_1 \cap b_1) \times (a_2 \cap b_{20}) \cap M, (a_1 \cap b_1) \times (a_2 \cap b_2) \cap M) \tag{3.7}$$

is the central quantity by which the directed action is completely described. In fact by (3.7), by $\lambda_{\mathscr{Q}_1}$, $\lambda_{\mathscr{Q}_2}$ and the function μ_1 defined by

$$\mu_1(a_1, (b_{10}, b_1)) = \lambda_{\mathscr{S}_1}(a_1 \cap b_{10}, a_1 \cap b_1),$$

due to T 2.8 and (3.6) all of λ_{12} is determined!

One could designate (3.7) as the central structure of experimentation. That it does not seem so central is only due to the great generality of (3.7). But actually the laboratories of experimental physicists are filled with devices in which systems x_1 exert a directed action on systems x_2 and the probabilities (3.7) are read off experimentally. In order to indicate the range of possible applications of (3.7), let us mention quite different examples: A rifle (x_1) to make a hole in a target (x_2); a transmitter (x_1) which acts on a receiver (x_2) (e.g. produces a television image); a piece of uranium (x_1) which blackens a photo plate (x_2); an accelerator plus a target (x_1) which cause traces in a bubble chamber (x_2). The readers should clarify for themselves the physical meaning of (3.7) by such examples.

The expression (3.7) suggests to consider, besides the SP structure $\mathscr{S}_1$ (generated by $\{a_1 \cap b_1 | a_1 \in \mathscr{Q}_1, b \in \mathscr{R}_1\}$ over M_1) the following SP structures over M_2: the structure $\mathscr{S}_2$ generated by $\{a_2 \cap b_2 | a_2 \in \mathscr{Q}_2, b_2 \in \mathscr{R}_2\}$ and the $\mathscr{S}_{20}$ generated by $\{a_2 \cap b_{20} | a_2 \in \mathscr{Q}_2, b_{20} \in \mathscr{R}_{20}\}$. We see immediately that $\mathscr{S}_{20} \subset \mathscr{S}_2$. Then (3.7) is only a particular case of the general expression

$$\lambda_{12}(c_1 \times c_{20} \cap M, c_1 \times c_2 \cap M) \tag{3.8}$$

for $c_1 \in \mathscr{S}_1$, $c_{20} \in \mathscr{S}_{20}$, $c_2 \in \mathscr{S}_2$ and $c_2 \subset c_{20}$. Here (3.8) is defined whenever $c_1 \times c_{20} \cap M$ and $c_1 \times c_2 \cap M$ are elements of $\mathscr{S}_{12}$. In order to tell when they are elements of $\mathscr{S}_{12}$, we first formulate the definition

D 3.2 We say that a $c_1 \in \mathscr{S}_1$ with $c_1 \neq \emptyset$ may be combined with a $c_{20} \in \mathscr{S}_{20}$ if $\emptyset \neq c_1' \subset c_1$, $c_1' \in \mathscr{S}_1$ and $\emptyset \neq c_{20}' \subset c_{20}$, $c_{20}' \in \mathscr{S}_{20}$ imply $c_1' \times c_{20}' \cap M \neq \emptyset$.

The two definitions D 3.1 and D 3.2 are intimately connected as shown by the theorem

T 3.7 c_1 being combinable with c_{20} (in the sense of D 3.2) is equivalent to the existence of an $a_2 \in \mathscr{Q}_2$ and a $b_{20} \in \mathscr{R}_{20}$ with $c_{20} \subset a_2 \cap b_{20}$ such that c_1 and a_2 may be combined (in the sense of D 3.1).

Proof. Let $c_{20} \subset a_2 \cap b_{20}$, suppose c_1 and a_2 may be combined, and assume $\emptyset \neq c_{20}' \subset c_{20}$, $\emptyset \neq c_1' \subset c$. From Th 4.5.1 in [2] II we easily get

$$c_{20} = \bigcup_{(i,k) \in A} (a_2^{(i)} \cap b_{20}^{(k)}) \quad \text{and} \quad c_{20}' = \bigcup_{(i,k) \in B} (a_2^{(i)} \cap b_{20}^{(k)}), \tag{3.9}$$

with $a_2^{(i)} \subset a_2$, $b_{20}^{(k)} \subset b_{20}$ and B a subset of A.

From this follows

$$c_1' \times c_{20}' \cap M = \bigcup_{(i,k) \in B} [c_1' \times (a_2^{(i)} \cap b_{20}^{(k)}) \cap M].$$

Because of $a_2^{(i)} \subset a_2$ we find c_1 and hence also c_1' combinable with $a_2^{(i)}$. Therefore T 3.5 yields $c_1' \times (a_2^{(i)} \cap b_{20}^{(k)}) \cap M \neq \emptyset$ and hence $c_1' \times c_{20}' \cap M \neq \emptyset$. Thus c_1 may be combined with c_{20}.

Conversely, if c_1 may be combined with c_{20}, choose in (3.9) in particular (as for T 3.3)

$$a_2 = \bigcup_i a_1^{(i)} \quad \text{and} \quad b_{20} = \bigcup_k b_{20}^{(k)}. \tag{3.10}$$

Moreover, for any pair $(i_0, k_0) \in A$ picking a particular $c_{20}' = a_2^{(i_0)} \cap b_{20}^{(k_0)}$, we obtain $c_1' \times (a_2^{(i_0)} \cap b_{20}^{(k_0)}) \cap M \neq \emptyset$. Likewise, $c_1' \times (a_2'' \cap b_{20}^{(k_0)}) \cap M \neq \emptyset$ and hence $c_1' \times a_2'' \cap M \neq \emptyset$ would follow for subsets $a_2'' \subset a_2^{(i_0)}$. Therefore c_1 may for all i be combined with $a_2^{(i)}$ and hence, by T 2.1 and (3.10), also with a_2. □

From T 3.4 and T 3.7 follows the important theorem

T 3.8 $\mathscr{S}_{12}$ is also the SP structure that is generated over M by

$A_3 = \{c_1 \times c_2 \cap M \mid c_1 \in \mathscr{S}_1, c_2 \in \mathscr{S}_2$ and there is a $c_{20} \in \mathscr{S}_{20}$ such that $c_2 \subset c_{20}$ and c_1 may be combined with $c_{20}\}$.

Proof. Consider $c_2 = a_2 \cap b_2$, where a_2 and c_1 may be combined. For this b_2 there is a $b_{20} \in \mathscr{R}_{20}$ with $b_2 \subset b_{20}$. Due to T 3.7, c_1 may be combined with $c_{20} = a_2 \cap b_{20}$ and $c_2 \subset c_{20}$. Consequently, A_1 from T 3.4 obeys $A_1 \subset A_3$.

From $c_2 \subset c_{20}$ with a $c_{20} \in \mathscr{S}_{20}$ combinable with c_1 follows by (3.9) that also $a_2^{(i)} \cap b_{20}^{(k)}$ may be combined with c_1. Then T 3.7 says that all $a_2^{(i)}$ may be combined with c_1. Applying Th 4.5.1 from [2] II to $\mathscr{S}_2$ we find that c_2 (with $c_2 \subset c_{20}$) can be represented as

$$c_2 = \bigcup_{(i,k) \in B} (a_2^{(i)} \cap b_2^{(k)}). \tag{3.11}$$

Here one can choose the $a_2^{(i)}$ to be the same as in (3.9) and assume $b_2^{(k)} \subset b_{20}^{(k)}$. Since all $a_2^{(i)}$ may be combined with c_1 and we have $a_2^{(i)} \cap b_2^{(k)} \subset a_2^{(i)} \cap b_{20}^{(k)}$, we conclude $c_1 \times (a_2^{(i)} \cap b_2^{(k)}) \cap M \in A_1$. From $c_2 \subset c_{20} \subset a_2 \cap b_{20}$ and because $a_2 \cap b_{20}$ may be combined with c_1, it then follows that $c_1 \times c_2 \cap M$ is an element of the SP structure generated by A_1, such that $A_3 \subset \mathscr{S}_{12}$. □

With c_{20} as in (3.9), from (3.5) follows that also

$$\lambda_{\mathscr{S}_1}(c_1, c_1') = \lambda_{12}(c_1 \times c_{20} \cap M, c_1' \times c_{20} \cap M) \tag{3.12}$$

holds whenever the right side is defined, i.e. whenever c_1 may be combined with c_{20}.

T 3.9 A probability function $\lambda_{\mathscr{S}_{20}}$ for $\mathscr{S}_{20}$ can be introduced uniquely by the requirements

$$\lambda_{\mathscr{S}_{20}}(a_2 \cap b_{20}, a_2' \cap b_{20}) = \lambda_{\mathscr{D}_2}(a_2, a_2'),$$
$$\lambda_{\mathscr{S}_{20}}(a_2 \cap b_{20}, a_2 \cap b_{20}') = \lambda_{\mathscr{R}_{20}}(b_{20}, b_{20}').$$

This $\lambda_{\mathscr{S}_{20}}$ makes $\mathscr{S}_{20}$ into an SSP over M_2.

Proof. Since $a_2 \in \mathscr{D}_2'$ and $b_{20} \in \mathscr{R}_{20}'$ imply $a_2 \cap b_{20} \neq \emptyset$ (see the remarks after APSZ 5.1), the proof of T 2.5 can at once be transferred. □

T 3.10 Besides (3.12), for λ_{12} holds

$$\lambda_{12}(c_1 \times c_{20} \cap M, c_1 \times c'_{20} \cap M) = \lambda_{\mathcal{S}_{20}}(c_{20}, c'_{20}) \tag{3.13}$$

with $c_{20}, c'_{20} \in \mathcal{S}_{20}$.

Proof. As in the proof of T 3.6, with (3.3) follows

$$\lambda_{12}(c_1 \times c_{20} \cap M, c_1 \times c'_{20} \cap M) = \frac{\sum\limits_{(i,k) \in A} \lambda_{12}((a_1 \cap b_1) \times c_{20} \cap M, (a_1^{(i)} \cap b_1^{(k)}) \times c'_{20} \cap M)}{\sum\limits_{(i,k) \in A} \lambda_{12}((a_1 \cap b_1) \times c_{20} \cap M, (a_1^{(i)} \cap b_1^{(k)}) \times c_{20} \cap M)}.$$

Furthermore, (3.12) yields

$$\lambda_{12}((a_1 \cap b_1) \times c_{20} \cap M, (a_1^{(i)} \cap b_1^{(k)}) \times c_{20} \cap M) = \lambda_{\mathcal{S}_1}(a_1 \cap b_1, a_1^{(i)} \cap b_1^{(k)})$$

and

$$\begin{aligned} &\lambda_{12}((a_1 \cap b_1) \times c_{20} \cap M, (a_1^{(i)} \cap b_1^{(k)}) \times c'_{20} \cap M) \\ &\quad = \lambda_{12}((a_1 \cap b_1) \times c_{20} \cap M, (a_1 \cap b_1) \times c'_{20} \cap M) \\ &\qquad \cdot \lambda_{12}((a_1 \cap b_1) \times c'_{20} \cap M, (a_1^{(i)} \cap b_1^{(k)}) \times c'_{20} \cap M) \\ &\quad = \lambda_{12}((a_1 \cap b_1) \times c_{20} \cap M, (a_1 \cap b_1) \times c'_{20} \cap M) \\ &\qquad \cdot \lambda_{\mathcal{S}_1}(a_1 \cap b_1, a_1^{(i)} \cap b_1^{(k)}). \end{aligned}$$

If $\lambda_{12}((a_1 \cap b_1) \times c_{20} \cap M, (a_1 \cap b_1) \times c'_{20} \cap M) = \lambda_{\mathcal{S}_{20}}(c_{20}, c'_{20})$, (3.13) then follows; hence it suffices to prove (3.13) for $c_1 = a_1 \cap b_1$.

With (3.9) and $c_{20} \subset a_2 \cap b_{20}$ follows

$$\begin{aligned} &\lambda_{12}((a_1 \cap b_1) \times c_{20} \cap M, (a_1 \cap b_1) \times c'_{20} \cap M) \\ &\quad = \frac{\sum\limits_{(i,k) \in B} \lambda_{12}((a_1 \cap b_1) \times (a_2 \cap b_{20}) \cap M, (a_1 \cap b_1) \times (a_2^{(i)} \times b_{20}^{(k)}) \cap M)}{\sum\limits_{(i,k) \in A} \lambda_{12}((a_1 \cap b_1) \times (a_2 \cap b_{20}) \cap M, (a_1 \cap b_1) \times (a_2^{(i)} \times b_{20}^{(k)}) \cap M)}. \end{aligned}$$

If $a'_2 \subset a_2$, $b'_{20} \subset b_{20}$ satisfy

$$\begin{aligned} &\lambda_{12}((a_1 \cap b_1) \times (a_2 \cap b_{20}) \cap M, (a_1 \cap b_1) \times (a'_2 \cap b'_{20}) \cap M) \\ &\quad = \lambda_{\mathcal{S}_{20}}(a_2 \cap b_{20}, a'_2 \cap b'_{20}) = \lambda_{\mathcal{D}_2}(a_2, a'_2)\, \lambda_{\mathcal{R}_{20}}(b_{20}, b'_{20}), \end{aligned} \tag{3.14}$$

we obtain

$$\begin{aligned} &\lambda_{12}((a_1 \cap b_1) \times c_{20} \cap M, (a_1 \cap b_1) \times c'_{20} \cap M) \\ &\quad = \frac{\sum\limits_{(i,k) \in B} \lambda_{\mathcal{S}_{20}}(a_2 \cap b_{20}, a_2^{(i)} \cap b_{20}^{(k)})}{\sum\limits_{(i,k) \in A} \lambda_{\mathcal{S}_{20}}(a_2 \cap b_{20}, a_2^{(i)} \cap b_{20}^{(k)})} = \lambda_{\mathcal{S}_{20}}(c_{20}, c'_{20}); \end{aligned}$$

thus it suffices to prove (3.14).

For $b_{10} \in \mathcal{R}_{10}$ with $b_{10} \supset b_1$ follows

$$\begin{aligned} &\lambda_{12}((a_1 \cap b_1) \times (a_2 \cap b_{20}) \cap M, (a_1 \cap b_1) \times (a'_2 \cap b'_{20}) \cap M) \\ &\quad = \frac{\lambda_{12}((a_1 \cap b_{10}) \times (a_2 \cap b_{20}) \cap M, (a_1 \cap b_1) \times (a'_2 \cap b'_{20}) \cap M)}{\lambda_{12}((a_1 \cap b_{10}) \times (a_1 \cap b_{20}) \cap M, (a_1 \cap b_1) \times (a_2 \cap b_{20}) \cap M)}. \end{aligned}$$

Furthermore, the numerator equals

$$\lambda_{12}((a_1 \cap b_{10}) \times (a_2 \cap b_{20}) \cap M, (a_1 \cap b_{10}) \times (a_2' \cap b_{20}') \cap M)$$
$$\cdot \lambda_{12}((a_1 \cap b_{10}) \times (a_2' \cap b_{20}') \cap M, (a_1 \cap b_1) \times (a_2' \cap b_{20}') \cap M);$$

thus one obtains

$$\lambda_{12}((a_1 \cap b_1) \times (a_2 \cap b_{20}) \cap M, (a_1 \cap b_1) \times (a_2' \cap b_{20}') \cap M)$$
$$= \frac{\lambda_{12}((a_1 \cap b_{10}) \times (a_2 \cap b_{20}) \cap M, (a_1 \cap b_{10}) \times (a_2' \cap b_{20}') \cap M)\, \lambda_{\mathscr{S}_1}(a_1 \cap b_{10}, a_1 \cap b_1)}{\lambda_{\mathscr{S}_1}(a_1 \cap b_{10}, a_1 \cap b_1)}$$

$$= \lambda_{12}(a_1 \times a_2 \cap M \cap b_{10} \times b_{20}, a_1 \times a_2' \cap M \cap b_{10} \times b_{20}')$$
$$= \lambda_{12}(a_1 \times a_2 \cap M \cap b_{10} \times b_{20}, a_1 \times a_2 \cap M \cap b_{10} \times b_{20}')$$
$$\cdot \lambda_{12}(a_1 \times a_2 \cap M \cap b_{10} \times b_{20}', a_1 \times a_2' \cap M \cap b_{10} \times b_{20}').$$

By APSZ 7 this becomes

$$\lambda_{\mathscr{R}_{120}}(b_{10} \times b_{20} \cap M, b_{10} \times b_{20}' \cap M)\, \lambda_{\mathscr{Q}_{12}}(a_1 \times a_2 \cap M, a_1 \times a_2' \cap M)$$
$$= \lambda_{\mathscr{R}_2}(b_{20}, b_{20}')\, \lambda_{\mathscr{Q}_2}(a_2, a_2'),$$

which completes the proof of (3.14). □

T 3.11 For each $c_2 \in \mathscr{S}_2$ there is a $c_{20} \in \mathscr{S}_{20}$ with $c_{20} \supset c_2$.

Proof. According to Th 4.5.1 in [2] II, for each $c_2 \in \mathscr{S}_2$ there exist an $a_2 \in \mathscr{Q}_2$ and a $b_2 \in \mathscr{R}_2$ with $c_2 \subset a_2 \cap b_2$. For $b_2 \in \mathscr{R}_2$ there exists a $b_{20} \in \mathscr{R}_{20}$ with $b_2 \subset b_{20}$. Hence $c_{20} \supset c_2$ holds with $c_{20} = a_2 \cap b_{20} \in \mathscr{S}_{20}$. □

In APSZ 5.2 we only required that for each $a_1 \in \mathscr{Q}_1'$ there be an $a_2 \in \mathscr{Q}_2'$ combinable with a_1. In the proof of T 3.6 we saw that from this more generally follows that for each $c_1 \in \mathscr{S}_1$ with $c_1 \neq \emptyset$ there is an $a_2 \in \mathscr{Q}_2'$ combinable with c_1.

But from T 3.7 then immediately follows

T 3.12 For each $c_1 \in \mathscr{S}_1$, $c_2 \neq \emptyset$, there is a $c_{20} \in \mathscr{S}_{20}'$ so that c_1 and c_{20} may be combined.

This among other things asserts $c_1 \times c_{20} \cap M \neq \emptyset$. But the derivation of T 3.12 from APSZ 5.2 was possible only because we presumed APSZ 9. Therefore it is understandably not sufficient to introduce an axiom analogous to APSZ 5.2 for interchanged $\mathscr{Q}_1, \mathscr{Q}_2$ in order to prove a relation analogous to T 3.12. As an axiom we therefore demand

APSZ 5.3 For each $a_2 \in \mathscr{Q}_2'$ and each $a_2' \cap b_2 \neq \emptyset$ with $a_2' \subset a_2$, $a_2' \in \mathscr{Q}_2'$, $b_2 \in \mathscr{R}_2$, there is an $a_1 \in \mathscr{Q}_1'$ that may be combined with a_2, so that $a_1 \times (a_2' \cap b_2) \cap M \neq \emptyset$. We assume $\lambda_{12}((a_1 \cap b_{10}) \times (a_2 \cap b_{20}) \cap M, (a_1 \cap b_{10}) \times (a_2 \cap b_2) \cap M)$ to be independent of b_{10} (see the remarks at the end of § 2).

From this follows

T 3.13 For each $c_2 \in \mathscr{S}_2$ with $c_2 \neq \emptyset$ there is a $c_{20} \in \mathscr{S}_{20}$ with $c_{20} \supset c_2$ and a $c_1 \in \mathscr{S}_1$ combinable with c_{20}, where $c_1 \times c_2 \cap M \neq \emptyset$.

Proof. By T 3.11 there is a $c_{20} \supset c_2$, which can be assumed of the particular form $a_2 \cap b_{20}$. Then c_2 is a union of elements of the form $a_2' \cap b_2$ with $a_2' \subset a_2$ and $b_2 \subset b_{20}$ (a fact already used many times). Because $c_2 \neq \emptyset$, there consequently exists such an $a_2' \cap b_2 \neq \emptyset$. According to APSZ 5.3, there is an a_1 combinable with a_2, so that $a_1 \times (a_2' \cap b_2) \cap M \neq \emptyset$.

If for a $b_{10} \in \mathscr{R}_{10}'$ and a $b_{20} \supset b_2$ the probability

$$\lambda_{12}\left((a_1 \cap b_{10}) \times (a_2' \cap b_{20}) \cap M,\ (a_1 \cap b_{10}) \times (a_2' \cap b_2) \cap M\right)$$

where zero, this would hold for all $b_{10} \in \mathscr{R}_{10}'$, since that λ_{12} is independent of b_{10}. It would imply

$$(a_1 \cap b_{10}) \times (a_2' \cap b_2) \cap M = \emptyset \quad \text{for all} \quad b_{10} \in \mathscr{R}_{10}'$$

and hence, because of (2.13), also $a_1 \times (a_2' \cap b_2) \cap M = \emptyset$, in contradiction to APSZ 5.3. Thus we get

$$(a_1 \cap b_{10}) \times (a_2' \cap b_2) \cap M \neq \emptyset$$

and *a fortiori*

$$(a_1 \cap b_{10}) \times c_2 \cap M \neq \emptyset.$$

The proof is completed by taking $c_1 = a_1 \cap b_{10}$. □

§ 4 Action Carriers

The structures introduced in the preceding §§ 1–3, describing the directed interaction of coupled systems, appear at first to have nothing to do with what till now is intuitively understood by microsystems. But how were microsystems such as electrons, atoms, nuclei, or molecules discovered? Surely, the hypothesis of atoms stood at the forefront in the historical development. By this hypothesis one wanted to "explain" the properties of macroscopic processes, such as the behavior of gases and the laws of chemistry. The positivists rightly pointed to the still "soft" spot, that the "reality" of atoms is not yet "proven" by the consequences of the atom hypothesis. But then the physicists learned to experiment with single atoms, single electrons, etc. They invented devices on which these microsystems evoked actions, and what "acted" that had to be "real". It is precisely this experimenting in the discovery of microsystems that we wish to describe by a mathematical picture. We assert that we have in §§ 1–3 introduced the fundamental structures for this mathematical picture, so that we can step by step begin with the "discovery" of microsystems.

The base sets introduced were M_1, M_2, the structure terms were M, then $\mathscr{Q}_1$, $\mathscr{R}_{10}$, $\mathscr{R}_1$ with $\lambda_{\mathscr{Q}_1}$, $\lambda_{\mathscr{R}_{10}}$ and $\mathscr{Q}_2$, $\mathscr{R}_{20}$, $\mathscr{R}_2$ with $\lambda_{\mathscr{Q}_2}$, $\lambda_{\mathscr{R}_{20}}$ and finally λ_{12}.

These structures (terms and axioms) enabled us to equip the set M with a structure $\mathscr{Q}_{12}$, $\mathscr{R}_{120}$, $\mathscr{R}_{12}$ with $\lambda_{\mathscr{Q}_{12}}$, $\lambda_{\mathscr{R}_{120}}$ and λ_{12} so that M can be called a set of physical systems, prepared by the procedures from $\mathcal{Q}_{12}$ and registered by those from $\mathcal{R}_{12}$. We called M endowed with $\mathcal{Q}_{12}, \mathcal{R}_{120}, \mathcal{R}_{12}, \lambda_{\mathcal{Q}_{12}}, \lambda_{\mathcal{R}_{120}}, \lambda_{12}$ the set of composite systems.

Of course one could "forget" how this set M of composite systems was derived, and start a mathematical theory from M as the base set and structures $\mathcal{Q}_{12}, \mathcal{R}_{120}, \mathcal{R}_{12}, \lambda_{\mathcal{Q}_{12}}, \lambda_{\mathcal{R}_{120}}, \lambda_{12}$, subject to the "axioms"

$\mathcal{Q}_{12}$ is an SSP structure.

$\mathcal{R}_{120}$ is an SSP structure.

$\mathcal{R}_{12}$ is an SP structure.

$\mathcal{R}_{120} \subset \mathcal{R}_{12}$.

For each $b \in \mathcal{R}_{12}$ there is a $b_0 \in \mathcal{R}_{120}$ with $b \subset b_0$. $C_{12} = \mathcal{Q}'_{12} \times \mathcal{R}'_{120}$.

$\mathcal{S}_{12}$ is an SSP structure.

For $a, a' \in \mathcal{Q}_{12}$ and $b_0, b'_0 \in \mathcal{R}_{120}$ we have

$$\lambda_{12}(a \cap b_0, a' \cap b_0) = \lambda_{\mathcal{Q}_{12}}(a, a'),$$

$$\lambda_{12}(a \cap b_0, a \cap b'_0) = \lambda_{\mathcal{R}_{120}}(b_0, b'_0).$$

$$M = \bigcup_{a \in \mathcal{Q}_{12}} a = \bigcup_{b \in \mathcal{R}_{12}} b.$$

In §§ 2 and 3 we proved these "axioms" as theorems. If one "forgets" the derivation and starts from M with $\mathcal{Q}_{12}, \mathcal{R}_{120}, \ldots$ and the given axioms, then M is a set of physical systems (just as M_m in II § 2) of which one "forgot" that they are "composite".

But the theory erected in §§ 1–3 enables one to endow M with yet *another* structure, by which M gets *another* physical meaning, as said in [3] § 10.5. Since we shall not presume these general considerations let us make some clarifying remarks on this method.

The difficulty in comprehending is due to the fact that one cannot "imagine" that one and *the same* set M can describe *different* physical facts. Despite the hint that the elements of M are only "pictures" of facts, in the "imagination" one illegally identifies these elements with the facts. Insisting that M is initially nothing but a kind of "index set" for real facts can perhaps avoid one's getting that false idea. Just this "indexing" has been described in [3] § 5 as "marking" to record the axioms $\hookleftarrow_r$.

But if one endowes M with a physically interpreted structure, its elements get a further meaning than of mere indices. This likewise happens in mathematics: A set M is first of all only a set; but if one endowes it with a group structure then M becomes a group.

Therefore in the following one must not worry when we endow M with another structure (likewise physically real; see [3] § 10.5), and then characterize it by another physical concept. But whoever still has qualms with this can imagine an auxiliary set $\tilde{M}$ and a bijective mapping $M \leftrightarrow \tilde{M}$ and interpret all the structures introduced below as structures of $\tilde{M}$ (due to $M \leftrightarrow \tilde{M}$). Then one will certainly no longer have trouble in viewing $\tilde{M}$ with these structures as the image of other physical facts.

As new structure terms over M we introduce

$$\mathscr{Q} = \{c_1 \times M_2 \cap M \mid c_1 \in \mathscr{S}_1\}, \tag{4.1}$$

$$\mathscr{R}_0 = \{M_1 \times c_{20} \cap M \mid c_{20} \in \mathscr{S}_{20}\}, \tag{4.2}$$

$$\mathscr{R} = \{M_1 \times c_2 \cap M \mid c_2 \in \mathscr{S}_2\}. \tag{4.3}$$

Therefore, $\mathscr{Q}$, $\mathscr{R}_0$, $\mathscr{R}$ are countable since $\mathscr{Q}_1$, $\mathscr{Q}_2$, $\mathscr{R}_1$, $\mathscr{R}_2$ are. Due to T2.2, the selection procedures $\mathscr{Q}$, $\mathscr{S}_1$ are isomorphic, and so are $\mathscr{R}_0$, $\mathscr{S}_{20}$ and likewise $\mathscr{R}$ and $\mathscr{S}_2$. By these isomorphisms one can transfer $\lambda_{\mathscr{S}_1}$ from $\mathscr{S}_1$ to $\mathscr{Q}$ (as $\lambda_{\mathscr{Q}}$) and likewise $\lambda_{\mathscr{S}_{20}}$ from $\mathscr{S}_{20}$ to $\mathscr{R}_0$ (as $\lambda_{\mathscr{R}_0}$). Thus the following relations hold as theorems (!):

APS 1 $\mathscr{Q}$ is an SSP structure.

APS 2 $\mathscr{R}$ is an SP structure.

APS 3 $\mathscr{R}_0$ is an SSP structure.

From $\mathscr{S}_{20} \subset \mathscr{S}_2$ follows immediately

APS 4.1 $\mathscr{R}_0 \subset \mathscr{R}$.

From T 3.11 and the above isomorphisms follows

APS 4.2 For each $b \in \mathscr{R}$ there is a $b_0 \in \mathscr{R}_0$ with $b \subset b_0$.

In a way entirely analogous to the definitions of combinability given in §§ 2 and 3, we adopt

D 4.1 We say that an $a \in \mathscr{Q}'$ and a $b_0 \in \mathscr{R}_0'$ may be combined if $a' \in \mathscr{Q}'$, $b_0' \in \mathscr{R}_0'$ and $a' \subset a$, $b' \subset b$ always imply $a' \cap b' \neq \emptyset$.

With D 3.2 and the above isomorphisms follows immediately that $a = c_1 \times M_2 \cap M$ may precisely then be combined with $b_0 = M \times c_{20} \cap M$ when c_1 may be combined with c_{20}. From this follows that

$$\Theta = \{a \cap b \mid a \in \mathscr{Q}',\ b \in \mathscr{R} \text{ and there is a } b_0 \in \mathscr{R}_0 \text{ with } b_0 \supset b \text{ and } a \text{ combinable with } b_0\}$$

is identical to the set A_3 in T 3.8. From T 3.8 thus follows that $\mathscr{S}_{12}$ is the SP structure generated over M by Θ. We now omit the indices on $\mathscr{S}_{12}$ and simply write $\mathscr{S}$. We write the probability λ_{12} accordingly as $\lambda_{\mathscr{S}}$. As theorem (!) therefore follows

APS 6 $\mathscr{S}$ is an SSP structure.

From T 3.12 follows

APS 5.1.1 For each $a \in \mathscr{Q}'$ there is a $b_0 \in \mathscr{R}_0'$ combinable with a.

From T 3.13 follows

APS 5.2 For each $b \in \mathscr{R}$ with $b \neq \emptyset$ there exist an $a \in \mathscr{Q}$ and a $b_0 \in \mathscr{R}_0$ with $b_0 \supset b$, $a \cap b \neq \emptyset$ and a, b_0 are combinable.

The next relations follow from (3.12) and (3.13):

APS 7 For a_1, $a_2 \in \mathscr{D}$ with $a_2 \subset a_1$ and b_{01}, $b_{02} \in \mathscr{R}_0$ with $b_{02} \subset b_{01}$ and a_1 combinable with b_{01} we obtain

7.1 $$\lambda_{\mathscr{S}}(a_1 \cap b_{01}, a_2 \cap b_{01}) = \lambda_{\mathscr{D}}(a_1, a_2),$$

7.2 $$\lambda_{\mathscr{S}}(a_1 \cap b_{01}, a_1 \cap b_{02}) = \lambda_{\mathscr{R}_0}(b_{01}, b_{02}).$$

From (2.9) and (2.13) follows

$$\bigcup_{a \in \mathscr{D}} a = \bigcup_{\substack{a_1 \in \mathscr{D}_1 \\ b_1 \in \mathscr{R}_1}} [(a_1 \cap b_1) \times M_2 \cap M] = \left[\left(\bigcup_{a_1 \in \mathscr{D}_1} a_1\right) \cap \left(\bigcup_{b_1 \in \mathscr{R}} b_1\right)\right]$$
$$= (M_1 \cap M_1) \times M_2 \cap M = M;$$

likewise (2.10) and (2.14) imply

$$\bigcup_{b \in \mathscr{R}} b = M_1 \times \left[\left(\bigcup_{a_2 \in \mathscr{D}_2} a_2\right) \cap \left(\bigcup_{b_2 \in \mathscr{R}_2} b_2\right)\right] \cap M = M_1 \times (M_2 \cap M_2) \cap M = M.$$

Consequently we find

APS 8 $$M = \bigcup_{a \in \mathscr{D}} a = \bigcup_{b \in \mathscr{R}} b.$$

D 4.2 We denote M, endowed with the structure $\{\mathscr{D}, \mathscr{R}_0, \mathscr{R}, \lambda_{\mathscr{D}}, \lambda_{\mathscr{R}_0}, \lambda_{\mathscr{S}}\}$ obeying APS 1 through APS 8, as *the set of action carriers.*

This notation should connote that one can interpret $\mathscr{D}$ as the set of preparation procedures for the action carriers, $\mathscr{R}_0$ as the set registration methods and $\mathscr{R}$ as the set of registration procedures. Precisely in this sense one designates the interaction between the systems (x_1, x_2) as "caused by action carriers going from x_1 to x_2''. Such a statement in everyday language must always be regarded as abbreviation for an entire complex relationship. In the above formulation there naturally enters an ontological notion whose basis in $\mathscr{PT}$ is the "physical reality" of M, $\mathscr{D}$, $\mathscr{R}_0$, $\mathscr{R}$, $\lambda_{\mathscr{D}}$, $\lambda_{\mathscr{R}_0}$, $\lambda_{\mathscr{S}}$; one can easily prove this reality by the methods from [3] §§ 10.4 and 10.5.

If one "forgets" the derivation of M, of the structure $\{\mathscr{D}, \mathscr{R}_0, \mathscr{R}, \lambda_{\mathscr{D}}, \lambda_{\mathscr{R}_0}, \lambda_{\mathscr{S}}\}$ and of APS 1 through APS 8 as theorems, one can start a theory from M as base set and APS 1 through APS 8 as axioms. Quantum mechanics was developed in just this way in [2] (see [2] II). The theorems derived there are thus also theorems in the theory presented here (some to be stated below).

If one denotes the theory presented here by $\mathscr{PT}_1$ and the theory presented in [2] II by $\mathscr{PT}_2$, then $\mathscr{PT}_2$ is a *restriction* of $\mathscr{PT}_1$ (in the sense of [3] § 8, [48] and XIII § 3 and provable with the methods presented there). Strictly speaking (in the notation from [3] § 8, [48] and XIII § 3), the relation

$$\mathscr{PT}_1 \rightarrow \mathscr{PT}_1' \rightsquigarrow \mathscr{PT}_2 \tag{4.4}$$

holds. Here the embedding $\mathscr{PT}_1' \rightsquigarrow \mathscr{PT}_2$ is trivial, since the M from $\mathscr{PT}_1'$ (i.e. from $\mathscr{PT}_1$) need only be identified (by a bijection) with the M from $\mathscr{PT}_2$ in such a way that also $\mathscr{D}, \mathscr{R}_0, \ldots$ from $\mathscr{PT}_1$ are identified with $\mathscr{D}, \mathscr{R}_0, \ldots$ from $\mathscr{PT}_2$.

D 4.2 need not entice one to additional ideas about these action carriers (for instance the whereabouts in space and time). By the axioms from §§ 1–3 it is not even guaranteed that one can view the action carriers as physical systems (see § 5). Still many axioms, i.e. natural laws, must be added to characterize M as the set of microsystems.

The following theorems can be deduced from APS 1 through APS 8.

T 4.1 The function $\lambda_{\mathscr{S}}$ for $\mathscr{S}$ is uniquely determined by $\lambda_{\mathscr{Q}}$ and by the special values

$$\lambda\,(a \cap b_0,\ a \cap b)$$

for a, b_0 combinable and $b \in \mathscr{R}$ with $b \subset b_0$.

Proof. See [2] II Th 4.5.2.

To formulate further theorems we adopt the definitions:

D 4.3 $\mathscr{F} = \{(b_0, b) \mid b_0 \in \mathscr{R}_0',\ b \in \mathscr{R},\ b \subset b_0\}$ be called the set of "effect processes" (or of "questions").

Since $\mathscr{R}$ is countable, $\mathscr{F}$ is so.

D 4.4 $\mathscr{E} = \{(a, f) \mid f = (b_0, b) \in \mathscr{F}$ and a, b_0 combinable$\}$.

D 4.5 On $\mathscr{E}$ we define the real-valued function μ by $\mu\,(a, f) = \mu\,(a, (b_0, b)) = \lambda_{\mathscr{S}}\,(a \cap b_0,\ a \cap b)$.

Then T 4.1 says that $\lambda_{\mathscr{S}}$ is completely determined by $\lambda_{\mathscr{Q}}$ and μ.

The μ defined by D 4.5 plays a central role in the statistical description of action carriers. Therefore the reader should review the definition of the sets $\mathscr{Q}$, $\mathscr{R}_0$, $\mathscr{R}$ (given in § 4) on the basis of $\mathscr{Q}_1, \mathscr{Q}_2, \mathscr{R}_{10}, \mathscr{R}_1, \mathscr{R}_{20}, \mathscr{R}_2$, which are the experiments to test μ.

Furthermore we adopt

D 4.6 $(b_0, b_0) = \mathbf{1}_{b_0}$ and $(b_0, \emptyset) = \mathbf{0}_{b_0}$, so that $\mathbf{1}_{b_0}$ and $\mathbf{0}_{b_0}$ are elements of $\mathscr{F}$.

D 4.7 For a decomposition of a $b_0 \in \mathscr{R}_0$ of the form $b_0 = \bigcup_{i=1}^{n} b_i$, with $b_i \cap b_k = \emptyset$ for $i \neq k$ $(b_i \in \mathscr{R})$ we call the $f_i = (b_0, b_i)$ a disjoint decomposition of $\mathbf{1}_{b_0}$ and write $\mathbf{1}_{b_0} = \dot{\bigcup}_{i=1}^{n} f_i$.

For each $f = (b_0, b) \in \mathscr{F}$ there is an $f' \in \mathscr{F}$ with $\mathbf{1}_{b_0} = f \,\dot{\cup}\, f'$, namely $f' = (b_0, b_0 \backslash b)$.

T 4.2 The function μ obeys

(i) $0 \leqq \mu\,(a, b) \leqq 1$.
(ii) For each $a \in Q'$ there is an $f_0 \in \mathscr{F}$ with $\mu\,(a, f_0) = 0$.
(iii) For each $a \in Q'$ there is an $f_1 \in \mathscr{F}$ with $\mu\,(a, f_1) = 1$.

(iv) Under a decomposition $a = \bigcup_{i=1}^{n} a_i$ (with $a_i \cap a_j = \emptyset$ for $i \neq j$) for all f with $(a, f) \in \mathscr{C}$ follows

$$\mu\left(\bigcup_{i=1}^{n} a_i, f\right) = \sum_{i=1}^{n} \lambda_i \, \mu\,(a_i, f)$$

with $0 < \lambda_i = \lambda_{\mathscr{Q}}\,(a, a_i) \leqq 1$ and $\sum_{i=1}^{n} \lambda_i = 1$.

(v) For $b_{01}, b_{02} \in \mathscr{R}_0'$ and $b_{01} \supset b_{02} \supset b$ and for $f_1 = (b_{01}, b)$, $f_2 = (b_{02}, b)$ every a with $(a, f_1) \in \mathscr{C}$ yields

$$\mu\,(a, f_1) = \lambda_{\mathscr{R}_0}(b_{01}, b_{02})\; \mu\,(a, f_2)\,.$$

(vi) For a, b_0 combinable, $\mathbf{1}_{b_0} = \dot{\bigcup}_{i=1}^{n} f_i$ implies $\sum_{i=1}^{n} \mu\,(a, f_i) = 1$.

(For each $f \in \mathscr{F}$ there is in particular an $f' \in \mathscr{F}$ with $\mu\,(a, f) + \mu\,(a, f') = 1$.)

(vii) For $(a, f) \in \mathscr{C}$ with $f = (b_0, b)$, we find $\mu\,(a, f) = 0 \Leftrightarrow a \cap b = \emptyset$.

(viii) If $b_0 = \bigcup_{i=1}^{n} b_{0i}$ with $b_{0i} \in \mathscr{R}_0'$ and $b_{0i} \cap b_{0k} = \emptyset$ for $i \neq k$ $\left(\text{i.e. } \bigcup_{i=1}^{n} b_{0i} \text{ a decomposition of } b_0 \text{ in } \mathscr{R}_0\right)$, for $f = (b_0, b)$ with $f_i = (b_{0i}, b_{0i} \cap b)$ and for each $a \in \mathscr{Q}'$ with $(a, f) \in \mathscr{C}$ follows

$$\mu\,(a, b) = \sum_{i=1}^{n} \lambda_{\mathscr{R}_0}(b_0, b_{0i})\; \mu\,(a, f_i)$$

and

$$\sum_{i=1}^{n} \lambda_{\mathscr{R}_0}(b_0, b_{0i}) = 1\,.$$

(In this case we call the f_i the decomposition of f induced by that of b_0.)

Proof. See [2] II Th 4.5.3.

T 4.3 If for prescribed $\lambda_{\mathscr{Q}}$ a function $\mu\,(a, f)$ on $\mathscr{C}$ satisfies (i), (iv), (v), (vi) and (vii) from T 4.2, for $\mathscr{S}$ there is one and only one probability $\lambda_{\mathscr{S}}$ with $\lambda_{\mathscr{S}}(a \cap b_0, a \cap b) = \mu\,(a, f)$. The $\lambda_{\mathscr{S}}$ thus determined fulfills APS 7 and the condition (viii) from T 4.2, while $\lambda_{\mathscr{R}_0}$ is uniquely determined by μ.

Proof. See [2] II Th 4.5.4.

T 4.3 shows that in experiments it suffices to test the function μ. Due to the meaning of μ as picture of a frequency, the conditions (i), (vi) and (vii) are almost trivially fulfilled in experiments. Condition (iv) expresses the fact that the preparation of the action carriers is not influenced by their registration (the initially imposed directedness of the action). In fact one *must* experiment in just this way; otherwise one has not performed an experiment suitable for the test. Condition (v) expresses the fact that a refinement of the registration method is statistically independent of the preparation (one also *must* experiment in this way).

Therefore, the conditions (i), (iv), (v), (vi) and (vii) "control" a "correct" experimentation rather than stating something about the investigated action carriers. The entire "information about the carriers" therefore resides in the function μ over $\mathscr{C}$. This shows us the path to be followed if we want to learn the "most possible" about the action carriers themselves regardless of the employed devices. We took a first step in this direction in § 4, considering only M with the structure $\mathscr{Q}$, $\mathscr{R}_0$, $\mathscr{R}$, μ and "forgetting" how M with this structure was obtained from the "pretheory" of the coupled macrosystems. In § 5 we shall take a further step in this direction.

§ 5 Ensembles and Effects

Into the concept and the description of the action carriers only that entered in § 4 what in § 3 we called the directed interaction between two macrosystems. Therefore, quite understandably one still misses a certain "independence" in the concept of action carriers. In fact, M with the structure $\mathscr{Q}$, $\mathscr{R}_0$, $\mathscr{R}$, μ is nothing but a description of the "interaction" under the utmost abstraction of all individual structures of the interacting macrosystems. But now we shall introduce further axioms for those carriers, which signify a first step in the direction of autonomy for action carriers.

§ 5.1 The Problem of Combining Preparation and Registration Procedures

We got a bearing on the combinability problem just before definition D 2.2. Already there we pointed out that the two systems $x_1 \in M_1$ and $x_2 \in M_2$ not only must be put side by side. Rather the combinability also "establishes a prescription", saying what should be viewed as experiments belonging to the "fundamental domain".

What is the aim we are pursuing by prescriptions for the selection of "correct" experiments?

There are two such objects which condition each other. First of all, from the fundamental domain we try to eliminate two complicated problems. One is the production of the action carriers by the preparing device; the other is the action of the carriers on the registration device. In the fundamental domain, we thus want only to keep the facts that the carriers are produced and are acting, without theoretically describing how this happens inside the devices. Secondly, we want to select the experiments in such a way that they inform us about the action carriers and not about the structure of the devices.

Intuitively the second intent is in the minds of all experimental physicists. For planning their experiments, they imagine the carriers to spread (after leaving the preparation device), and the registration devices to test this spreading. This imagination is very obvious for scattering experiments.

During the historical development of the theory for microsystems one has adopted many additional imaginations for the structure of the carriers, which one later had to abandon (e.g. the carriers as particles that fly through space). Our task is it to avoid such additional imaginations when we introduce axioms for the allowed combination of preparation and registration methods.

Thus eliminating the detailed description of the processes inside the preparation and registration devices, we strongly simplify the intended theory. This elimination is a first step in the direction of describing the action carriers as they are.

These two objects in our minds can be illustrated by two examples.

The first example is a rifle as preparation device (1) and a target (sandbag) as registration device (2). Obviously the action of (1) on (2) is directed.

Our first object intended is to include in the fundamental domain neither the physical and chemical processes inside the rifle not the burst of the bullet into the sandbag. In the same direction goes our second intention, namely to describe *only* the flight of the bullet (as action carrier). We have to accomplish both intentions by introducing axioms for combining preparations and registrations.

As the second example let us take a radio transmitter as preparation device (1) and a reciever as registration device (2). First we want to eliminate (from the fundamental domain) the description of how the transmitter and reciever are functioning. We see immediately that this elimination strongly simplifies the problem of how the action spreads from the transmitter to the receiver. In this example we know how to describe the action carriers (electro-magnetic waves), namely by the homogenous (!) Maxwell equations. How can we for the combination of transmitters and receivers introduce such axioms that in the fundamental domain the propagation processes need only be described by the homogeneous (!) Maxwell equations?

The desired reduction of the fundamental domain shall be described (of course not only for the above examples) by the axioms APSZ 5.4, APS 5.1.3, APS 5.1.4. How can these axioms be made plausible? The second example is best suited to guide us to them.

Let the transmitter operate during the time interval from t_1 to t_2 and occupy the space region V. Obviously we must forbid to register at a $t < t_1$ because there is no electromagnetic wave at such a time. If a registration before t_1 were allowed, one could estimate the time about which the wave was emitted. But this emission cannot be described by the homogeneous Maxwell equations. We also must forbid to register within the region V (inside the transmitter). We can only allow to register at times and in regions, where the waves can be described by the homogeneous equations, hence sufficiently late and outside V. When we imagine subsets $\mathscr{Q}_{2\alpha}$ of $\mathscr{Q}_2$ with a parameter α (to describe the time after which and the region outside which one registers), it is plausible to introduce the following axiom for the set Δ_2 of all $\mathscr{Q}_{2\alpha}$:

APSZ 5.4 There is a downward directed set $\Delta_2 \subset \mathscr{P}(\mathscr{Q}_2)$ with $\emptyset \notin \Delta_2$ so that for each $a_1 \in \mathscr{Q}_1$ there is at least one element $\mathscr{Q}_{2\alpha} \in \Delta_2$ such that a_1 is combinable with all $a_2 \in \mathscr{Q}_{2\alpha}$.

The next relation easily follows (as a theorem) from APSZ 5.4 and the theorems in § 3.

APS 5.1.2 There is a downward directed set $\Gamma \subset \mathscr{P}(\mathscr{R}_0')$ with $\emptyset \notin \Gamma$ such that for each $a \in \mathscr{Q}'$ there is at least one element $\mathscr{R}_{0\alpha}' \in \Gamma$ such that a is combinable with all $b_0 \in \mathscr{R}_{0\alpha}'$.

This APS 5.1.2 has in [2] III § 1 been used as an axiom.

The demand of APSZ 5.4 and its implication APS 5.1.2 are weak; they do not essentially restrict the fundamental domain. In particular, APSZ 5.4 does not indicate the physical significance of the index α. There are two ways of strengthening the prescriptions for the combination. One possibility would be to introduce in $\mathscr{M}\mathscr{T}_\Sigma$ relations between the index α and space-time-regions. The second possibility is to introduce additional axioms for the set Γ, having in mind the physical significance of the index α without introducing mathematical relations for this significance. We choose the second possibility. This does not mean that we do not attach importance to describing the relations between space-time-regions and preparation or registration procedures occuring in these regions (for such relations see [46]). We do not introduce such relations for the sole reason that we shall have no opportunity to use them in other contexts within this book.

To strengthen APS 5.1.2 let us ask, what information about the carriers the registration methods of $\mathscr{R}_{0\alpha}$ can give. It might turn out that the time specified by an index α is "too late".

Nothing about the actions proceeding from the $x_1 \in a \in \mathscr{Q}'$ (for fixed a) can be learned any more from the $b_0 \in \mathscr{R}_{0\alpha}'$ because these actions have meanwhile petered out. That just this does not happen in the intended fundamental domain of the microsystems (not *only* there) will be expressed by an axiom. This axiom APS 5.1.3 will assert that the structure of the action carriers is not *lost* in time even if perhaps it is temporally variable. This shows the first sign of a certain autonomy of the action carriers. We formulate this "not getting lost" so that it suffices to test the elements of $\mathscr{Q}'$ by means of the procedures b_0 from anyone of the elements $\mathscr{R}_{0\alpha}' \in \Gamma$. Since, according to APS 5.1.2, for any $a_1, a_2 \in \mathscr{Q}'$ there is a common $\mathscr{R}_{0\alpha}' \in \Gamma$ such that all $b_0 \in \mathscr{R}_{0\alpha}'$ are combinable as well with a_1 as with a_2, we formulate

APS 5.1.3 The set Γ from APS 5.1.2 can be so chosen that the following holds: For *each* element $\mathscr{R}_{0\alpha}' \in \Gamma$, with a_1, b_0 and a_2, b_0 combinable for all $b_0 \in \mathscr{R}_{0\alpha}'$, from

$$\mu(a_1, (b_0, b)) = \mu(a_2, (b_0, b))$$

for all $b_0 \in \mathscr{R}_{0\beta}'$ and all $b \subset b_0$ follows

$$\mu(a_1, f) = \mu(a_2, f)$$

for all $f \in \mathscr{F}$, where (a_1, f) and $(a_2, f) \in \mathscr{C}$.

This axiom asserts that two preparation procedures for action carriers do not differ at all in their probability distributions if they cannot be distinguished by the registration methods b_0 of a single $\mathscr{R}_{0\alpha}' \in \Gamma$. Hence nothing of the structure of the action carriers gets *lost* if one registers after any length of time.

In our example of the electromagnetic waves as action carriers, APS 5.1.3 holds because any solution of the homogeneous Maxwell equations is determined by the values of the field at a particular time.

The next theorem follows from APS 5.1.3:

T 5.1.1 The relation $\sim$ defined by

$$a_1 \sim a_2 \colon \mu(a_1, f) = \mu(a_2, f) \text{ for all } f \text{ with } (a_1, f), (a_2, f) \in \mathscr{C}$$

is an equivalence relation.

Proof. See [2] III Th 1.2.

We now adopt

D 5.1.1 Let $\mathscr{K}$ be the set of classes of $\mathscr{Q}'$ relative to the equivalence relation from T 5.1.1. Let φ denote the canonical mapping: $\mathscr{Q}' \xrightarrow{\varphi} \mathscr{K}$.

The set $\mathscr{K}$ is countable since $\mathscr{Q}$ is countable.

From T 5.1.1 follows immediately that

$$\hat{\mu}(\varphi(a), f) = \mu(a, f) \tag{5.1.1}$$

defines a function $\hat{\mu}$ on a subset of $\mathscr{K} \times \mathscr{F}$, formed by all those pairs (w, f) for which an $a \in w$ exists with $(a, f) \in \mathscr{C}$.

The intuitive notions having led to APS 5.1.3 lead us to formulate a further axiom. We have imagined that for any $a \in \mathscr{Q}'$ there exists a set $\mathscr{R}'_{0\alpha} \in \Gamma$ with whose elements a is combinable. If we now compare the different a which belong to the same class w, one can perhaps find, for an $a' \in w$ not yet combinable with all $b_0 \in \mathscr{R}'_{0\alpha}$ (since the preparation procedure a' of the action carriers is not "terminated" by the time corresponding to α), an $a \in w$ for which the preparation has already terminated at an earlier time. One could thus require that in each w and for any α one finds an $a \in w$ combinable with all $b_0 \in \mathscr{R}'_{0\alpha}$. Also vice versa, for fixed $a \in \mathscr{Q}'$ and any $f \in \mathscr{F}$ with $(a, f) \notin \mathscr{C}$, there should exist an f' to be registered later (with $(a, f) \in \mathscr{C}$), which yields the same probabilities. These intuitive considerations suggest

APS 5.1.4 For each $b_0 \in \mathscr{R}'_0$ there is in each class $w \in \mathscr{K}$ an $a \in w$ such that a and b_0 are combinable. For any pair $a \in \mathscr{Q}'$, $f \in \mathscr{F}$ there is an $f' \in \mathscr{F}$ with $(a, f') \in \mathscr{C}$, making $\hat{\mu}(w, f) = \hat{\mu}(w, f')$ wherever defined.

As the intuitive path to axioms APS 5.1.3 and APS 5.1.4 has shown, these are by no means trivial. It certainly is thinkable that these axioms so severely restrict the fundamental domain of the theory that it becomes useless just for the intended interaction situations. Therefore, if difficulties arise (for example, for a "relativistic quantum theory") one must not assume that the axioms APS 5.1.3 and APS 5.1.4 cannot be responsible.

From APS 5.1.4 follows

T 5.1.2 $\hat{\mu}$ is defined on all of $\mathscr{K} \times \mathscr{F}$.

Proof: See [2] III Th 1.3.

T 5.1.3 The relation $\sim$ defined by

$$f_1 \sim f_2 \colon \hat{\mu}(w, f_1) = \hat{\mu}(w, f_2) \text{ for all } w \in \mathscr{K}$$

is an equivalence relation. For every pair $a \in \mathscr{Q}'$, $f \in \mathscr{F}$ there is an $f' \in \mathscr{F}$ with $f \sim f'$ and $(a, f') \in \mathscr{C}$.

D 5.1.2 Let $\mathscr{L}$ be the set of classes of $\mathscr{F}$ relative to the equivalence relation in T 5.1.3. Let ψ be the canonical mapping $\mathscr{F} \xrightarrow{\psi} \mathscr{L}$.

The set $\mathscr{L}$ is countable since $\mathscr{F}$ is countable.

From T 4.2 easily follows

T 5.1.4 A real-valued function $\tilde{\mu}$ is defined on $\mathscr{K} \times \mathscr{L}$ by $\tilde{\mu}(w, \psi(f)) = \hat{\mu}(w, f)$ with $w \in \mathscr{K}, f \in \mathscr{F}$; it satisfies

(i) $0 \leqq \tilde{\mu}(w, g) \leqq 1$ for $g \in \mathscr{L}$.
(ii) $\tilde{\mu}(w_1, g) = \tilde{\mu}(w_2, g)$ for all $g \in \mathscr{L}$ implies $w_1 = w_2$.
(iii) $\tilde{\mu}(w, g_1) = \tilde{\mu}(w, g_2)$ for all $w \in \mathscr{K}$ implies $g_1 = g_2$.
(iv) There is a $g_0 \in \mathscr{L}$ with $\mu(w, g_0) = 0$ for all $w \in \mathscr{K}$.
(v) There is a $g_1 \in \mathscr{L}$ with $\mu(w, g_1) = 1$ for all $w \in \mathscr{K}$.
(vi) For each $g \in \mathscr{L}$ there is a $g' \in \mathscr{L}$ with $\mu(w, g) + \mu(w, g') = 1$ for all $w \in \mathscr{K}$.

Later on, instead of g_0 we simply write **0** and instead of g_1 simply **1**. If no error is possible, we shall simply write μ instead of $\tilde{\mu}$ so that $\mu(\varphi(a), \psi(f)) = \mu(a, f)$ holds (in a not quite correct notation).

One should recall that the partition of $\mathscr{Q}'$ and $\mathscr{F}$ into classes depends essentially on the combinable pairs (a, b_0). If one had declared the pair (a, b_0) combinable if and only if the corresponding devices are not in each others way, one would have obtained (for example for $\mathscr{Q}'$) an essentially more refined partition into classes. The classes from $\mathscr{K}$ would split into finer ones because the $a \in \mathscr{Q}'$ could be distinguished, for instance, by the space-time region in which they prepare. But just by choosing the coarser partition we believe we can dig out the structures of the "interaction carriers". They are structures which no longer contain the special form of the devices belonging to the $a \in \mathscr{Q}'$.

Let it further be noted that APS 5.1.1, APS 5.1.2, APS 5.1.3 are not mutually independent: APS 5.1.3 is formulated so that it comprises APS 5.1.2 from which APS 5.1.1 follows as a theorem. Whereas APS 5.1.1, APS 5.1.2 and APS 5.2 are easily reduced (see §§ 1 through 4) to physically not very deep statements about composite systems, APS 5.1.3 and APS 5.1.4 thus present important "selection axioms" for "correct" experiments. The "aim" of this selection is to introduce the equivalence relations in $\mathscr{Q}'$ and $\mathscr{F}$. (A similarity of this procedure with the introduction of selection norms in protophysics is unmistakable; see [42].) Therefore, the axioms APS 5.1.3 and APS 5.1.4 represent essential statements about the interaction, declaring it possible to select such a fundamental domain that this selection axiom can be satisfied. Of course, even with APS 5.1.3 and APS 5.1.4 there are in fact larger domains of the theory than only the microsystems.

The consequence APS 5.1.2 of APSZ 5.4 and the axioms APS 5.1.3, APS 5.1.4 do not explicitly describe the relation between the elements of Γ and space-time-regions. But these relations will in § 7 be important in formulating the transport of the preparation and registration devices relative to each other.

In applying quantum mechanics, one very often disregards that APS 5.1.2 to APS 5.1.4 (together with the structures of § 7) actually reduce the fundamental domain in such a way that the discription of physical processes inside the preparation and registration devices is eliminated. To say it very clearly: By introducing

APS 5.1.2 to APS 5.1.4, we deprived ourselves from the beginning of the possibility to describe the processes inside the preparing and registering devices by quantum mechanics alone. Or to say it in a familiar form: Quantum mechanics is from the beginning not suited to explain the "measuring process" (the registration). The attempt to derive the measuring process from quantum mechanics resembles that of deriving the inhomogeneous Maxwell equation from the homogeneous one.

Therefore it is amusing to watch the various philosophical gymnastics by which one hopes to escape from this hard fact. It is not possible, simultaneously to get a "simple" theory of the microsystems as action carriers (in the form of quantum mechanics) *and* a theory of the preparation and registration devices.

For instance it is essentially simpler to describe the structure of light (as action carrier) than to describe the production of this light by a star.

This example of the spreading of light may seduce us to some errors.

(1) One may assume that the emission of a photon by an atom (describable by quantum mechanics) were a preparation process for light. This is not the case. One must take a *macroscopic* device to establish a preparation procedure. It is possible to construct a preparation device producing single excited atoms and to register the emitted photons (see [2] XI).
(2) It seems wrong that we cannot measure the position of a light source in space time by registrations far away from the source. Indeed we observe the stars and also other light sources in this way. But the registrations *alone* cannot tell us where the sources are. It is not difficult, by optical methods with lenses and mirrors to procedure such light beams that the light seems to come from "virtual" sources. To infer "real" sources is only possible when one has additional information. This is not available from registrations far away from the sources (registrations combinable with the preparations). To conclude: Quantum mechanics alone is not suited to describe the preparation and registration devices.

Is this a reason, not to be interested in the processes inside the preparation and registration devices? On the contrary! But for this purpose we have to develop a more comprehensive theory than quantum mechanics (similarly as the inhomogeneous Maxwell equation yields a more comprehensive theory than the homogeneous one does). How such a more comprehensive theory may look will be seen in XI.

§ 5.2 Physical Systems

Let us express the importance of the axioms APS 5.1.3 and APS 5.1.4 by a new concept, which sharpens the definition D 4.2.

D 5.2.1 We shall designate M, endowed with the structure $\mathcal{Q}$, $\mathcal{R}_0$, $\mathcal{R}$, $\lambda_{\mathcal{Q}}$, $\lambda_{\mathcal{R}_0}$, $\lambda_{\mathcal{S}}$ which satisfies APS 1 through APS 4, APS 6 through APS 8 and APS 5.1.3, APS 5.1.4, APS 5.2, as a *set of physical systems.*

Therefore, one can for physical systems introduce the sets $\mathcal{K}$ and $\mathcal{L}$ defined in D 5.1.1 and D 5.1.2. In view of the importance of these sets, we likewise introduce two new concepts:

D 5.2.2 The elements of $\mathcal{K}$ will be called *ensembles* or *states*, those of $\mathcal{L}$ will be calles *effects*.

Using the methods of [3] §§ 10.4 and 10.5, one can (by means of the mappings φ and ψ) easily show that $\mathcal{K}$ and $\mathcal{L}$ are sets of real situations. Our intention to recognize the structure of the physical systems from M suggests, in forgetting the devices to go a step farther, namely also to forget $\mathcal{Q}$, $\mathcal{R}_0$, $\mathcal{R}$, $\lambda_{\mathcal{Q}}$, $\lambda_{\mathcal{R}_0}$, $\lambda_{\mathcal{L}}$, considering only $\mathcal{K}$, $\mathcal{L}$ with the structure $\mathcal{K} \times \mathcal{L} \xrightarrow{\mu} [0, 1]$. This is a very customary way of presenting quantum mechanics.

One can surely construct a $\mathcal{PT}_3$ so that the axiomatic basis $\mathcal{MT}_{\Sigma_3}$ is replaced by a structure species Σ_3, There $\mathcal{K}$, $\mathcal{L}$ are taken as the base sets and $\mathcal{K} \times \mathcal{L} \xrightarrow{\mu} [0, 1]$ as the structure term with the relations from T 5.1.4 as axioms. But one buys such a presentation (that seemingly relates purely to the microsystems) at the expense of serious difficulties in the interpretation as we will now make clear.

According to our derivation, $\mathcal{PT}_3$ would be a restriction (with trivial embedding) of $\mathcal{PT}_1$, the theory we have started to erect in § 1. Therefore, one could continue the scheme (4.4) in the sense of [3] § 8 and [48] to

$$\mathcal{PT}_1 \rightarrow \mathcal{PT}_1' \rightsquigarrow \mathcal{PT}_2 \rightarrow \mathcal{PT}_2' \rightsquigarrow \mathcal{PT}_3 . \qquad (5.2.1)$$

In this sense also the mapping principles of $\mathcal{PT}_3$ are fixed as shown in [3] § 8 and [48]. All the theories presented (in § 1 through § 5.1) "before" $\mathcal{PT}_3$ can then be counted among the pretheories of $\mathcal{PT}_3$ (see [48]), whereby one obtains the mapping principles of $\mathcal{PT}_3$ from the pretheories (as hinted in [3] § 10.5 and [48]). But also without knowledge of the general considerations from [3] or [48] is it obvious what the sets $\mathcal{K}$ and $\mathcal{L}$ mean since they were *derived* from the interpreted sets M_1, M_2, $\mathcal{Q}_1$, $\mathcal{Q}_2$, $\mathcal{R}_1$, $\mathcal{R}_2$, etc. If one abandons all this in order to interpret the elements of $\mathcal{K}$, $\mathcal{L}$ and μ, one must try to explain the mapping principles in everyday language. The explanations in [1] XI § 1.7, for instance, show how difficult and obscure such explanations are.

Still worse than obscurities are the erroneous interpretations that can occur if for $\mathcal{PT}_3$ one foregoes the "pretheories" from § 1 through § 5.1. Such erroneous interpretations can easily be suggested by intuitive concepts. In classical mechanics, an ensemble is characterized by a density $\varrho(x)$ in the phase space Γ. One can regard this $\varrho(x)$ as realized by a very large number of systems whose swarm of points in Γ approximates ϱ. In this sense the structure of ϱ seems to determine the partition of a large swarm of systems *uniquely*. Whenever $\varrho_1 = \varrho_2$, a swarm of systems characterizing ϱ_1 is in no way physically distinguishable from a swarm of systems characterizing ϱ_2. Such notions carried over to quantum mechanics can easily lead to misjudgments. Only so is it understable that one could perceive the so-called "Einstein-Podolski-Rosen paradox" as a paradox although it is none. Subsets a_1, a_2 of microsystems ($a_1, a_2 \in \mathcal{Q}'$) can very well be physically distinguishable although $\varphi(a_1) = \varphi(a_2)$. Yes, it can even happen that two preparation procedures a_1 and a_2 in spite of $\varphi(a_1) = \varphi(a_2)$ differ so strongly that *necessarily* $a_1 \cap a_2 = \emptyset$ (see V § 7 and VII T 5.3.2). Similar remarks hold for the effects.

Evidently something in the structure of the physical systems which transfer the action is lost when one "forgets" everything up to ($\mathcal{K}$, $\mathcal{L}$, μ). We shall in V compensate this loss by introducing the concepts "preparator" and "observable".

The suspicion that $(\mathcal{K}, \mathcal{L}, \mu)$ does *not* contain everything that can be said about the structure of the microsystems is already suggested by the fact that without the preparation and registration procedures it is no longer possible to express mathematically a relation between M and the sets $\mathcal{K}$, $\mathcal{L}$. To the structure $(\mathcal{K}, \mathcal{L}, \mu)$, the concepts of preparators and observables attach pictures for preparation and registration procedures. These pictures appear in mathematically idealized form, namely as abstract Boolean rings. Therefore, one *cannot completely understand* quantum mechanics if one *only* starts from $\mathcal{K}$, $\mathcal{L}$, μ.

The only reason why in definition D 5.2.1 we have used the phrase "physical systems" and not "physical objects" is that we reserve the word "objects" for special physical systems (see V § 10).

§ 5.3 Mixing and De-mixing of Ensembles and Effects

An important theorem for μ (i.e. for $\tilde{\mu}$) follows from T 4.2 (iv):

T 5.3.1 From a de-mixing $a = \bigcup_{i=1}^{n} a_i$ of a preparation procedure a, for all $g \in \mathcal{L}$ follows

$$\mu(\varphi(a), g) = \sum_{i=1}^{n} \lambda_i \, \mu(\varphi(a_i), g), \tag{5.3.1}$$

where $\lambda_i = \lambda_{\mathcal{Q}}(a, a_i)$, $0 < \lambda_i \leqq 1$ and $\sum_{i=1}^{n} \lambda_i = 1$.

Proof. By [2] III Th 2.1, it results from T 4.2 (iv) and T 5.1.3.

D 5.3.1 If for a $w \in \mathcal{K}$ there is a set of real numbers λ_i with $0 < \lambda_i \leqq 1$, $\sum_{i=1}^{n} \lambda_i = 1$, and a set $w_i \in \mathcal{K}$ so that

$$\mu(w, g) = \sum_{i=1}^{n} \lambda_i \, \mu(w_i, g) \tag{5.3.2}$$

holds for all $g \in \mathcal{L}$, then one calls (5.3.2) a de-mixing of w relative to the w_i with weights λ_i.

By the construction of preparation devices, used to select systems, a further structure of preparation procedures is suggested. It reverses the de-mixing of preparation procedures and is therefore called the mixing of preparation procedures:

If one has constructed two selection procedures $c', c'' \in \mathcal{S}_1$, one can evidently construct another selection procedure $c \in \mathcal{S}_1$ as follows: One takes a random generator B with two indications (+) and (−), where (+) appears with the frequency α (hence (−) with the frequency $1 - \alpha$). To the device B one attaches devices selected by c' resp. c'', so that upon appearance of (+) the device selected due to c' is used for the action on systems from M_2 and upon appearance of (−) the device selected due to c'' is used. The new "large" device thus composed from B and the devices selected due to c' resp. c'' is again a system $x_1 \in M_1$ with directed interaction. One thus obtains from B, c', c'' a new selection procedure $c \in \mathcal{S}$. How can one express this technical construction possibility of c from c', c'' and B mathematically?

The new procedure c is de-mixed into c_+ and c_- according to (+) or (−) having appeared on B. Therefore, $c = c_+ \cup c_-$ and $c_+ \cap c_- = \emptyset$.

One would be tempted to simply set $c_+ = c'$ and $c_- = c''$. But this is false since the selection procedure c' concerns only the selection due to c' and not a selection of the "large" systems comprizing the device B. Nevertheless, c_+ evidently is very similar to c'. This similarity can be expressed best with the aid of the mapping φ from D 5.1.1.

According to the definition of $\mathscr{Q}$ in § 4, to each element of $\mathscr{S}_1$ there corresponds in a one-to-one fashion an element of $\mathscr{Q}$: $c' \leftrightarrow a'$, $c'' \leftrightarrow a''$, $c_+ \leftrightarrow a_+$, $c_- \leftrightarrow a_-$. Then we can express the similarity of c' with c_+ simply by $\varphi(a') = \varphi(a_+)$. Because of the isomorphism of $\mathscr{Q}$ with $\mathscr{S}_1$, the axiom to describe the possibility of constructing c shall be formulated directly by means of the elements of $\mathscr{Q}$. Let us first introduce two definitions.

D 5.3.2 Two preparation procedures a, $\tilde{a} \in \mathscr{Q}'$ are called *isomorphic* (this is what we intuitively designated above as being similar) if there is an isomorphic mapping i of the Boolean ring $\mathscr{Q}(a)$ ($\mathscr{Q}(a) = [\emptyset, a] \subset \mathscr{Q}$) onto the Boolean ring $\mathscr{Q}(\tilde{a})$ such that

$$\varphi(i\,a') = \varphi(a') \quad \text{and} \quad \lambda_{\mathscr{Q}}(a, a') = \lambda_{\mathscr{Q}}(i\,a, i\,a') \tag{5.3.3}$$

hold and if (a, b_0) combinable is equivalent to $(\tilde{a}, b_0)$ combinable.

D 5.3.3 A preparation procedure a is called a direct *mixture* of the preparation procedures a_1 and a_2 if these are isomorphic to two preparation procedures a_1', a_2' with $a_1' \cap a_2' = \emptyset$, $a = a_j' \cup a_2'$. One calls $\alpha = \lambda_{\mathscr{Q}}(a, a_1')$ resp. $(1-\alpha) = \lambda_{\mathscr{Q}}(a, a_2')$ the weight of a_1 respectively a_2 in the direct mixture a.

Due to the above intuitive considerations for the construction of devices let us now require:

AP 1 For every pair $a_1, a_2 \in \mathscr{Q}'$ and for every rational number α with $0 < \alpha < 1$ there is a direct mixture $a \in \mathscr{Q}'$ of a_1 and a_2 with α the weight of a_1 in a.

The phrase "there is" in this axiom due to [3] § 10.4 asserts that a direct mixture a of the stated form is "physically possible". Just this was to be suggested by the above considerations about possibilities of constructing devices.

The "natural law" AP 1 is no indication for "correct" experimentation. It is an assertion about what is doable (physically possible, according to [3], in the sense of "at one's disposal"). However, AP 1 contains no suggestion for realizing a direct mixture. Just before introducing the axiom AP 1 we have described such a suggestion; perhaps one can think of yet other possibilities.

From AP 1 follows

T 5.3.2 For each sequence $w_i \in \mathscr{K}$ $(i = 1, \ldots, n)$ and each sequence of rational numbers $\lambda_i > 0$ $(i = 1, \ldots, n)$ with $\sum_i \lambda_i = 1$ there is an $a \in \mathscr{Q}'$ and a de-mixing $a = \bigcup_{i=1}^{n} a_i$ with $\varphi(a_i) = w_i$ and $\lambda_{\mathscr{Q}}(a, a_i) = \lambda_i$. With $w = \varphi(a)$ we have:

$$\mu(w, g) = \sum_{i=1}^{n} \lambda_i\, \mu(w_i, g) \quad \text{for all} \quad g \in \mathscr{L}.$$

Proof. See [2] III Th 2.2.

That we required AP 1 only for rational α corresponds to the fact that $\mathscr{Q}$ is countable (a consequence from [3] § 9).

In regard to AP 1 let the reader be warned of the wrong inference that for each de-mixing $a = \bigcup_{i=1}^{n} a_i$ the device A belonging to a *must* consist of a random generator B, which triggers other subdevices A_i belonging to the a_i. Actually, the indications on the device A need *not* relate to a random generator. They can be so intimately connected with the processes within the device A that it is *impossible* to define subdevices A_i of A which function according to the occurrence of the indications of some random generator. It will be quite decisive for understanding quantum mechancis that there really exist demixings $a = \bigcup_{i=1}^{n} a_i$ which do not correspond to a decomposition of a device A into subdevices A_i switched by a random generator.

A contradiction to quantum mechanics would arise if one rashly would sharpen the axiom AP 1 to the following requirement:

If there is given a de-mixing of w into the w_i according to D 5.3.1 and if $w = \varphi(a)$, then a has a de-mixing $a = \bigcup_{i=1}^{n} a_i$, with $\varphi(a_i) = w_i$ and $\lambda_{\mathscr{Q}}(a, a_i) = \lambda_i$.

In VII, § 5.3 we shall see that there must be selection procedures a with $\varphi(a) = w$ for which there is no de-mixing $a = \bigcup_{i=1}^{n} a_i$ with $\varphi(a_i) = w_i$ and $\lambda_{\mathscr{Q}}(a, a_i) = \lambda_i$ although $w = \varphi(a)$ obeys (5.3.2). Of course, AP 1 then postulates *another* selection procedure a' with $\varphi(a') = w$ which allows a de-mixing $a' = \bigcup_{i=1}^{n} a'_i$ with $\varphi(a'_i) = w_i$ and $\lambda_{\mathscr{Q}}(a', a'_i) = \lambda_i$.

These considerations about the preparation procedures can easily be transferred in a slightly modified form to the effect procedures.

To begin with, from T 4.2 (viii) follows

T 5.3.3 If $b_0 = \bigcup_{i=1}^{n} b_{0i}$ is a de-mixing of the registration method b_0 into the b_{0i} with weight λ_i, an effect procedure $f = (b_0, b)$ with $f_i = (b_{0i}, b_{0i} \cap b)$ satisfies

$$\mu(w, \psi(f)) = \sum_{i=1}^{n} \lambda_i \mu(w, \psi(f_i)) \tag{5.3.4}$$

for all $w \in \mathscr{K}$.

Proof. See [2] III Th 2.3.

D 5.3.4 If for a $g \in \mathscr{L}$ there is a set of real numbers λ_i with $0 < \lambda_i \leqq 1$, $\sum_{i=1}^{n} \lambda_i = 1$ and a set $g_i \in \mathscr{L}$ so that

$$\mu(w, g) = \sum_{i=1}^{n} \lambda_i \mu(w, g_i) \tag{5.3.5}$$

holds for all $w \in \mathscr{K}$, one calls (5.3.5) a de-mixing of the effect g into the effects g_i with the weights λ_i.

Thus (5.3.4) presents a de-mixing of $g = \psi(b_0, b)$ into the $g_i = \psi(b_{0i}, b_{0i} \cap b)$ with the weights $\lambda_i = \lambda_{\mathscr{R}_0}(b_0, b_{0i})$.

The above prescription for constructing a selection procedure $c \in \mathscr{S}_1$ from two selection procedures $c', c'' \in \mathscr{S}_1$ and a random generator can at once be transferred to the elements of $\mathscr{S}_{20}$. By the isomorphic correspondence of $\mathscr{R}_0$ with $\mathscr{S}_{20}$ thus follows the construction of a "direct mixture" $b_0 \in \mathscr{R}_0$ from two $b_{01}, b_{02} \in \mathscr{R}_0$. In order to formulate this, correspondingly to D 5.3.2, D 5.3.3 we adopt

D 5.3.5 Two registration procedures b_0 and b_0' are called *isomorphic* if there is an isomorphic mapping i of the Boolean ring $\mathscr{R}(b_0)$ onto the Boolean ring $\mathscr{R}(b_0')$ such that $\psi(i\,b_0, i\,b) = \psi(b_0, b)$ holds and i maps the subring $\mathscr{R}_0(b_0)$ isomorphically onto $\mathscr{R}_0(b_0')$, and if (a, b_0) combinable is equivalent to (a, b_0') combinable.

D 5.3.6 A registration procedure b_0 is said to be a *direct mixture* of the registration procedures b_{01}, b_{02} if these are isomorphic to some b_{01}' and b_{02}' with $b_{01}' \cap b_{02}' = \emptyset$ and $b_0 = b_{01}' \cup b_{02}'$. One calls $\alpha = \lambda_{\mathscr{R}_0}(b_0, b_{01}')$ resp. $1 - \alpha = \lambda_{\mathscr{R}_0}(b_0, b_{02}')$ the weight of b_{01} resp. b_{02} in the direct mixture b_0.

As axiom let us require

AR 1 For every pair $b_{01}, b_{02} \in \mathscr{R}_0$ and every rational number α with $0 < \alpha < 1$ there is a direct mixture $b_0 \in \mathscr{R}_0'$ of b_{01}, b_{02}, where b_{01}, has the weight α in b_0.

From AR 1 follows

T 5.3.4 For every sequence $g_i \in \mathscr{S}$ $(i = 1, \ldots, n)$ and every sequence of rational numbers $\lambda_i > 0$ $(i = 1, \ldots, n)$ with $\sum \lambda_i = 1$, there exist a $b_0 \in \mathscr{R}_0'$ and a de-mixing $b_0 = \bigcup_{i=1}^{n} b_{0i}$ with $b_{0i} \in \mathscr{R}_0'$, and a $b \in \mathscr{R}$ with $b \subset b_0$ such that $\psi(b_{0i}, b_{0i} \cap b) = g_i$ and $\lambda_{\mathscr{R}_0}(b_0, b_{0i}) = \lambda_i$. With $\psi(b_0, b) = g$ then follows

$$\mu(w, g) = \sum_{i=1}^{n} \lambda_i \, \mu(w, g_i) \quad \text{for all} \quad w \in \mathscr{K}.$$

Proof. See [2] III Th 2.4.

A special case of T 5.3.2 is

APK For every sequence $w_i \in \mathscr{K}$ $(i = 1, \ldots, n)$ and every sequence of rational numbers $\lambda_i > 0$ $(i = 1, \ldots, n)$ with $\sum_i \lambda_i = 1$, there is $w \in \mathscr{K}$ with

$$\mu(w, g) = \sum_{i=1}^{n} \lambda_i \, \mu(w_i, g) \quad \text{for all} \quad g \in \mathscr{S}.$$

Likewise, T 5.3.4 contains the special case

ARK For every sequence $g_i \in \mathscr{S}$ $(i = 1, \ldots, n)$ and every sequence of rational numbers $\lambda_i > 0$ $(i = 1, \ldots, n)$ with $\sum_i \lambda_i = 1$, there is a $g \in \mathscr{S}$ such that

$$\mu(w, g) = \sum_{i=1}^{n} \lambda_i \, \mu(w, g_i) \quad \text{for all} \quad w \in \mathscr{K}.$$

If one wants to forget $\mathcal{D}$, $\mathcal{R}_0$, $\mathcal{R}$, $\lambda_{\mathcal{S}}$ and use only $\mathcal{K}$, $\mathcal{L}$, μ as the basis for a theory, then one must require APK, ARK as axioms. One does this mostly by describing *all* that in § 5.3 has been depicted and mathematically derived, only in everyday language. Of course, we also have thus described a part, namely that up to the statement of the axioms AP 1 and AR 1.

§ 5.4 Re-elimination of the Action Carrier

Since one very often uses only the method of describing physical systems by the sets $\mathcal{K}$, $\mathcal{L}$ and the function μ, the question arises how to present in a clearer form the production of ensembles and the registration of effects.

Therefore, we now start backwards with the set $\mathcal{K}$ of ensembles, the set $\mathcal{L}$ of effects and the probability μ which satisfies (i) through (vi) of T 5.1.4.

In order to give the concept "ensemble" a meaning, we must introduce a set M of systems to which this concept refers. Likewise the concept "effect" must be explained. Therefore, over M we introduce the structure terms $\mathcal{D}$, $\mathcal{R}_0$, $\mathcal{R}$, with $\mathcal{D}$ the set of preparation procedures, $\mathcal{R}_0$ the set of registration methods, and $\mathcal{R}$ the set of registration procedures.

With $\mathcal{F}$ as in D 4.3 we then obtain $\mathcal{K} = \varphi\mathcal{D}'$, $\mathcal{L} = \psi\mathcal{F}$ and

$$\tilde{\mu}\,(\varphi\,(a), \psi\,(b_0, g)) = \lambda_{\mathcal{S}}\,(a \cap b_0, a \cap b)\,.$$

This equation defines $\lambda_{\mathcal{S}}$ if one furthermore prescribes $\lambda_{\mathcal{D}}$. But how are the elements of $\mathcal{D}$, $\mathcal{R}_0$, $\mathcal{R}$ defined by the devices with which these procedures are performed?

To begin with, to every $x \in M$ one must assign a preparation device and a registration device. Due to AZ 1, AZ 2 a mapping

$$M_1 \xrightarrow{\pi} M \tag{5.4.1}$$

is defined by $\pi(x_1) = (x_1, x_2)$. This $\pi(x_1)$ is called the action carrier prepared by the device x_1. By (4.1) there is defined an isomorphic mapping

$$\mathcal{S}_1 \xrightarrow{h} \mathcal{D} \tag{5.4.2}$$

(see the remarks after (4.1) through (4.3)). We can transfer (5.4.1) canonically to:

$$\mathcal{P}(M_1) \xrightarrow{\pi} \mathcal{P}(M)\,. \tag{5.4.3}$$

One easily recognizes that the h in (5.4.2) is just the mapping π from (5.4.3) restricted to $\mathcal{S}_1$. For (5.4.2) we can thus write

$$\mathcal{S}_1 \xrightarrow{\pi} \mathcal{D}. \tag{5.4.4}$$

Hence $\varphi\,\pi(c_1)$ with $c_1 \in \mathcal{S}_1$ is just the "ensemble prepared by the devices $x_1 \in c_1$".

Similarly to (5.4.1), a mapping

$$M_2 \xrightarrow{\varrho} M \tag{5.4.5}$$

is defined by $\varrho(x_2) = (x_1, x_2)$. This $\varrho(x_2)$ is called the action carrier registered by the device x_2. Canonically extending (5.4.5) one obtains a mapping

$$\mathcal{P}(M_2) \xrightarrow{\varrho} \mathcal{P}(M)\,, \tag{5.4.6}$$

which reproduces the isomorphisms defined by (4.2), (4.3):

$$\mathcal{S}_{20} \xrightarrow{\varrho} \mathcal{R}_0,$$
$$\mathcal{S}_2 \xrightarrow{\varrho} \mathcal{R}. \tag{5.4.7}$$

As a subset of $\mathcal{R}_{20} \times \mathcal{R}_2$ we introduce

$$\mathcal{N}_2 = \{(c_{20}, c_2) \mid c_{20} \in \mathcal{S}_{20}, c_2 \in \mathcal{S}_2, c_2 \subset c_{20}\}. \tag{5.4.8}$$

Then (5.4.7) canonically yields the bijective mapping

$$\mathcal{N}_2 \xrightarrow{\varrho} \mathcal{F}. \tag{5.4.9}$$

Thus $\psi\varrho(c_{20}, c_2)$ with $(c_{20}, c_2) \in \mathcal{N}_2$ is the effect defined by the indications c_2 on the registration devices from c_{20}.

If now the action carriers $x \in \pi(c_1)$ generated by the devices $x_1 \in c_1$ are registered by devices $x_2 \in c_{20}$, then the probability of an indication c_2 equals

$$\tilde{\mu}(\varphi\,\pi(c_1), \psi\,\varrho(c_{20}, c_2)) = \lambda_{12}(c_1 \times c_{20} \cap M, c_1 \times c_2 \cap M). \tag{5.4.10}$$

According to T 2.8 it suffices to choose in particular $c_1 = a_1 \cap b_1$, $c_{20} = a_2 \cap b_{20}$, $c_2 = a_2 \cap b_2$ with $b_2 \in \mathcal{R}_2(b_{20})$. Then (5.4.10) reads

$$\begin{aligned}\tilde{\mu}(\varphi\,\pi(a_1 \cap b_1, \psi\,\varrho(a_2 \cap b_{20}, a_2 \cap b_2)) \\ = \lambda_{12}((a_1 \cap b_1) \times (a_2 \cap b_{20}) \cap M, (a_1 \cap b_1) \times (a_2 \cap b_2) \cap M).\end{aligned} \tag{5.4.11}$$

The left side describes as $\tilde{\mu}(w, g)$ the probability that action carriers of an ensemble $w = \varphi\,\pi(a_1 \cap b_1)$ cause an effect $g = \psi\,\varrho(a_2 \cap b_{20}, a_2 \cap b_2)$. If one "forgets" how w and g have arisen, the left side of (5.4.11) thus gives a description in which only the action carriers occur as physical systems. But the right side says what is really meant, namely a macroscopically describable experiment in which "at first" nothing is seen of physical systems as action carriers.

The $\tilde{\mu}$ in (5.4.11) due to (3.7) determines λ_{12}, if also $\lambda_{\mathcal{D}_1}$ and $\lambda_{\mathcal{D}_2}$ are given together with the $\mu_1(a_1, (b_{10}, b_1))$ which describes the macroscopic behavior of the preparing device. The left side of (5.4.11) therefore completely describes the action of system 1 on system 2, i.e. the directed interaction of the systems 1, 2. Therefore one can call (5.4.11) the mathematical "description of the interaction of system 1 and 2 by action carriers". These are emitted from the systems 1 (as the set $\pi(a_1 \cap b_1)$ which belongs to the ensemble $\varphi\,\pi(a_1 \cap b_1)$) and cause the "effects" $\psi\,\varrho(a_2 \cap b_{20}, a_2 \cap b_2)$ on the systems 2. This is just the mode in which one uses to describe the experiments with microsystems (e.g. scattering experiments; see [2] XVI).

Before we investigate the sets $\mathcal{K}, \mathcal{L}$, and to give them more structure by further axioms, let us in the following § 6 finish the description (sketched only briefly in § 1) of composite systems in state spaces and trajectory spaces.

§ 6 Objectivating Method of Describing Experiments

Here let us connect the base terms introduced in § 1 with the sets $\mathcal{K}$, $\mathcal{L}$ and the function $\mathcal{K} \times \mathcal{L} \xrightarrow{\mu} [0, 1]$. Before doing this, let us point out the remarkable fact that in § 2 through § 5 we did not really use the objectivating description of the macrosystems $x_1 \in M_1, x_2 \in M_2, (x_1, x_2) \in M$.

§ 6.1 The Method of Describing Composite Macrosystems in the Trajectory Space

Already in § 1 we introduced the state spaces Z_1, Z_2 defined by pretheories and assumed them equipped with uniform structures g and p. The description of the systems in M as composite systems then means that the states of $x=(x_1, x_2)$ are described by the state space $Z = Z_1 \times Z_2$, where one must set $Z_g = Z_{1g} \times Z_{2g}$ and $Z_p = Z_{1p} \times Z_{2p}$. Thus one must choose the uniform structures g and p in Z to be the product uniform structures, as appears physically meaningful.

Constructing (as in II § 1) the trajectory space Y_1 for Z_1, the trajectory space Y_2 for Z_2, and the trajectory space Y for $Z = Z_1 \times Z_2$, one immediately finds $Y = Y_1 \times Y_2$ and $Y_p = Y_{1p} \times Y_{2p}$ and thus $\hat{Y} = \hat{Y}_1 \times \hat{Y}_2$.

Therefore, on the basis of the axiom AT 1, the sets $Y = Y_1 \times Y_2$ and $\hat{Y} = \hat{Y}_1 \times \hat{Y}_2$ can be derived in the extended axiomatic basis with Y_1, Y_2 as base sets.

§ 6.2 Trajectory Effects of the Composite Systems

In § 2 we introduced the sets $\mathscr{R}_1, \mathscr{R}_{10}; \mathscr{R}_2, \mathscr{R}_{20}$ of registration procedures. In order to interpret these (as in II § 3.1) as trajectory registration procedures, we first introduce (analogously to II § 2.3) instead of $\mathscr{R}_1, \mathscr{R}_{10}, \mathscr{R}_2, \mathscr{R}_{20}$ the base terms $\tilde{\mathscr{R}}_1, \tilde{\mathscr{R}}_{10}, \tilde{\mathscr{R}}_2, \tilde{\mathscr{R}}_{20}$, and as axiom demand

AT 2 $\tilde{\mathscr{R}}_{10} \subset \tilde{\mathscr{R}}_1, \tilde{\mathscr{R}}_{20} \in \tilde{\mathscr{R}}_2$.
$\tilde{\mathscr{R}}_1, \tilde{\mathscr{R}}_{10}, \tilde{\mathscr{R}}_2, \tilde{\mathscr{R}}_{20}$ are "sets of procedures" in the sense of II § 2.1.

As picture relations we introduce (as in II, § 2.1) two relations $\alpha_i\,(i = 1, 2)$ with the interpretation: $(\varrho_i, x_i) \in \alpha_i \subset \tilde{\mathscr{R}}_i \times M_i$ is the picture of "the system x_i was selected due to the procedure ϱ_i", i.e. "for x_i the digital indication characterized by ϱ_i has occured" (see II § 2.3). Then two mappings $\tilde{\mathscr{R}}_i \xrightarrow{h_i} \mathscr{P}(M_i)$ are defined by

$$\varrho_i \xrightarrow{h_i} \{x_i \mid x_i \in M_i, (\varrho_i, x_i) \in \alpha_i\} \quad \text{with} \quad i = 1, 2.$$

By these one can replace α_1 and α_2, i.e. replace $(\varrho_i, x_i) \in \alpha_i$ by $x_i \in h_i(\varrho_i)$. Then defining $\mathscr{R}_i = h_i(\tilde{\mathscr{R}}_i)$, $\mathscr{R}_{i0} = h_i(\tilde{\mathscr{R}}_{i0})$ as axioms we introduce

AT 3 Equations analogous to II (2.1.1) and II (2.1.2) hold for h_1, h_2. For each $\varrho_i \in \tilde{\mathscr{R}}_i$ there is a $\varrho_{i0} \in \tilde{\mathscr{R}}_{i0}$ with $\varrho_i \leqq \varrho_{i0}\,(i = 1, 2)$.

AT 4 $\tilde{\mathscr{R}}_{i0} \xrightarrow{h_i} \mathscr{R}_{i0}\,(i = 1, 2)$ are isomorphic mappings.

AT 5 $\mathscr{R}_{i0}\,(i = 1, 2)$ are SSP.

These AT 4, AT 5 very briefly say that as further picture relations we have introduced probabilities over $\mathscr{R}_{10}$ and $\mathscr{R}_{20}$ with the corresponding axioms (also see II § 2.3).

According to II § 2.3, the relations APSZ 2, APSZ 3 and APSZ 4.1, 2 follow as theorems from AT 3, AT 4, AT 5.

We retain APSZ 1, APSZ 8.1, APSZ 5.1, APSZ 8.2, APSZ 6, APSZ 7 from § 2 as axioms.

Corresponding to II § 3.1, as further picture relations we introduce two probabilities $\lambda_{\text{Meas 1}}$ and $\lambda_{\text{Meas 2}}$, postulating the requirements from II § 3.1 as the axiom

AT 6 For $\lambda_{\text{Meas 1}}$ and $\lambda_{\text{Meas 2}}$ there hold relations analogous to AS 2 from II § 2.1 and equations analogous to II (3.1.1).

Then the considerations from II (3.1.8) with

$$\Phi_i = \{(\varrho_{i0}, \varrho_i) \mid \varrho_{i0} \in \hat{\mathscr{R}}_{i0}, \varrho_i \in \tilde{\mathscr{R}}_i, \varrho_{i0} \neq 0, \varrho_i \leqq \varrho_{i0}\}$$

lead to

$$\Phi_i \xrightarrow{\psi_i} L(\hat{Y}_i) \quad \text{for} \quad i = 1, 2. \tag{6.2.1}$$

The physical meaning of II (3.1.8) and thus of (6.2.1) was that, with $\psi_i(\varrho_{i0}, \varrho_i) = f(y_i)$, the probability $\lambda_{\text{Meas}\,i}(y_i; \varrho_{i0}, \varrho_i)$ for the occurrence of the indication ϱ_i equals $f(y_i)$.

Corresponding to II § 3.1, as further axiom we introduce

AT 7 la $\psi_i(\Phi_i)$ with $i = 1, 2$ is norm-dense in $L(\hat{Y}_i)$.

Let $\tilde{\mathscr{R}}_{12}$ be the system of procedures freely generated by the set $\tilde{\mathscr{R}}_1 \times \tilde{\mathscr{R}}_2$. One obtains $\tilde{\mathscr{R}}_{12}$ by a procedure that is entirely analogous to that from [2] II Th 2.2: The elements of $\tilde{\mathscr{R}}_{12}$ are of the form

$$\sum_{i,j} \dot{+} (\varrho_1^{(i)}, \varrho_2^{(j)}) \text{ with } \varrho_1^{(i)} \leqq \varrho_1 \in \tilde{\mathscr{R}}_1, \varrho_2^{(j)} \leqq \varrho_2 \in \tilde{\mathscr{R}}_2$$
$$\varrho_1^{(i)} \cap \varrho_1^{(k)} = 0 \text{ for } i \neq k,$$
$$\varrho_2^{(j)} \cap \varrho_2^{(l)} = 0 \text{ for } j \neq l,$$

with $\dot{+}$ understood as addition in the Boolean algebra $[0, (\varrho_1, \varrho_2)]$, freely generated by $[0, \varrho_1]$ and $[0, \varrho_2]$.

Corresponding to $\tilde{\mathscr{R}}_{12}$, let $\tilde{\mathscr{R}}_{120}$ be the system of procedures freely generated by the set $\tilde{\mathscr{R}}_{10} \times \tilde{\mathscr{R}}_{20}$. We inquire into the structure of $\lambda_{\text{Meas}}(y; \varrho_{120}, \varrho_{12})$ with $y \in \hat{Y}$, $\varrho_{120} \in \tilde{\mathscr{R}}_{120}$, $\varrho_{120} \neq 0$, $\varrho_{12} \in \tilde{\mathscr{R}}_{12}$, $\varrho_{12} < \varrho_{120}$. This $\lambda_{\text{Meas}}(y; \varrho_{120}, \varrho_{12})$ is determined when $\lambda_{\text{Meas}}(y; (\varrho_{10}, \varrho_{20}), (\varrho_1, \varrho_2))$ is known.

We now assume that the trajectories on the systems 1 and 2 are registered independently, i.e. that $y = (y_1, y_2)$ yields

$$\lambda_{\text{Meas}}(y; (\varrho_{10}, \varrho_{20}), (\varrho_1, \varrho_2)) = \lambda_{\text{Meas 1}}(y_1; \varrho_{10}, \varrho_1)\, \lambda_{\text{Meas 2}}(y_2; \varrho_{20}, \varrho_2). \tag{6.2.2}$$

Then, the probability for measuring trajectories, defined by

$$\Phi_{12} \xrightarrow{\psi_{12}} L(\hat{Y}) \tag{6.2.3}$$

with

$$\Phi_{12} = \{(\varrho_{120}, \varrho_{12}) \mid \varrho_{120} \in \tilde{\mathscr{R}}_{120}, \varrho_{12} \in \tilde{\mathscr{R}}_{12}, \varrho_{120} \neq 0, \varrho_{12} \leqq \varrho_{120}\}$$

is determined by (6.2.2), i.e. by

$$\psi_{12}((\varrho_{10}, \varrho_{20}), (\varrho_1, \varrho_2)) = \psi_1(\varrho_{10}, \varrho_1)\, \psi_2(\varrho_{20}, \varrho_2). \tag{6.2.4}$$

Since the la $\psi_i(\varrho_{i0}, \varrho_i)$ with $i = 1, 2$ are dense in the $L(\hat{Y}_i)$, we find la $\psi_{12}(\Phi_{12})$ dense in $L(\hat{Y})$. In fact, the linear subspace of $C(\hat{Y})$ spanned by all the products $f_1(y_1) f_2(y_2)$ and $|f_1(y_1) f_2(y_2)|$ with $f_i \in \psi_i(\Phi_i)$ is equal to the subspace spanned by all $f_1(y_1) f_2(y_2)$ with $f_i \in$ la $\psi_i(\Phi_i)$. Hence this subspace is dense in $C(\hat{Y})$.

The following sets will later be of importance: $\hat{\mathscr{R}}_{12}$ be the system of SP generated over $M_1 \times M_2$ by the set $\{b_1 \times b_2 \mid b_1 \in \mathscr{R}_1, b_2 \in \mathscr{R}_2\}$. Correspondingly, $\hat{\mathscr{R}}_{120}$ be the system of SP generated by the set $\{b_{10} \times b_{20} \mid b_{10} \in \mathscr{R}_{10}, b_{20} \in \mathscr{R}_{20}\}$ such that $\hat{\mathscr{R}}_{120} \subset \hat{\mathscr{R}}_{12}$.

A mapping $\mathscr{P}(M_1 \times M_2) \xrightarrow{r} \mathscr{P}(M)$ is defined by $d \subset M_1 \times M_2$ and $d \to d \cap M$. By the definition of $\mathscr{R}_{12}$ and $\mathscr{R}_{120}$ in § 2, one sees that

$$\hat{\mathscr{R}}_{12} \xrightarrow{r} \mathscr{R}_{12}, \hat{\mathscr{R}}_{120} \xrightarrow{r} \mathscr{R}_{120} \tag{6.2.5}$$

holds and that these mappings are surjective. Furthermore, it follows easily that r is a homomorphism of $\hat{\mathscr{R}}_{12}$ onto $\mathscr{R}_{12}$ while the mapping $\hat{\mathscr{R}}_{120} \xrightarrow{r} \mathscr{R}_{120}$ is an isomorphism (because $b_{10} \times b_{20} \neq \emptyset \Rightarrow b_{10} \times b_{20} \cap M \neq \emptyset$, see the remarks after APSZ 5.1).

The mapping

$$(\varrho_1, \varrho_2) \to h_1(\varrho_1) \times h_2(\varrho_2)$$

can be extended to all of $\tilde{\mathscr{R}}_{12}$ as the homomorphism

$$\tilde{\mathscr{R}}_{12} \xrightarrow{h_{12}} \hat{\mathscr{R}}_{12}. \tag{6.2.6}$$

Then a homomorphism

$$\tilde{\mathscr{R}}_{12} \xrightarrow{h} \mathscr{R}_{12} \tag{6.2.7}$$

is defined by $h = r\, h_{12}$, where

$$\tilde{\mathscr{R}}_{120} \xrightarrow{h} \mathscr{R}_{120} \tag{6.2.8}$$

holds and is an isomorphism.

The physical interpretation of h is given by: $h(\varrho_1, \varrho_2)$ is that subset of systems $x = (x_1, x_2) \in M$ for which x_i with $i = 1, 2$ triggers the indication ϱ_i.

In § 2 we have not introduced the sets $\tilde{\mathscr{R}}_1$, $\tilde{\mathscr{R}}_2$ to make it *a priori* clear that an axiomatic basis for quantum mechanics can be erected without them.

Below we shall see that we can forget $\tilde{\mathscr{R}}_1$, $\tilde{\mathscr{R}}_2$ even in grasping the objectivating method of describing composite systems.

§ 6.3 Trajectory Ensembles of the Composite Systems

We can immediately extend the considerations of II § 3.2 to Φ_{12} and to $\mathscr{F}_{12}$ defined by

$$\mathscr{F}_{12} = \{(b_{120}, b_{12}) \mid b_{120} \in \mathscr{R}_{120}, b_{12} \in \mathscr{R}_{12}(b_{120}), b_{120} \neq \emptyset, b_{12} \subset b_{120}\}. \tag{6.3.1}$$

In this connection, II (3.2.1) goes over into

$$\begin{array}{ccc} \Phi_{12} & \xrightarrow{\psi_{12}} & L(\hat{Y}) \\ h \downarrow & & \downarrow u_{a_{12}} \\ \mathscr{F}_{12} & \xrightarrow{\mu_{12}(a_{12}, \ldots)} & [0, 1] \end{array} \tag{6.3.2}$$

with h from (6.2.7), with $a_{12} \in \mathscr{D}'_{12}$, $u_{a_{12}} \in C'(Y)$ and (see (2.7))

$$\mu_{12}(a_{12}, (b_{120}, b_{12})) = \lambda_{12}(a_{12} \cap b_{120}, a_{12} \cap b_{12}) . \tag{6.3.3}$$

Corresponding to II § 3.2, as axiom we therefore require

AT 8 The diagram (6.3.2) holds and $u_{a_{12}}$ is an affine mapping.

Due to T 2.8 it suffices to restrict $\mathscr{F}_{12}$ to the subset

$$\tilde{\mathscr{F}}_{12} = \{(b_{120}, b_{12}) \mid b_{120} = b_{10} \times b_{20} \cap M,\, b_{12} = b_1 \times b_2 \cap M,\; b_{10} \in \mathscr{R}_{10},\, b_{20} \in \mathscr{R}_{20},\, b_1 \in \mathscr{R}_1(b_{10}),\, b_2 \in \mathscr{R}_2(b_{20})\}$$

and instead of $a_{12} \in \mathscr{D}'_{12}$ to choose in particular the preparation procedures $a_{12} = a_1 \times a_2 \cap M$ with $a_1 \times a_2 \cap M \in \Gamma_{12}$ to be used below.

Due to II (3.2.2) there is a mapping $\mathscr{D}'_{12} \xrightarrow{\varphi_{12}} C'(\hat{Y})$ with

$$\mu_{12}(a_{12}, (h(\varrho_{120}), h(\varrho_{12}))) = \langle \varphi_{12}(a_{12}), \psi_{12}(\varrho_{120}, \varrho_{12}) \rangle . \tag{6.3.4}$$

The remarks just made about T 2.8 show that φ_{12} is already determined by $\Gamma_{12} \xrightarrow{\varphi_{12}} C'(\hat{Y})$ and the equation

$$\begin{aligned} &\mu_{12}(a_1 \times a_2 \cap M, (h_1(\varrho_{10}) \times h_2(\varrho_{20}) \cap M,\, h_1(\varrho_1) \times h_2(\varrho_2) \cap M)) \\ &\quad = \lambda_{12}(a_1 \times a_2 \cap M \cap h_1(\varrho_{10}) \times h_2(\varrho_{20}),\, a_1 \times a_2 \cap M \cap h_1(\varrho_1) \times h_2(\varrho_2)) \\ &\quad = \langle \varphi_{12}(a_1 \times a_2 \cap M), \psi_1(\varrho_{10}, \varrho_1)\, \psi_2(\varrho_{20}, \varrho_2) \rangle . \end{aligned} \tag{6.3.5}$$

With $f_1 = \mathbf{1} \in L(\hat{Y}_1)$, $f_2 \in L(\hat{Y}_2)$, $u \in K(\hat{Y})$ (where $f_1(y_1) f_2(y_2) = \mathbf{1} f_2(y_2) = f_2(y_2) \in K(\hat{Y})$!), a mapping $C'(\hat{Y}) \xrightarrow{R_2} C'(\hat{Y}_2)$ is determined by

$$\langle u, f_2 \rangle = \langle R_2 u, f_2 \rangle_2 ,$$

with $\langle u, f \rangle$ the canonical dual form of $C'(\hat{Y})$, $C(\hat{Y})$ and $\langle v, f \rangle_2$ that of $C'(\hat{Y}_2)$, $C(\hat{Y}_2)$. This mapping is dual to $C(\hat{Y}_2) \xrightarrow{S_2} C(\hat{Y})$ with $S_2 f_2 = \mathbf{1} f_2$; it easily implies $R_2 K(\hat{Y}) \subset K_2(\hat{Y}_2)$. One calls $R_2 u$ with $u \in K(\hat{Y})$ the contraction of u to $u_2 = R_2 u \in K_2(\hat{Y}_2)$. Likewise one calls R_2 the "contraction operator (or simply 'contractor') on the system 2". As a dual mapping, R_2 is not only norm continuous but also $\sigma(C'(\hat{Y}), C(\hat{Y}))$, $\sigma(C'(\hat{Y}_2), C(\hat{Y}_2))$-continuous.

In an entirely analogous way one can define R_1 as the contractor on the system 1.

With $\overline{\text{co}}(A)$ as the norm closure of the convex set generated by A, analogous to $K_m(\hat{Y})$ in II § 3.2 we define

$$K_{12m}(\hat{Y}) = \overline{\text{co}}\, \varphi_{12}(\mathscr{D}'_{12}) . \tag{6.3.6}$$

Due to the definition of $\mathscr{D}_{12}$ as the SP structure generated by Γ_{12}, we easily obtain $\text{co}\, \varphi_{12}(\mathscr{D}'_{12}) = \text{co}\, \varphi_{12}(\Gamma_{12})$ and thus

$$K_{12m}(\hat{Y}) = \overline{\text{co}}\, \varphi_{12}(\Gamma_{12}) . \tag{6.3.7}$$

We define

$$K_{1m}(\hat{Y}_1) = R_1 K_{12m}(\hat{Y}) , \qquad K_{2m}(\hat{Y}_2) = R_2 K_{12m}(\hat{Y}) . \tag{6.3.8}$$

In analogy to II § 3.3, we can define $\hat{S}_1$ as the support of $K_{1m}(\hat{Y}_1)$ and $\hat{S}_2$ as the support of $K_{2m}(\hat{Y}_2)$. With $\hat{S}$ as the support of $K_{12m}(\hat{Y})$, we easily find

$$\hat{S}_{12} \overset{\text{def}}{=} \hat{S}_1 \times \hat{S}_2 \supset \hat{S} . \tag{6.3.9}$$

If one applies (6.3.5) with $\varrho_1 = \varrho_{10}$, in analogy to II (3.3.3) a function ψ_{2S} is uniquely defined by $\mathscr{F}_2 = \{(b_{20}, b_2) \mid b_{20} \in \mathscr{R}_{20}, b_2 \in \boldsymbol{R}_2(b_{20})\}$ and the diagram

$$\begin{array}{ccccc} & & L(\hat{Y}_2) & & \\ & \overset{\psi_2}{\nearrow} & & \searrow & \\ \Phi_2 & & & & L(\hat{S}_2)\,; \\ & \underset{h_2}{\searrow} & & \underset{\psi_{2S}}{\nearrow} & \\ & & \mathscr{F}_2 & & \end{array} \tag{6.3.10}$$

the mapping $\mathscr{F}_1 \xrightarrow{\psi_{1S}} L(\hat{S}_1)$ follows correspondingly. The two mappings ψ_{iS} $(i = 1, 2)$ enable us to "forget" the sets $\tilde{\mathscr{R}}_i$ if one replaces $\hat{Y}_i$ by $\hat{S}_i$. Then we can supplement the discussion from § 2 by the "objectivating description" to follow.

Because of (6.3.9), one can identify $K_{12m}(\hat{Y})$ with a subset $K_{12m}(\hat{S}_{12})$ of $K_{12}(\hat{S}_{12}) \subset C'(\hat{S}_{12})$ (see analogous considerations in II § 3.3) and thus regard φ_{12} as the mapping $\mathscr{D}_{12} \xrightarrow{\varphi_{12}} C'(\hat{S}_{12})$. In particular, (6.3.5) then goes over into

$$\begin{aligned} &\mu_{12}(a_1 \times a_2 \cap M, (b_{10} \times b_{20} \cap M, b_1 \times b_2 \cap M)) \\ &\quad = \lambda_{12}(a_1 \times a_2 \cap M \cap b_{10} \times b_{20}, a_1 \times a_2 \cap M \cap b_1 \times b_2) \\ &\quad = \langle \varphi_{12}(a_1 \times a_2 \cap M), \psi_{1S}(b_{10}, b_1)\, \psi_{2S}(b_{20}, b_2) \rangle \end{aligned} \tag{6.3.11}$$

where $\langle \ldots, \ldots \rangle$ is the canonical bilinear form of the dual pair $C'(\hat{S}_{12})$, $C(\hat{S}_{12})$. According to T 2.8, all of λ_{12} is thus determined by $\lambda_{\mathscr{D}_1}$, $\lambda_{\mathscr{D}_2}$, and

$$\langle \varphi_{12}(a_1 \times a_2 \cap M), \psi_{1S}(b_{10}, b_1)\, \psi_{2S}(b_{20}, b_2) \rangle .$$

The $\psi_{1S}(b_{10}, b_1)$ and $\psi_{2S}(b_{20}, b_2)$ have nothing to do with the interaction, but only describe the registration of the trajectories. The trajectory ensembles $\varphi_{12}(a_1 \times a_2 \cap M)$, i.e. the mapping

$$\Gamma_{12} \xrightarrow{\varphi_{12}} K(\hat{S}_{12})$$

(with Γ_{12} as in (2.1)), describe the interaction of the systems 1 and 2 *completely*.

In particular, from (6.3.11) with $b_1 = b_{10}$ follows

$$\begin{aligned} &\lambda_{12}(a_1 \times a_2 \cap M \cap b_{10} \times b_{20}, a_1 \times a_2 \cap M \cap b_{10} \times b_2) \\ &\quad = \langle \varphi_{12}(a_1 \times a_2 \cap M), \psi_{2S}(b_{02}, b_2) \rangle , \end{aligned}$$

such that

$$\lambda_{12}(a_1 \times a_2 \cap M \cap b_{10} \times b_{20}, a_1 \times a_2 \cap M \cap b_{10} \times b_2)$$

does not depend on b_{10}. Likewise follows that

$$\lambda_{12}(a_1 \times a_2 \cap M \cap b_{10} \times b_{20}, a_1 \times a_2 \cap M \cap b_1 \times b_{20})$$

does not depend on b_{20}. Hence, the additional requirement in APSZ 9 and APSZ 5.3 is not necessary whenever on uses the connections between the registrations and the trajectories (also see the end of § 2) described here.

With

$$\mathscr{F}_{12} = \{(b_{120}, b_{12}) \mid b_{120} \in \hat{\mathscr{R}}_{120}, b_{12} \in \hat{\mathscr{R}}_{12}, b_{12} \subset b_{120} \neq \emptyset\} ,$$

with the mapping r from (6.2.5), h_{12} from (6.2.6) and their canonical extensions, from (6.3.10) and a corresponding diagram with the index 1 we obtain

$$\begin{array}{ccccc} L(\hat{Y}) & \longrightarrow & L(\hat{S}_{12}) & \longrightarrow & L(\hat{S}) \\ \uparrow \psi_{12} & & \uparrow \hat{\psi}_{12S} & & \uparrow \psi_{12S} \\ \Phi_{12} & \xrightarrow{h_{12}} & \hat{\mathscr{F}}_{12} & \xrightarrow{r} & \mathscr{F}_{12}\,. \end{array} \qquad (h: \Phi_{12} \to \mathscr{F}_{12}) \tag{6.3.12}$$

Since $\hat{\psi}_{12S}$ is determined uniquely by

$$\hat{\psi}_{12S}(b_{10}\times b_{20}, b_1\times b_2) = \psi_{1S}(b_{10}, b_1)\,\psi_{2S}(b_{20}, b_2), \tag{6.3.13}$$

one can in diagram (6.3.12) forget the left part (which refers to $L(\hat{Y})$ and Φ_{12}) and thus forget $\tilde{\mathscr{R}}_1, \tilde{\mathscr{R}}_2$.

Thus we may call (6.3.11) the fundamental formula of the objectivating manner of describing the experiments with composite systems. But we are interested in a special situation, which in § 3 has been called the *directed* interaction.

§ 6.4 The Structure of the Trajectory Measures for Directed Action

Because of $\psi_{2S}(b_{20}, b_{20}) = \mathbf{1}$, the relation APSZ 9 with (6.3.11) says that

$$\begin{aligned} &\lambda_{12}(a_1\times a_2\cap M\cap b_{10}\times b_{20},\ a_1\times a_2\cap M\cap b_1\times b_{20}) \\ &= \langle \varphi_{12}(a_1\times a_2\cap M),\ \psi_{1S}(b_{10}, b_1)\rangle \end{aligned} \tag{6.4.1}$$

does not depend on a_2. That (3.1) does not depend on b_{20} is secured automatically by the objectivating description, i.e. it holds due to (6.3.11).

Using the contractor R_1 which maps $C'(\hat{S}_{12})$ on $C'(\hat{S}_1)$ one can write (6.4.1) as

$$\begin{aligned} &\lambda_{12}(a_1\times a_2\cap M\cap b_{10}\times b_{20},\ a_1\times a_2\cap M\cap b_1\times b_{20}) \\ &= \langle R_1\,\varphi_{12}(a_1\times a_2\cap M),\ \psi_{1S}(b_{10}, b_1)\rangle_1\,, \end{aligned} \tag{6.4.2}$$

where the last subscript indicates the canonical bilinear form for $C'(\hat{S}_1)$, $C(\hat{S}_1)$.

Since the $\psi_{1S}(b_{10}, b_1)$ separate (because la $\psi_1(\Phi_1)$ is dense in $L(\hat{S}_1)$), axiom APSZ 9 is equivalent to

AT 9 $R_1\,\varphi_{12}(a_1\times a_2\cap M)$ does not depend on a_2. (6.4.3)

This uniquely characterizes a directed interaction.

We retain the axioms APSZ 5.2 and APSZ 5.3 from § 3, where the last additional requirement can be dropped from APSZ 5.3 (see § 6.3).

Because of APSZ 5.2, a mapping $\mathscr{Q}_1 \xrightarrow{\varphi_1} K(\hat{S}_1)$ is defined by

$$a_1 \to R_1\,\varphi_{12}(a_1\times a_2\cap M)\,.$$

Using (6.3.8) and identifying $K_{1m}(\hat{Y}_1)$ with $K_{1m}(\hat{S}_1)$, we then find

$$\overline{\text{co}}\,\varphi_1(\mathscr{Q}_1) = K_{1m}(\hat{S}_1)\,.$$

By T 3.6 the probability (6.4.2) goes over into

$$\lambda_{\mathscr{S}_1}(a_1\cap b_{10}, a_1\cap b_1) = \mu_1(a_1, (b_{10}, b_1)) = \langle \varphi_1(a_1), \psi_{1S}(b_{10}, b_1)\rangle_1\,, \tag{6.4.4}$$

which holds for the trajectory effect $\psi_{1S}(b_{10}, b_1)$ on systems 1 prepared according to a_1. This simply expresses the expected fact that the dynamics of the systems 1 is independent of the coupled-on systems 2.

We also retain the further axioms APSZ 5.4, APS 5.1.3, APS 5.1.4 from § 5.1 so that the function μ from § 5.1 can be defined.

In the case of a directed interaction, we are interested in the probability function $\mu(a,f) = \tilde{\mu}(\varphi(a), \psi(f))$ from D 4.5 and T 5.1.4, which forms the basis for all further discussion in the following chapters. According to the definition of μ and the definition of $\mathscr{Q}$, $\mathscr{R}_0$, $\mathscr{R}$, we have

$$\mu(a, (b_0, b)) = \lambda_{\mathscr{S}}(a \cap b_0, a \cap b) = \lambda_{12}(c_1 \times c_{20} \cap M, c_1 \times c_2 \cap M)$$

with $c_1 \in \mathscr{S}_1$, $c_{20} \in \mathscr{S}_{20}$, $c_2 \in \mathscr{S}_2$. This value is determined by the particular values for $c_1 = a_1 \cap b_1$, $c_{20} = a_2 \cap b_{20}$, $c_2 = a_2 \cap b_2$, i.e. by

$$\lambda_{12}((a_1 \cap b_1) \times (a_2 \cap b_{20}) \cap M, (a_1 \cap b_1) \times (a_2 \cap b_2) \cap M).$$

By means of (3.6) and (6.3.11), with $a = a_1 \cap b_1 \times M_2 \cap M$, $b_0 = M_1 \times (a_2 \cap b_{20}) \cap M$, $b = M_1 \times (a_2 \cap b_2) \cap M$, follows

$$\langle \varphi_{12}(a_1 \times a_2 \cap M, \psi_{1S}(b_{10}, b_1)\, \psi_{2S}(b_{20}, b_2)\rangle = \lambda_{\mathscr{S}_1}(a_1 \cap b_{10}, a_1 \cap b_1)\, \mu(a, (b_0, b)). \tag{6.4.5}$$

The $\mathscr{Q}$ defined in § 4 yields

$$\lambda_{\mathscr{S}_1}(a_1 \cap b_{10}, a_1 \cap b_1) = \lambda_{\mathscr{Q}}(a_0, a), \tag{6.4.6}$$

with $a_0 = (a_1 \cap b_{10}) \times M_2 \cap M$. Hence (6.4.5) implies

$$\langle \varphi_{12}(a_1 \times a_2 \cap M), \psi_{1S}(b_{10}, b_1)\, \psi_{2S}(b_{20}, b_2)\rangle = \lambda_{\mathscr{Q}}(a_0, a)\, \tilde{\mu}(\varphi(a), \psi(b_0, b)), \tag{6.4.7}$$

while T 3.6 and (6.4.4) give

$$\lambda_{\mathscr{Q}}(a_0, a) = \lambda_{\mathscr{S}_1}(a_1 \cap b_{10}, a_1 \cap b_1) = \langle \varphi_1(a_1), \psi_{1S}(b_{10}, b_1)\rangle_1. \tag{6.4.8}$$

Using (5.4.11), from (6.4.7) we thus get

$$\tilde{\mu}(\varphi\pi(a_1 \cap b_1), \psi\varrho(a_2 \cap b_{20}, a_2 \cap b_2)) = \frac{\langle \varphi_{12}(a_1 \times a_2 \cap M, \psi_{1S}(b_{10}, b_1)\, \psi_{2S}(b_{20}, b_2)\rangle}{\langle \varphi_1(a_1), \psi_{1S}(b_{10}, b_1)\rangle_1}. \tag{6.4.9}$$

This directly connects the function $\tilde{\mu}$ with the trajectory effects and trajectory ensembles, thus emphasizing most clearly that $\tilde{\mu}(\varphi(a), \psi(b_0, b))$ is obtained from an objectivating description. From (6.4.9) with $b_2 = b_{20}$ follows that $R_1\varphi_{12}(a_1 \times a_2 \cap M) = \varphi_1(a_1)$, i.e. the directedness of the interaction. According to the remarks following (3.7), due to (6.4.8) the mapping $\Gamma_{12} \xrightarrow{\varphi_{12}} K(\hat{S}_{12})$ and thus the interaction are determined by $\tilde{\mu}$ and $\mathscr{Q}_1 \xrightarrow{\varphi_1} K(\hat{S}_1)$.

In order to rewrite (6.4.9), we consider for $C'(\hat{S}_{12})$, $C(\hat{S}_{12})$ the expression

$$\langle u, f_1(y_1)\, g(y_1, y_2)\rangle, \tag{6.4.10}$$

with $u \in K(\hat{S}_{12})$, $f_1(y_1) \in L(\hat{S}_1)$ and $g(y_1, y_2) \in L(\hat{S}_{12})$.

Multiplication of $g(y_1, y_2)$ with $f_1(y_1) \in L(\hat{S}_1)$ is a norm continuous mapping $C(\hat{S}_{12}) \to C(\hat{S}_{12})$. The dual mapping of $C'(\hat{S}_{12})$ into itself will be denoted f_1'. For $f_1 \in L(\hat{S}_1)$ one immediately finds $f_1' K(\hat{S}_{12}) \subset \check{K}(\hat{S}_{12})$, where $\check{K}(\hat{S}_{12})$ is the truncated cone $\bigcup_{0 \leq \lambda \leq 1} \lambda K(\hat{S}_{12})$. In this sense, from (6.4.10) follows

$$\langle u, f_1 g \rangle = \langle f_1' u, g \rangle;$$

hence (6.4.9) can be rewritten

$$\tilde{\mu}(\varphi\pi(a_1 \cap b_1), \psi\varrho(a_2 \cap b_{20}, a_2 \cap b_2)) = \frac{\langle R_2 \psi_{1S}'(b_{10}, b_1)\, \varphi_{12}(a_1 \times a_2 \cap M), \psi_{2S}(b_{20}, b_2)\rangle_2}{\langle \varphi_1(a_1), \psi_{1S}(b_{10}, b_1)\rangle_1}. \tag{6.4.11}$$

Here we have

$$\frac{R_2 \psi_{1S}'(b_{10}, b_1)\, \varphi_{12}(a_1 \times a_2 \cap M)}{\langle \varphi_1(a_1), \psi_{1S}(b_{10}, b_1)\rangle_1} \in K(\hat{S}_2), \tag{6.4.12}$$

because $\psi_{2S}(b_{20}, b_{20}) = 1$ implies $\psi(b_0, b_0) = \mathbf{1}$ and thus $\tilde{\mu}(\varphi(a), \psi(b_0, b_0)) = 1$.

Entirely analogously to the derivation of (6.4.11), from (6.4.7) with $a_0 = (a_1 \cap b_{10}) \times M_2 \cap M$, $a = (a_1 \cap b_1) \times M_2 \cap M$ one obtains

$$\begin{aligned} &\lambda_{\mathscr{Q}}(a_0, a)\, \tilde{\mu}(\varphi(a), \psi\varrho(a_2 \cap b_{20}, a_2 \cap b_2)) \\ &= \lambda_{\mathscr{I}_1}(a_1 \cap b_{10}, a_1 \cap b_1)\, \tilde{\mu}(\varphi\pi(a_1 \cap b_1), \psi\varrho(a_2 \cap b_{20}, a_2 \cap b_2)) \\ &= \langle R_1 \psi_{2S}'(b_{20}, b_2)\, \varphi_{12}(a_1 \times a_2 \cap M), \psi_{1S}(b_{10}, b_1)\rangle_1. \end{aligned} \tag{6.4.13}$$

The formulas (6.4.9), (6.4.11), (6.4.13) show that their left sides do not depend directly on b_1, b_{20}, b_2, but only on the trajectory effects $k_i = \psi_{iS}(b_{i0}, b_i)$.

In order to exhibit this more distinctly, we can also write

$$\begin{aligned} w &= \varphi\pi(a_1 \cap b_1) = \alpha(a_1, k_1), \\ g &= \psi\varrho(a_2 \cap b_{20}, a_2 \cap b_2) = \beta(a_2, k_2) \end{aligned} \tag{6.4.14}$$

with mappings α, β given by

$$\begin{array}{cc} \mathscr{Q}_1' \times \psi_{1S}(\mathscr{F}_1) \subset \mathscr{Q}_1' \times L(\hat{S}_1), & \mathscr{Q}_2' \times \psi_{2S}(\mathscr{F}_2) \subset \mathscr{Q}_2' \times L(\hat{S}_2); \\ \alpha \downarrow & \beta \downarrow \\ \mathscr{K} & \mathscr{L} \end{array}$$

then (6.4.9) becomes

$$\langle \varphi_1(a_1), k_1\rangle_1\, \tilde{\mu}(\alpha(a_1, k_1), \beta(a_2, k_2)) = \langle \varphi_{12}(a_1 \times a_2 \cap M), k_1 k_2\rangle. \tag{6.4.15}$$

§ 6.5 Complete Description by Trajectories

$\varphi((a_1 \cap b_1) \times m_2 \cap M) = \alpha(a_1, k_1)$ is uniquely determined by (6.4.15), where the form of α shows that not the registration procedure b_1 but only the trajectory effect k_1 enters. On the other hand, the SP $a_1 \in \mathscr{Q}_1$ appears as a "production procedure" of the preparation devices $x_1 \in M_1$. We say that the systems $x_1 \in M_1$ are "completely" described by the trajectories from $\hat{S}_1$ whenever $\varphi_1(a_1) = \varphi_1(\tilde{a}_1)$ implies $\alpha(a_1, k_1) = \alpha(\tilde{a}_1, k_1)$. Then two preparation procedures a_1 and $\tilde{a}_1$, which are not distinguishable by the trajectories on a system 1, are also not distinguished by the action of the systems 1 on other systems 2. Equivalent to this is that the right side of (6.4.15) does not depend on a_1 except through $\varphi_1(a_1)$. Thus $\varphi_{12}(a_1 \times a_2 \cap M)$ depends only on $\varphi_1(a_1)$, so that a function γ_{12} is determined by the diagram

$$\begin{array}{l} \mathscr{D}_1 \times \mathscr{D}_2 \xrightarrow{\varphi_{12}} K_{12m}(\tilde{S}_{12}) \\ \Big\downarrow{\scriptstyle \varphi_1 \times \mathbf{1}} \quad \Big\downarrow \quad \nearrow{\scriptstyle \gamma_{12}} \\ \varphi_1(\mathscr{D}_1) \times \mathscr{D}_2 \end{array}$$

The fact that $\varphi_{12}(a_1 \times a_2 \cap M)$ depends only on $\varphi_1(a_1)$ must for physical reasons be extended to the fact that γ_{12} is norm continuous in its first variable. Then one can extend γ_{12} to the norm closure $K_{1m}(S_1)$ of $\varphi_1(\mathscr{D}_1)$:

$$K_{1m}(\hat{S}_1) \times \mathscr{D}_2 \xrightarrow{\gamma_{12}} K_{12m}(\hat{S}_{12}) . \tag{6.5.1}$$

Therefore (6.4.15) goes over into

$$\langle u_1, k_1 \rangle_1 \tilde{\mu}(\alpha_1(u_1, k_1), \beta(a_2, k_2)) = \langle \gamma_{12}(u_1, a_2), k_1 k_2 \rangle \tag{6.5.2}$$

with $u_1 \in \varphi_1(\mathscr{D}_1)$, $k_1 \in \psi_{1S}(\mathscr{F}_1)$ where α_1 is determined by the diagram

$$\begin{array}{l} \mathscr{D}_1 \times \psi_{1S}(\mathscr{F}_1) \xrightarrow{\alpha} \mathscr{K} \\ \Big\downarrow{\scriptstyle \varphi_1 \times \mathbf{1}} \quad \Big\downarrow \quad \nearrow{\scriptstyle \alpha_1} \\ \varphi_1(\mathscr{D}_1) \times \psi_{1S}(\mathscr{F}_1) \end{array} \tag{6.5.3}$$

The "complete describability" of the systems 1 by trajectories is not self-evident. But most often one assumes that for a sufficiently fine state space the complete description of the systems 1 by trajectories is possible. In order to obtain a similar statement for the systems 2, one must know how these behave without "external influence". In IX § 1 we shall obtain this description by choosing a_1 so that the systems $x_1 \in a_1$ prepare "vacuum". Here let us simply presume it possible experimentally to investigate the systems 2 without external influence (what experimenters do for testing their registration devices). Thus a mapping

$$\mathscr{D}_2 \xrightarrow{\varphi_2^{(0)}} K_{2m}(\hat{S}_2)$$

is so defined that the probabilities for the trajectory effects k_2 on the systems prepared due to a_2 are given by

$$\langle \varphi_2^{(0)}(a_2), k_2 \rangle_2$$

in case the systems are not externally influenced.

Let it be emphasized that the support of $\varphi_2^{(0)}(\mathscr{D}_2)$ need not be all of $\hat{S}_2$: For, *with* external influence from the systems 1, trajectories can by all means appear which are otherwise impossible (i.e. physically excluded; see [3] § 10).

The systems 2 are called completely describable by trajectories if $\varphi_{12}(a_1 \times a_2 \cap M)$ does not explicitly depend on a_2 but only on $\varphi_2^{(0)}(a_2)$ and that norm continuously.

If the systems 1 as well as the systems 2 are completely describable by trajectories, then a norm continuous bilinear function δ_{12} is determined by the diagram

$$\begin{array}{l} \mathscr{D}_1 \times \mathscr{D}_2 \xrightarrow{\varphi_{12}} K_{12m}(\hat{S}_{12}) \\ \Big\downarrow{\scriptstyle \varphi_1 \times \varphi_2^{(0)}} \Big\downarrow \quad \Big\uparrow{\scriptstyle \delta_{12}} \\ K_{1m}(\hat{S}_1) \times K_{2m}^{(0)}(\hat{S}_2) \end{array} \tag{6.5.4}$$

where $K_{2m}^{(0)}(\hat{S}_2)$ is the norm closure of the set $\varphi_2^{(0)}(\mathcal{D}_2')$. Then (6.5.2) can be written

$$\langle u_1, k_1 \rangle_1 \tilde{\mu}(\alpha_1(u_1, k_1), \beta_2(u_2^{(0)}, k_2)) = \langle \delta_{12}(u_1, u_2^{(0)}), k_1 k_2 \rangle \tag{6.5.5}$$

with $u_2^{(0)} \in \varphi_2^{(0)}(\mathcal{D}_2)$, where β_2 is determined by the diagram

$$\begin{array}{ccc} \mathcal{D}_2 \times \psi_{2S}(\mathcal{F}_2) & \xrightarrow{\beta} & \mathcal{S} \\ \downarrow \varphi_2^{(0)} \times \mathbf{1} \quad \downarrow & \nearrow \beta_2 & \\ \varphi_2^{(0)}(\mathcal{D}_2) \times \psi_{2S}(\mathcal{F}_2) & & \end{array} \tag{6.5.6}$$

In (6.5.4) one can identify $K_{1m}(\hat{S}_1) \times K_{2m}^{(0)}(\hat{S}_2)$ with a subset of $K_{12m}(\hat{S}_{12})$ and thus extend δ_{12} to the norm closed subset of $K_{12m}(\hat{S}_{12})$ generated by $K_{1m}(\hat{S}_1) \times K_{2m}^{(0)}(\hat{S}_2)$:

$$K_{12m}^{(0)}(\hat{S}_{12}) \overset{\text{def}}{=} \overline{\text{co}}\,(K_{1m}(\hat{S}_1) \times K_{2m}^{(0)}(\hat{S}_2)) \,.$$

The elements of $K_{12m}^{(0)}(\hat{S}_{12})$ describe the system pairs (x_1, x_2) "as if no interaction were present". In this sense δ_{12} can be called the "interaction operator"; $\delta_{12} K_{12m}^{(0)}(\hat{S}_{12})$ must be norm dense in $K_{12m}(\hat{S}_{12})$.

With $\mathcal{B}_{12m}^{(0)}$ as the Banach subspace generated by the $K_{12m}^{(0)}(\hat{S}_{12})$ in $C'(\hat{S}_{12})$, one can extend δ_{12} to all of $\mathcal{B}_{12m}^{(0)}$, so that the mapping δ_{12}' dual to δ_{12} maps the space $C(\hat{S}_{12})$ into $\mathcal{B}_{12m}^{(0)\prime}(\hat{S}_{12})$. One can thus rewrite the right side of (6.5.5) as

$$\langle \delta_{12}(u_1, u_2^{(0)}), k_1 k_2 \rangle = \langle u_1 u_2^{(0)}, \delta_{12}'(k_1 k_2) \rangle^{(0)} \,, \tag{6.5.7}$$

where $\langle \ldots, \ldots \rangle^{(0)}$ is the canonical bilinear form of $\mathcal{B}_{12m}^{(0)}(\hat{S}_{12})$, $\mathcal{B}_{12m}^{(0)\prime}(\hat{S}_{12})$.

For fixed $u_2^{(0)}$, k_1, k_2, by (6.5.7) a norm continuous linear form over u_1 is defined, i.e. an element $\chi_1 \in \mathcal{B}_{1m}'(\hat{S}_1)$ with $\mathcal{B}_{1m}(\hat{S}_1)$ as the Banach subspace of $C'(\hat{S}_1)$ generated by $K_{1m}(\hat{S}_1)$. With j_1 as the embedding $\mathcal{B}_{1m}(\hat{S}_1) \xrightarrow{j_1} C'(S_1)$ and j_1' its dual that maps $C(\hat{S}_1)$ into $\mathcal{B}_{1m}'(\hat{S}_1)$, we find $j_1' C(\hat{S}_1)$ dense in $\mathcal{B}_{1m}'(\hat{S}_1)$ under the $\sigma(\mathcal{B}_{1m}'(\hat{S}_1), \mathcal{B}_{1m}(\hat{S}_1))$-topology, since $C(\hat{S}_1)$ separates the $u \in K_{1m}(\hat{S}_1)$. In this sense χ_1 represents a "generalized trajectory effect" on the systems 1. On physical grounds, one must expect that χ_1 is not only approximated by elements of $j_1' C(S_1)$ but that even $\chi_1 \in j_1' C(S_1)$ holds. Then there is a "proper" trajectory effect $l_1 \in C(\hat{S}_1)$ with $\chi_1 = j_1' l_1$, i.e. with

$$\langle \delta_{12}(u_1, u_2^{(0)}), k_1 k_2 \rangle = \langle u_1, l_1 \rangle_1 \,. \tag{6.5.8}$$

We can apply a quite analogous reasoning to the systems 2. For fixed u_1, k_1, k_2, an element $\chi_2 \in \mathcal{B}_{2m}^{(0)\prime}(\hat{S}_2)$ is defined by (6.5.7). With j_2 as injection $\mathcal{B}_{2m}^{(0)}(\hat{S}_2) \to C'(S_2)$, we find $j_2' C(\hat{S}_2)$ is dense in $\mathcal{B}_{2m}^{(0)\prime}(\hat{S}_2)$. On physical grounds, one can expect $\chi_2 \in j_2' C(\hat{S}_2)$, i.e. that there is an $l_2 \in C(\hat{S}_2)$ with

$$\langle \delta_{12}(u_1, u_2^{(0)}), k_1 k_2 \rangle = \langle u_2^{(0)}, l_2 \rangle_2 \,. \tag{6.5.9}$$

The complete description of the systems 1 and 2 by the trajectories therefore implies that one can obtain "no new information" beyond the trajectories from any interaction of the systems 1 with the systems 2. In the following § 6.6 we shall describe this in more detail by extending the registration procedures.

§ 6.6 Use of the Interaction for the Registration of Macrosystems

Let us first consider the systems 1! Proceeding from the sets $\mathscr{Q}_1$ and $\mathscr{R}_1$ of preparation and registration procedures, we have first considered (in the sense of § 6.2) only trajectory registration procedures. We will now extend $\mathscr{R}_{10}$ and $\mathscr{R}_1$.

The mapping π defined in (5.4.1) is bijective, with $\pi^{-1}(x_1, x_2) = x_1$. (We could directly have defind the mapping $(x_1, x_2) \to x_1$ of M into M_1 but want to clarify the connection with (5.4.1) by the notation π^{-1}.)

With $\bar{\mathscr{R}}_{10}$ as the system of SP generated by the two sets $\mathscr{R}_{10}$ and

$$\{\bar{b}_{10} \mid \bar{b}_{10} = \pi^{-1}(b_{10} \times (a_2 \cap b_{20}) \cap M),\ b_{10} \in \mathscr{R}_{10},\ a_2 \in \mathscr{Q}_2,\ b_{20} \in \mathscr{R}_{20}\}, \tag{6.6.1}$$

and $\bar{\mathscr{R}}_1$ as the system of SP generated by $\mathscr{R}_1$ and

$$\{\bar{b}_1 \mid \bar{b}_1 = \pi^{-1}(b_1 \times (a_2 \cap b_2) \cap M),\ b_1 \in \mathscr{R}_1,\ a_2 \in \mathscr{Q}_2,\ b_2 \in \mathscr{R}_2\}. \tag{6.6.2}$$

$\bar{\mathscr{R}}_{10}$ and $\bar{\mathscr{R}}_1$ form an extended registration structure for the systems 1, namely extended by the "registrations of the systems 1 by means of their action on the systems 2". Of course, also the registration of trajectories of the systems 1 is for physics an action of the systems 1 on the devices to register trajectories. For these trajectories we have presumed and explained in II § 3.1, that the trajectory registration can be described *by pretheories*. But the "new" registrations indicated in (6.6.1) and (6.6.2) are describable by their probability functions only in a theory of the coupling between different macrosystems, which provides

$$\begin{aligned}
&\bar{\lambda}_1(a_1 \cap \pi^{-1}(b_{10} \times (a_2 \cap b_{20}) \cap M),\ a_1 \cap \pi^{-1}(b_1 \times (a_2 \cap b_2) \cap M)\\
&\quad \overset{\text{def}}{=} \lambda_{12}((a_1 \cap b_{10}) \times (a_2 \cap b_{20}) \cap M,\ (a_1 \cap b_1) \times (a_2 \cap b_2) \cap M)\\
&\quad = \langle \varphi_{12}(a_1 \times a_2 \cap M),\ \psi_{1S}(b_{10}, b_1)\ \psi_{2S}(b_{20}, b_2)\rangle .
\end{aligned} \tag{6.6.3}$$

If the systems 1 are completely describable by the trajectories as presented in § 6.5, then (6.5.8) yields

$$\langle \varphi_{12}(a_1 \times a_2 \cap M),\ \psi_{1S}(b_{10}, b_1)\ \psi_{2S}(b_{20}, b_2)\rangle = \langle \varphi_1(a_1), l_1\rangle_1 \tag{6.6.4}$$

with an $l_1 \in C(\hat{S}_1)$; i.e. the new registration procedures yield no new information about the systems 1.

But in physics often just the opposite happens; one checks by (6.6.3) whether one has already attained a complete description by trajectories. If it has not yet been attained one tries by an extension (a refinement of the state space) to achieve a complete description by trajectories.

One can proceed quite analogously with the systems 2. By means of the sets

$$\{\bar{b}_{20} \mid \bar{b}_{20} = \varrho^{-1}((a_1 \cap b_{10}) \times b_{20} \cap M),\ a_1 \in \mathscr{Q}_1,\ b_{10} \in \mathscr{R}_{10},\ b_{20} \in \mathscr{R}_{20}\} \tag{6.6.5}$$

and

$$\{\bar{b}_2 \mid \bar{b}_2 = \varrho^{-1}((a_1 \cap b_1) \times b_2 \cap M),\ a_1 \in \mathscr{Q}_1,\ b_1 \in \mathscr{R}_1,\ b_2 \in \mathscr{R}_2\}, \tag{6.6.6}$$

with ϱ^{-1} as in (5.4.5) one extends $\mathscr{R}_{20}, \mathscr{R}_2$ to $\bar{\mathscr{R}}_{20}, \bar{\mathscr{R}}_2$ with the probability function

$$\begin{aligned}
&\bar{\lambda}_2(a_2 \cap \varrho^{-1}((a_1 \cap b_{10}) \times b_{20} \cap M),\ a_2 \cap \varrho^{-1}((a_1 \cap b_1) \times b_2 \cap M))\\
&\quad \overset{\text{def}}{=} \lambda_{12}((a_1 \cap b_{10}) \times (a_2 \cap b_{20}) \cap M,\ (a_1 \cap b_1) \times (a_2 \cap b_2) \cap M)\\
&\quad = \langle \varphi_{12}(a_1 \times a_2 \cap M),\ \psi_{1S}(b_{10}, b_1)\ \psi_{2S}(b_{20}, b_2)\rangle .
\end{aligned} \tag{6.6.7}$$

If a complete description by trajectories holds, then (6.5.9) implies

$$\langle \varphi_{12}(a_1 \times a_2 \cap M),\, \psi_{1S}(b_{10}, b_1)\, \psi_{2S}(b_{20}, b_2)\rangle = \langle \varphi_2^{(0)}(a_2), l_2\rangle_2 . \tag{6.6.8}$$

Thus from the additional registrations (due to the influence possibilities of the systems 1 on the systems 2) one does not obtain any new information.

The very extensive investigations of macroscopic systems that go beyond any description by trajectories or at least beyond the "till now known" trajectories rest on experiments with two auxiliary systems. The principle of such experiments is sketched in Fig. 3, where the arrows demonstrate interactions. One investigates the system (2, 2) by means of the systems 1 and (2, 3). If one combines the systems (2, 2) and (2, 3) into a system 2 (as in Fig. 3), one obtains that experimental structure from which in §§ 1–3 we started an axiomatic basis for quantum mechanics. For such an axiomatics it is not necessary to introduce more detailed structures of the registration devices 2, e.g. that 2 is composed from the two systems (2, 2) and (2, 3). On the other hand, it is not difficult in principle to *specialize* the considerations from §§ 1–3 to a more precise description of such situations, e.g. that in Fig. 3. In fact, that 2 is composed of (2, 2) and (2, 3) is just a special case of the general considerations from §§ 1–3. These hints must suffice here since we first want to develop an axiomatic basis. In XI and XII we shall return to consider the problem sketched in Fig. 3.

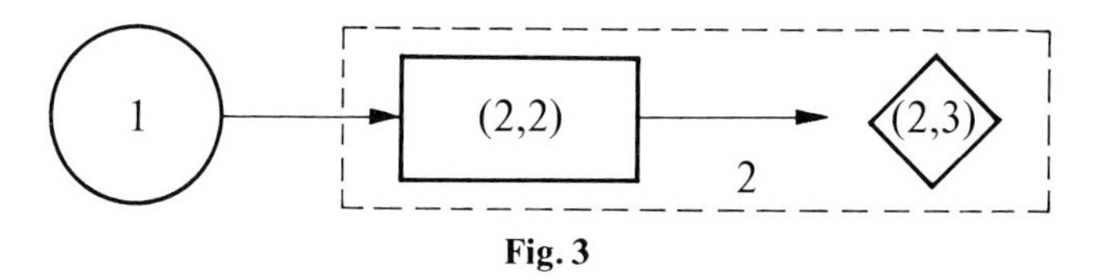

Fig. 3

In conclusion one thing must yet be emphasized: Till now there are no experiences or theoretical indications on the basis of the considerations presented in X and XI that an objectivating description of macrosystems would not be possible. Therefore, there cannot be a quantum mechanics of the macrosystems in the proper sense (i.e. including such structures as non-coexistent observables and non-coexistent decompositions; see V and X).

§ 6.7 The Relation Between the Two Forms of an Axiomatic Basis

In § 4 we denoted by $\mathscr{P}\mathscr{T}_1$ the theory with the axiomatic basis developed in § 1 through § 5 (i.e. without the base sets $Y_1, Y_2, \ldots$).

Let $\mathscr{P}\mathscr{T}_{1t}$ denote the extended axiomatic basis given in § 6.1–§ 6.6. The theorems presented there then show directly that a relation of the form

$$\mathscr{P}\mathscr{T}_{1t} \to \mathscr{P}\mathscr{T}_{1t}' \rightsquigarrow \mathscr{P}\mathscr{T}_1 \tag{6.7.1}$$

holds, where $\mathscr{P}\mathscr{T}_{1t}'$ arises from $\mathscr{P}\mathscr{T}_{1t}$ if one forgets the sets $Y_1, Y_2, \tilde{\mathscr{R}}_1, \tilde{\mathscr{R}}_{10}, \tilde{\mathscr{R}}_2, \tilde{\mathscr{R}}_{20}$. Then $\mathscr{P}\mathscr{T}_{1t}' \rightsquigarrow \mathscr{P}\mathscr{T}_1$ is the trivial embedding obtained by identifying the corresponding sets from $\mathscr{P}\mathscr{T}_{1t}$ and $\mathscr{P}\mathscr{T}_1$ (e.g. $h_1\, \tilde{\mathscr{R}}_1$ with $\mathscr{R}_1$), as already done throughout this § 6.

To a "theory of microsystems", $\mathscr{PT}_{1t}$ contributes nothing further than $\mathscr{PT}_1$, since in $\mathscr{PT}_{1t}$ only the devices are more precisely described than in $\mathscr{PT}_1$. But for investigating the compatibility problem (i.e. the measurement process in quantum mechanics) in XI, just $\mathscr{PT}_{1t}$ is the appropriate point of departure. The problem of the measurement process therefore cannot be solved solely within a quantum mechanics of the form $\mathscr{PT}_3$ (see § 5.2), but rather only within a more comprehensive theory, as discussed in XI.

§ 7 Transport of Systems Relative to Each Other

As already mentioned at the end of II § 4.3, in the case of composite systems (x_1, x_2) one can transport the systems x_2 (leaving the x_1 fixed). Such transports are spatial translations, time displacements, or the imparting of a uniform velocity. Thus the transports are in a Newtonian space-time called Galileo transformations, in a relativistic space-time Poincaré transformations. It is important to note that the physical interpretation of these transformations (of macrosystems relative to the laboratory coordinate system) is assumed known from pretheories. Not any unknown microsystems are transformed in ways not recognizable by pretheories. Therefore, in quantum mechanics, first of all place and time are defined by pretheories solely for the macroscopic devices, defined as objective properties and *not* (!) as observables of the action carriers, as to be introduced in V.

In order to introduce a mathematical structure for the transports interpretable by pretheories, we think of a countable subgroup Δ specified in the group $\Delta_{\mathscr{G}}$ of Galileo transformations (we shall only speak of these since everything carries over analogously to Poincaré transformations) where Δ is dense in $\Delta_{\mathscr{G}}$.

As an example of an element from Δ, let us consider the spatial translation by a vector $\boldsymbol{a}$. What does it mean that the system x_2' is translated by $\boldsymbol{a}$ relative to the system x_2? Evidently that the position of x_2' is made to differ just by $\boldsymbol{a}$ from that of x_2. Positioning the devices 2 in the laboratory system, however, belongs to preparation. When a_2 is the preparation procedure used to fix x_2, to fix x_2' we must use just that procedure a_2' which prepares systems shifted by $\boldsymbol{a}$ relative to those from a_2.

Thus we arrive at representing the real relation of transporting the systems 2 mathematically (by a structure species) as follows:

To every $\delta \in \Delta$ not containing a translation in the time, we assign an automorphism $\mathscr{Q}_2 \xrightarrow{o} \mathscr{Q}_2$, denoting it too by δ. For time translations δ_τ, each δ_τ as a mapping in $\mathscr{Q}_2$ coincides withe the mapping R_τ' introduced in II § 4.2, where its domain has in detail been discussed.

In a transport of the devices 2, however, let us not only include the positioning by the preparation procedures, but also the "co-transport" of the registration procedures. Hence we impose that to each $\delta \in \Delta$ (not containing a time translation) there corresponds an automorphism of $\mathscr{R}_{20}$ and $\mathscr{R}_2$. For time translations δ_τ, each δ_τ as a mapping in $\mathscr{R}_{20}$ and $\mathscr{R}_2$ shall coincide with the mapping R_τ defined in II § 4.2.

If we would desire to find, as we did in II § 4 for δ_τ, an explicit representation also for other elements of Δ, we ought to describe the structure of the phase space Z in more detail. Only then could we describe the changes of states under translations in space, under rotations and under imparting of uniform velocities. Here we do not want to do so, because in this book we shall only very broadly refer (in IX § 1) to the whole Galileo group as transforming the devices 2. Moreover, the example of the δ_τ shows us how to proceed.

Only in order to emphasize clearly the *physical meaning* of the transport of the registration devices, have we introduced the pertinent mathematical structures already in the theory of coupled macrosystems. Only *after* this interpretation (that lies purely in the macroworld) of the relative space-time position of the systems $x_1 \in M_1$ and $x_2 \in M_2$ is it possible later to give meaning to the "position observable" for microsystems as an only indirectly measurable quantity (as done in [2] II § 4).

For δ as automorphism of the sets $\mathscr{Q}_2, \mathscr{R}_{20}, \mathscr{R}_2$, we conclude that it also maps the sets $\mathscr{S}_{20}$ and $\mathscr{S}_2$ into themselves. On the basis of (4.2) and (4.3), we therefore can also define δ as an automorphism of $\mathscr{R}_0$ and $\mathscr{R}$.

Then this automorphism may canonically be extended to $\mathscr{F} \xrightarrow{\delta} \mathscr{F}$ with $\mathscr{F}$ as in D 4.3. Just these mappings for the $\delta \in \Delta$ form the point of departure for representing the Galileo group in quantum mechanics. We have in [2] VII presented this in detail and shall briefly scetch it in IX § 1.

In order to emphasize more precisely the interpretation of δ in the macroworld, let us in particular for a time translation δ_τ use the macroscopic description of the probability $\tilde{\mu}$, as obtained in § 6.4 and § 6.5:

From (6.4.9) we get

$$\tilde{\mu}(\varphi\,\pi(a_1 \cap b_1), \psi(\delta_\tau f)) = \frac{\langle \varphi_{12}(a_1 \times R'_\tau a_2 \cap M), \psi_{1S}(b_{10}, b_1)\, V_\tau\, \psi_{2S}(b_{20}, b_2)\rangle}{\langle \varphi_1(a_1), \psi_{1S}(b_{10}, b_1)\rangle_1} \tag{7.1}$$

with $f = \varrho(a_2 \cap b_{20}, a_2 \cap b_2) \in \mathscr{F}$, which by (6.4.15) implies

$$\langle \varphi_1(a_1), k_1\rangle_1\, \tilde{\mu}(\alpha(a_1, k_1), \beta(R'_\tau a_2, V_\tau k_2)) = \langle \varphi_{12}(a_1 \times R'_\tau a_2 \cap M), k_1\, V_\tau k_2\rangle. \tag{7.2}$$

By (6.5.2) this becomes

$$\langle u_1, k_1\rangle_1\, \tilde{\mu}(\alpha_1(u_1, k_1), \beta(R'_\tau a_2, V_\tau k_2)) = \langle \gamma_{12}(u_1, R'_\tau a_2), k_1\, V_\tau k_2\rangle, \tag{7.3}$$

so that (6.5.5) yields

$$\langle u_1, k_1\rangle_1\, \tilde{\mu}(\alpha_1(u_1, k_1), \beta_2(V'_{-\tau} u_2^{(0)}, V_\tau k_2)) = \langle \delta_{12}(u_1, V'_{-\tau} u_2^{(0)}), k_1\, V_\tau k_2\rangle. \tag{7.4}$$

Here the right side most clearly shows, how $\psi(\delta_\tau f) = \beta_2(V'_{-\tau} u_2^{(0)}, V_\tau k_2)$ is determined by the purely kinematic transformation V_τ of the trajectory effects and by the mapping δ_{12} which describes the interaction.

Similarly as the devices 2 can for fixed devices 1 be transformed by elements δ of Δ, one can also transform the devices 1 for fixed devices 2. To the reader it may be left to elaborate this in detail. δ as transformation of the devices 2 must affect the probabilities just as δ^{-1} as transformation of the devices 1. This fact has been employed in [2] VII.

IV Embedding of Ensembles and Effect Sets in Topological Vector Spaces

No new physical laws will arise in this chapter. We shall in fact forget all considerations from III except the countable sets $\mathcal{K}, \mathcal{L}$ with the mapping $\mathcal{K} \times \mathcal{L} \xrightarrow{\mu} [0, 1]$ for which III T 5.1.4 and the relations APK and ARK (from III §§ 5.3) hold. Hence we shall not gain new physical insights, but rather arrange the mathematical framework so flexibly that it is comfortable (!) to formulate further axioms and to prove theorems. Nevertheless, we shall use some general considerations from [3] §§ 6 and 9 in order to clarify the meaning of new concepts.

Let it be emphasized again, in order to avoid misunderstanding, that the vector spaces introduced will not change the structure of $\mathcal{M}\mathcal{T}_\Sigma$ in any way. Everything deals *only* (!) with structures derived from $\mathcal{K}$, $\mathcal{L}$, μ, i.e. with inner terms with respect to $\mathcal{K}, \mathcal{L}, \mu$ in the sense of [3] § 7.2, hence only with concepts for a simpler style of expression. Notwithstanding this *general* warning, in the following sections we shall often point out possible misunderstandings.

§ 1 Embedding of $\mathcal{K}$, $\mathcal{L}$ in a Dual Pair of Vector Spaces

We begin with a definition

D 1.1 Let **D** be the set of all functions $\mathcal{K} \xrightarrow{y} \mathbf{R}$ that can be represented as

$$y(w) = \sum_{i=1}^{n} \alpha_i \mu(w, g_i) \tag{1.1}$$

with real numbers α_i and the elements $g_i \in \mathcal{L}$. Let **B** be the set of all functions $\mathcal{L} \xrightarrow{x} \mathbf{R}$ that can be represented as

$$x(g) = \sum_{i=1}^{n} \beta_i \mu(w_i, g) \tag{1.2}$$

with real numbers β_i and the elements $w_i \in \mathcal{K}$.

It follows immediately that **D** and **B** are linear, real vector spaces.

From III T 5.1.4 (ii) and (iii) follows that $w \to \mu(w, g)$ and $g \to \mu(w, g)$ are injective mappings $\mathcal{K} \to \mathbf{B}$ resp. $\mathcal{L} \to \mathbf{D}$. Thus, one can identify $\mathcal{K}$ with a subset of **B** and $\mathcal{L}$ with a subset of **D**. In the sequal, we shall always adopt this identification. From (1.1) resp. (1.2) follows directly that $\mathcal{L}$ resp. $\mathcal{K}$ linearly spans the space **D** resp. **B**. (Note that initially we use neither the relations (iv) and (v) from III T 5.1.4 nor APK and ARK!)

T 1.1 A bilinear from $\mathbf{B} \times \mathbf{D} \to \mathbf{R}$, which coincides with μ when restricted to $\mathscr{K} \times \mathscr{L}$, is uniquely defined by $\mathscr{K} \times \mathscr{L} \xrightarrow{\mu} [0, 1]$. Thus $\langle \mathbf{B}, \mathbf{D} \rangle$ becomes a dual pair, where **D** separates **B** and **B** separates **D**.

Proof. For $w \in \mathscr{K}$ and $g \in \mathscr{L}$, with $\langle \ldots, \ldots \rangle$ as bilinear form, we must have

$$\langle w, g \rangle = \mu(w, g).$$

With (1.1) this implies

$$\langle w, y \rangle = \sum_{i=1}^{n} \alpha_i \, \mu(w, g_i). \tag{1.3}$$

But (1.3) as the definition of $\langle w, y \rangle$ is only meaningful if

$$y^{(1)}(w) = \sum_{i=1}^{n} \alpha_i^{(1)} \, \psi(w, g_i^{(1)}), \quad y^{(2)}(w) = \sum_{i=1}^{n} \alpha_i^{(2)} \, \mu(w, g_i^{(2)})$$

and $y^{(1)} = y^{(2)}$ also make $\langle w, y_1 \rangle = \langle w, y_2 \rangle$, which is immediately clear.

As easily follows that $\langle w, y \rangle$ is linear in y.

Since $\langle x, y \rangle$ must be bilinear, with (1.2) in the form

$$\langle x, g \rangle = \sum_{i=1}^{n} \beta_i \langle w_i, g \rangle \tag{1.4}$$

and (1.3) follows

$$\langle x, y \rangle = \sum_{i=1}^{n} \beta_i \langle w_i, y \rangle. \tag{1.5}$$

Equation (1.4) is meaningful if

$$x^{(1)}(g) = \sum_{i=1}^{n} \beta_i^{(1)} \, \mu(w_i^{(1)}, g), \quad x^{(2)}(g) = \sum_{i=1}^{n} \beta_i^{(2)} \, \mu(w_i^{(2)}, g)$$

and $x^{(1)} = x^{(2)}$ imply $\langle x^{(1)}, y \rangle = \langle x^{(2)}, y \rangle$; but this follows from (1.5) and (1.3). The linearity of $\langle x, y \rangle$ in x follows easily.

From $\langle x_1, g \rangle = \langle x_2, g \rangle$ for all $g \in \mathscr{L}$, with (1.2) follows $x_1 = x_2$. From $\langle w, y_1 \rangle = \langle w, y_2 \rangle$ for all $w \in \mathscr{K}$, with (1.1) follows $y_1 = y_2$. Therefore, **D** separates **B** and **B** separates **D**. □

Because of the connection of $\langle x, y \rangle$ with μ, in the sequel we write $\langle x, y \rangle = \mu(x, y)$.

Therefore, the embedding of $\mathscr{K}, \mathscr{L}$ in the dual pair **B, D** with $\mu(x, y)$ as canonical bilinear form is possible for purely mathematical reasons (without any further law of nature); it follows from III T 5.1.4 (i), (ii), (iii). From III T 5.1.4 (iv) follows the zero element of **D** is an element of $\mathscr{L}$: $\mathbf{0} \in \mathscr{L}$. From III T 5.1.4 (v) follows $\mathbf{1} \in \mathscr{L}$ (we denote the element g_1 by **1**) and

$$\mu(w, \mathbf{1}) = 1 \quad \text{for all} \quad w \in \mathscr{K}.$$

From III T 5.1.4 (vi) follows that $g \in \mathscr{L}$ implies $\mathbf{1} - g \in \mathscr{L}$.

D 1.2 A subset A of **B** and **D** is called *convex* if $x_1, x_2 \in A$ implies

$$\lambda x_1 + (1-\lambda) x_2 \in A \quad \text{for} \quad 0 < \lambda < 1. \tag{1.6}$$

A is called *rational* convex if (1.6) holds at least for all rational λ.

By means of (1.3), (1.4) one immediately sees:

T 1.2 Relation APK resp. ARK from III § 5.3 is equivalent to $\mathscr{K}$ resp. $\mathscr{L}$ being rational convex.

Let us collect all the properties obtained from III T 5.1.4 and APK, ARK.

$\mathscr{K}$ and $\mathscr{L}$ can be embedded in a dual pair of vector spaces **B**, **D** so that $\mathscr{K}$ linearly spans **B**, $\mathscr{L}$ linearly spans **D** and so that $\mathscr{K} \times \mathscr{L} \xrightarrow{\mu} [0, 1]$ is the restriction of the canonical bilinear form of $\langle \mathbf{B}, \mathbf{D} \rangle$. The following holds:

(a) $\mathbf{0} \in \mathscr{L}$,
(b) there is an element $\mathbf{1} \in \mathscr{L}$ with $\mu(w, \mathbf{1}) = 1$ for all $w \in \mathscr{K}$,
(c) $g \in \mathscr{L}$ implies $\mathbf{1} - g \in \mathscr{L}$,
(d) $\mathscr{K}$ and $\mathscr{L}$ are rational convex sets.

Of course, for $\mathscr{K} \times \mathscr{L} \xrightarrow{\mu} [0, 1]$ the properties from III T 5.1.4 conversely follow from (a) through (d) for the sets $\mathscr{K}$, $\mathscr{L}$ and the mapping $\mathscr{K} \times \mathscr{L} \xrightarrow{\mu} [0, 1]$.

We stress here once more that the spaces **B**, **D** are only auxiliary constructions. They enable us to formulate the properties (a) through (d) for $\mathscr{K}$, $\mathscr{L}$ and the mapping $\mathscr{K} \times \mathscr{L} \xrightarrow{\mu} [0, 1]$ in a mathematically convenient way.

§ 2 Uniform Structures of the Physical Imprecision on $\mathscr{K}$ and $\mathscr{L}$

The sets $\mathscr{K}$ and $\mathscr{L}$ were obtained as partition classes of the sets $\mathscr{Q}'$ and $\mathscr{F}$. Two $a_1, a_2 \in \mathscr{Q}'$ belong to the same class provided $\mu(a_1, f) = \mu(a_2, f)$ holds for all $f \in \mathscr{F}$ (as far as $\mu(a_1, f)$ and $\mu(a_2, f)$ are defined). Likewise, two $f_1, f_2 \in \mathscr{F}$ belong to the same class provided $\mu(a, f_1) = \mu(a, f_2)$ holds with all $a \in \mathscr{Q}'$ for which $\mu(a, f_1)$ and $\mu(a, f_2)$ are defined. Thus, the elements of $\mathscr{K}$ are distinguished by the effect procedures, hence by the elements of $\mathscr{L}$. Vice versa, these elements of $\mathscr{L}$ are distinguished by those of $\mathscr{K}$. Physically, an equation, for example of the form $\mu(a_1, f) = \mu(a_2, f)$, is of course an idealization, in as much as these probabilities can be compared with experience only my means of imprecision sets (see [3] § 11.6). Therefore, also the elements of $\mathscr{K}$ and $\mathscr{L}$ can only imprecisely be separated from each other. These considerations lead to uniform structures of physical imprecision on $\mathscr{K}$ and $\mathscr{L}$ (as described in [3] §§ 6 and 9 and [40]).

From these general considerations and the meaning of the elements of $\mathscr{K}$, as a uniform structure of the physical imprecision on $\mathscr{K}$ there follows the initial uniform structure (see [3] § 9) associated with the mappings $\mathscr{K} \xrightarrow{\mu(w,g)} [0, 1]$ for all $g \in \mathscr{L}$. Likewise, as the uniform structure of the physical imprecision on $\mathscr{L}$ one obtains the initial uniform structure associated with the mappings $\mathscr{L} \xrightarrow{\mu(w,g)} [0, 1]$ for all $w \in \mathscr{K}$. As mentioned in [3] § 9, $\mathscr{K}$ and $\mathscr{L}$ are precompact relative to these uniform structures, since [0, 1] is a bounded set of real numbers. Since the sets $\mathscr{K}$ and $\mathscr{L}$ are countable, such uniform structures are also metrizable (see [3] § 9).

These uniform structures of the physical imprecision can be expressed especially simply if one regards $\mathscr{K}$ and $\mathscr{L}$ as subsets of **B** resp. **D** (as in § 1). It is customary to denote by $\sigma(\mathbf{B}, A)$ the initial topology generated by the mappings $\mathbf{B} \xrightarrow{\mu(w,y)} \mathbf{R}$ for the elements y of a subset A of **D**. Likewise one denotes by $\sigma(\mathbf{D}, A)$ the topology generated on **D** by a subset A of **B**. A vector space topology determines uniquely a uniform structure (see [7] I § 1.4), i.e. the continuous linear functionals always are also uniformly continuous. Thus it suffices for vector spaces to prescribe the "topology" of the physical imprecision (see [3] § 9). Therefore, the physical imprecision on $\mathscr{K}$ is to be described by the topology $\sigma(\mathbf{B}, \mathscr{L})$ restricted to $\mathscr{K}$, the imprecision on $\mathscr{L}$ by the topology $\sigma(\mathbf{D}, \mathscr{K})$ restricted to $\mathscr{L}$.

As described in [3] § 9, one can now complete the sets $\mathscr{K}$ and $\mathscr{L}$ relative to $\sigma(\mathbf{B}, \mathscr{L})$ resp. $\sigma(\mathbf{D}, \mathscr{K})$ to compact sets $\hat{\mathscr{K}}, \hat{\mathscr{L}}$. Due to the physical interpretation of uniform structures of physical imprecision (see [3] §§ 6 and 9), it is physically meaningless, in a precisely way to single out the set $\mathscr{K}$ as a subset of $\hat{\mathscr{K}}$ and $\mathscr{L}$ as a subset of $\hat{\mathscr{L}}$. Of course, one also cannot choose an arbitrary countable subset of $\hat{\mathscr{K}}$ as the set $\mathscr{K}$, if one requires that the uniform structure generated on $\mathscr{L}$ by $(\mathbf{D}, \mathscr{K})$ does not change. Thus, the question arises for which subsets $\tilde{\mathscr{K}}$ of $\hat{\mathscr{K}}$ are the uniform structures generated on $\mathscr{L}$ by $\sigma(\mathbf{D}, \tilde{\mathscr{K}})$ equal to those generated by $\sigma(\mathbf{D}, \mathscr{K})$. In order for $\sigma(\mathbf{D}, \tilde{\mathscr{K}})$ and $\sigma(\mathbf{D}, \mathscr{K})$ to lead to the same uniform structure on $\mathscr{L}$, the mapping $\mathscr{L} \xrightarrow{f_x} [0, 1]$ must for $x \in \tilde{\mathscr{K}}$ be uniformly continuous with $f_x(g)$ as the extension of the mapping $\mathscr{K} \xrightarrow{\mu(w,g)} [0, 1]$ to $\hat{\mathscr{K}} \to [0, 1]$. From this follows that there is a largest subset $\bar{\mathscr{K}}$ of $\hat{\mathscr{K}}$ which contains $\mathscr{K}$ and for which $\sigma(\mathbf{D}, \bar{\mathscr{K}})$ and $\sigma(\mathbf{D}, \mathscr{K})$ generate the same uniform structure on $\mathscr{L}$, namely: $\bar{\mathscr{K}} = \{x \mid x \in \hat{\mathscr{K}}$ and $f_x(g)$ is uniformly continuous on $\mathscr{L}$ relative to $\sigma(\mathbf{D}, \mathscr{K})\}$. Correspondingly one can define $\bar{\mathscr{L}}$. The mapping $\mathscr{K} \times \mathscr{L} \xrightarrow{\mu} [0, 1]$ can be extended to $\bar{\mathscr{K}} \times \bar{\mathscr{L}} \xrightarrow{\mu} [0, 1]$ (but *not* to a mapping $\hat{\mathscr{K}} \times \hat{\mathscr{L}} \to [0, 1]$!).

Therefore, every countable subset of $\mathscr{K}$ which does not generate a weaker initial uniform structure on $\mathscr{L}$ and hence on $\bar{\mathscr{L}}$ is physically undistinguishable from $\mathscr{K}$; the analogous result holds for $\mathscr{L}$. Thus it appears meaningful to consider the sets $\bar{\mathscr{K}}, \bar{\mathscr{L}}$ and $\hat{\mathscr{K}}, \hat{\mathscr{L}}$ further, where $\hat{\mathscr{K}}, \hat{\mathscr{L}}$ are also the completions of $\bar{\mathscr{K}}, \bar{\mathscr{L}}$ relative to $\sigma(\mathbf{B}, \bar{\mathscr{L}})$, $\sigma(\mathbf{B}, \mathscr{L})$ resp. $\sigma(\mathbf{D}, \bar{\mathscr{K}})$, $\sigma(\mathbf{D}, \mathscr{K})$.

In order to describe the sets $\bar{\mathscr{K}}, \bar{\mathscr{L}}, \hat{\mathscr{K}}, \hat{\mathscr{L}}$, it is again practical (not necessary!) to return to the vector spaces **B**, **D** and to complete them topologically. We will do so in the next section.

§ 3 Embedding of $\mathscr{K}$ and $\mathscr{L}$ in Topologically Complete Vector Spaces

Since **B** is linearly spanned by $\mathscr{K}$, we have $\sigma(\mathbf{D}, \mathscr{K}) = \sigma(\mathbf{D}, \mathbf{B})$. For fixed $x \in \mathbf{B}$ the linear functional $\mu(x, y)$ is $\sigma(\mathbf{D}, \mathbf{B})$-continuous on **D** and can thus be uniquely extended to the $\sigma(\mathbf{D}, \mathbf{B})$-completion of **D** which one can identify with $\mathbf{B}^*$ (the set of *all* linear functionals on **B**; see [7] IV § 1.3). Therefore, one can think of $\mu(x, y)$ as the restriction to $\langle \mathbf{B}, \mathbf{D} \rangle$ of the canonical bilinear form of $\langle \mathbf{B}, \mathbf{B}^* \rangle$ for which we shall again write $\mu(x, y)$. Therefore, $\hat{\mathscr{L}}$ is equal to the $\sigma(\mathbf{B}^*, \mathbf{B})$-closure of $\mathscr{L}$ in $\mathbf{B}^*$. The set $\bar{\mathscr{L}}$ was to consist of all $y \in \hat{\mathscr{L}}$ for which (with x in $\mathscr{K}$) $\mu(x, y)$ is $\sigma(\mathbf{B}, \mathscr{L})$-

uniformly continuous. Since **D** is linearly spanned by $\mathscr{S}$, we have $\sigma(\mathbf{B}, \mathscr{S}) = \sigma(\mathbf{B}, \mathbf{D})$. This suggests to look for the set of all $y \in \mathbf{B}^*$ for which $\mu(w, y)$ is $\sigma(\mathbf{B}, \mathbf{D})$-uniformly continuous on $\mathscr{K}$. This set is obviously a linear vector space $\tilde{\mathscr{D}} \subset \mathbf{B}^*$ and yields $\bar{\mathscr{S}} = \hat{\mathscr{S}} \cap \tilde{\mathscr{D}}$.

We can define a norm in $\tilde{\mathscr{D}}$ by

$$\|y\| = \sup_{w \in \mathscr{K}} |\mu(w, y)|, \tag{3.1}$$

since $\mu(w, y)$ as a uniformly continuous function on the precompact set $\mathscr{K}$ must be bounded.

T 3.1 With the norm (3.1), $\tilde{\mathscr{D}}$ is a Banach space.

Proof. This follows from known general theorems, but can also be easily shown directly. If $y_n \in \tilde{\mathscr{D}}$ is a Cauchy sequence, $\mu(x, y_n)$ is a Cauchy sequence for each $x \in \tilde{\mathscr{D}}$. Thus, there is a linear functional $f(x)$ on **B** which is bounded on $\mathscr{K}$ and yields

$$\sup_{w \in \mathscr{K}} |\mu(w, y_n) - f(w)| \underset{n}{\rightarrow} 0.$$

Hence f is uniformly continuous on $\mathscr{K}$ because

$$\begin{aligned} |f(w_1) - f(w_2)| \leqq& |f(w_1) - \mu(w_1, y_n)| \\ &+ |\mu(w_1, y_n) - \mu(w_2, y_n)| + |\mu(w_2, y_n) - f(w_2)|. \quad \square \end{aligned}$$

D 3.1 We denote by R the subset of all $y \in \mathbf{B}^*$ for which $|\mu(w, y)|$ is bounded on $\mathscr{K}$.

It follows directly that R is a subspace of $\mathbf{B}^*$. From T 3.1 also follows $\tilde{\mathscr{D}} \subset R$. The same norm (3.1) can be introduced immediately in R. Then R is a Banach space, which follows even more easily than in the proof of T 3.1.

D 3.2 Let $\mathscr{D}$ denote the norm closure of **D** in R.

Therefore $\mathscr{D}$ is also a Banach space. From $\mathbf{D} \subset \tilde{\mathscr{D}}$ and T 3.1 follows $\mathscr{D} \subset \tilde{\mathscr{D}}$, hence the three Banach spaces $\mathscr{D}$, $\tilde{\mathscr{D}}$, R fulfill

$$\mathscr{D} \subset \tilde{\mathscr{D}} \subset R. \tag{3.2}$$

Form $\bar{\mathscr{S}} = \hat{\mathscr{S}} \cap \tilde{\mathscr{D}}$ follows $\hat{\mathscr{S}} \cap \mathscr{D} \subset \bar{\mathscr{S}}$. Later we shall find $\bar{\mathscr{S}} = \hat{\mathscr{S}} \cap \mathscr{D}$. The aim of the next considerations is a similar result for $\bar{\mathscr{K}}$. To this end, we start from the dual pair $\langle \mathbf{B}, R \rangle$.

T 3.2 The Banach space R' dual to R makes $\mathbf{B} \subset R'$.

Proof. With $x = \sum_{i=1}^{n} a_i w_i \, (w_i \in \mathscr{K})$ follows

$$|\mu(x, y)| = \left| \sum_{i=1}^{n} a_i \mu(w_i, y) \right| \leqq \sum_{i=1}^{n} |a_i| \, |\mu(w_i, y)| \leqq \|y\| \sum_{i=1}^{n} |a_i|. \quad \square$$

D 3.3 For a subset $\tilde{R}$ with $\mathbf{D} \subset \tilde{R} \subset R$, we define a norm in $\mathbf{B}$ by

$$\| x \|_{\tilde{R}} = \sup_{\substack{y \in \tilde{R} \\ y \neq 0}} \frac{|\mu(x, y)|}{\| y \|}. \tag{3.3}$$

Because of $\mathbf{D} \subset \tilde{R}$, one easily sees that $\| x \|_{\tilde{R}} = 0$ implies $x = \mathbf{0}$; hence (3.3) is meaningful. From $\tilde{R}_1 \subset \tilde{R}_2$ follows $\| x \|_{\tilde{R}_1} \leqq \| x \|_{\tilde{R}_2}$, thus in particular $\| x \|_{\mathbf{D}} \leqq \| x \|_{\tilde{R}} \leqq \| x \|_R$, where $\| x \|_R$ is the norm in the Banach space R' restricted to $\mathbf{B}$. Let $B_{\tilde{R}}$ denote the Banach space generated by $\mathbf{B}$ with the norm (3.3). For $y \in \tilde{R}$ the $\mu(x, y)$ are linear functionals on $\mathbf{B}$; they are continuous in the norm (3.3) and hence can be continued to $B_{\tilde{R}}$. In this way, $\mu(x, y)$ can be extended to a function on $B_{\tilde{R}} \times \tilde{R}$.

Using (b) (end of § 1), we obtain

T 3.3 The following holds: $\| w \|_{\tilde{R}} = 1$ for all $w \in \mathscr{K}$; $\| g \| \leqq 1$ for all $g \in \mathscr{L}$; $\| y \| = \sup \{ |\mu(x, y)| \mid \| x \|_{\tilde{R}} = 1, x \in B_{\tilde{R}} \}$ for all $y \in \tilde{R}$.

Proof. From (3.1) follows $|\mu(w, y)| \leqq \| y \|$ for all $y \in R$, which together with (3.3) yields $\| w \|_{\tilde{R}} \leqq 1$. For $g \in \mathscr{L}$ we have $|\mu(w, g)| \leqq 1$ so that (3.1) gives $\| g \| \leqq 1$. For $g = \mathbf{1}$, thereby follows $\| \mathbf{1} \| = 1$ and this by (3.3) implies

$$\| w \|_{\tilde{R}} \geqq \frac{\mu(w, \mathbf{1})}{\| \mathbf{1} \|} = 1,$$

hence we get $\| w \|_{\tilde{R}} = 1$.

From $\mathscr{K} \subset B_{\tilde{R}}$ follows

$$\sup \{ |\mu(x, y)| \mid \| x \|_{\tilde{R}} \leqq 1, x \in B_{\tilde{R}} \} \geqq \sup \{ |\mu(w, y)| \mid w \in \mathscr{K} \} = \| y \|$$

for $y \in \tilde{R}$, because $\| w \| = 1$ holds for $w \in \mathscr{K}$. From (3.3) follows

$$|\mu(x, y)| \leqq \| x \|_{\tilde{R}} \| y \|$$

and

$$\sup \{ |\mu(x, y)| \mid \| x \|_{\tilde{R}} \leqq 1, x \in B_{\tilde{R}} \} \leqq \| y \|. \quad \square$$

Therefore, if $f(x)$ is a linear functional form $B'_{\tilde{R}}$ (the Banach space dual to $B_{\tilde{R}}$), then $f(x)$ restricted to $\mathbf{B}$ is an element of $\mathbf{B}^*$. Because $|f(x)| \leqq C \| x \|$ for a suitable C on $\mathscr{K}$, we have $\| f(w) \| \leqq C$ (because $\| w \|_{\tilde{R}} = 1$ for $w \in \mathscr{K}$) and hence $f(x)|_{\mathbf{B}} \in R$. If $f_1(x)|_{\mathbf{B}} = f_2(x)|_{\mathbf{B}}$, also $f_1(x) = f_2(x)$ holds for all $x \in B_{\tilde{R}}$ because f_1 and f_2 are continuous. Thus $B'_{\tilde{R}}$ can be identified with a subset of R, namely with the subset of all $y \in R$ for which $\mu(x, y)$ is continuous on $\mathbf{B}$ in the norm (3.3).

This proves the theorem

T 3.4 We have $\mathbf{D} \subset \tilde{R} \subset B'_{\tilde{R}} \subset R$. The norm in the dual Banach space $B'_{\tilde{R}}$ coincides with that of R *at least* on the subset $\tilde{R}$ (!) (see T 3.3).

D 3.4 A subspace $\tilde{R}$ with $\mathbf{D} \subset \tilde{R} \subset R$ is called *symmetric* if $\tilde{R} = B'_{\tilde{R}}$.

Symmetric subspaces exist, for T 3.4 with $\tilde{R} = R$ gives $B'_R = R$.

For a symmetric subspace $\tilde{R}$, by T 3.4 the norm of $B'_{\tilde{R}}$ coincides with that of R on all of $B'_{\tilde{R}}$.

T 3.5 A $\sigma(B'_{\tilde{R}}, B_{\tilde{R}})$-closed subspace T of a symmetric subspace $\tilde{R} = B'_{\tilde{R}}$ is a $\sigma(R, B_R)$-closed subspace of R.

Proof. The subspace T is $\sigma(R, B_R)$-closed if $T \cap R_{|1|}$ (with $R_{|1|}$ the unit ball of R) is $\sigma(R, B_R)$-closed (see [7] IV § 6.4). Since the norms of $B'_{\tilde{R}}$ and R coincide, for the unit ball $B'_{\tilde{R}|1|}$ of $B'_{\tilde{R}}$ we find

$$B'_{\tilde{R}|1|} = B'_{\tilde{R}} \cap R_{|1|}.$$

Because of $T \subset B_{\tilde{R}}$, we therefore have

$$T \cap R_{|1|} = T \cap B'_{\tilde{R}|1|}. \tag{3.4}$$

$B'_{\tilde{R}|1|}$ is $\sigma(B'_{\tilde{R}}, B_R)$-compact and hence (since T should be $\sigma(B'_{\tilde{R}}, B_{\tilde{R}})$-closed) $T \cap B'_{\tilde{R}|1|}$ is also $\sigma(B'_{\tilde{R}}, B_{\tilde{R}})$ compact.

By $\mu(x, y)$ a linear form on R is uniquely assigned to each $x \in B_R$. Let $\mu(x, y)|_{\tilde{R}}$ be this linear form restricted to $\tilde{R}$. We will show that $\mu(x, y)|_{\tilde{R}} \in B_{\tilde{R}}$ holds: If $x_n \in \mathbf{B}$ is a sequence that in the norm $\|\cdot\|_R$ converges to $x \in B_R$, due to $\|x\|_{\tilde{R}} \leqq \|x\|_R$ the x_n also form a Cauchy sequence in the $\|\cdot\|_{\tilde{R}}$ norm, so that $\mu(x, y)|_{\tilde{R}} \in B_{\tilde{R}}$ holds. On $\tilde{R} = B'_{\tilde{R}}$, therefore $\sigma(\tilde{R}, B_R)$ is coarser than $\sigma(\tilde{R}, B_{\tilde{R}})$. Since $\mathscr{K} \subset B_R$, the $s(\tilde{R}, B_R)$-topology separates. Since on a compact set a weaker separating topology coincides with the topology of the compact set (see [5] I § 9.4), $\sigma(\tilde{R}, B_R)$ and $\sigma(\tilde{R}, B_{\tilde{R}})$ coincide on $T \cap B'_{\tilde{R}|1|}$. Thus $T \cap B'_{\tilde{R}|1|}$ is $\sigma(R, B_R)$-closed (being compact), so that $T \cap R_{|1|}$ due to (3.4) is $\sigma(R, B_R)$-closed. □

T 3.6 Every $\sigma(R, B_R)$-closed subspace $\tilde{R}$ of R with $\mathbf{D} \subset \tilde{R}$ is symmetric.

Proof. Since $\tilde{R}$ is $\sigma(R, B_R)$-closed, $B_{\tilde{R}}$ is isomorphic to the factor space $B_R/\tilde{R}^{\perp}$ and $\tilde{R}$ is isomorphic to the Banach space dual to $B_R/\tilde{R}^{\perp}$ (see [15] § 22.3), hence $B_{\tilde{R}'} = \tilde{R}'$. The isomorphic mapping $B_{\tilde{R}} \leftrightarrow B_R/\tilde{R}^{\perp}$ is just given by $\mu(x, y) \leftrightarrow \{x' | x' \in B_R$ and $\mu(x', x)|_{\tilde{R}} = \mu(x, y)\}$ (with $x \in B_{\tilde{R}}$). □

The preceding theorems allow us to prove the important, central theorem

T 3.7 There is a subspace $\tilde{R}$ of R with

(1) $\mathbf{D} \subset \tilde{R}$;
(2) $\tilde{R}$ with the norm (3.1) is a Banach space;
(3) $\tilde{R} = B'_{\tilde{R}}$ ($B_{\tilde{R}}$ the completion of $\mathbf{B}$ relative to the norm (3.3));
(4) $\sigma(B_{\tilde{R}}, \mathscr{S})$ separates (i.e. $x \in B_{\tilde{R}}$, $\mu(x, g) = 0$ for all $g \in \mathscr{S}$ implies $x = 0$).

The conditions (1) to (4) uniquely determine $\tilde{R}$.

(5) $\tilde{R}$ is the intersection of all symmetric subspaces of R.
(6) $\tilde{R}$ is the $\sigma(R, B_R)$-closed subspace spanned by $\mathbf{D}$.
(7) $\mathscr{D} \subset \tilde{R}$.
(8) For $w \in \mathscr{K}$ we have $\|w\|_{\tilde{R}} = 1$; for $g \in \mathscr{S}$ we have $\|g\| \leqq 1$. (Since $\tilde{R} = B'_{\tilde{R}}$ we have, as is known, $\|y\| = \sup\{|\mu(x, y)| \mid x \in B_{\tilde{R}}$ and $\|x\|_{\tilde{R}} \leqq 1\}$ for all $y \in \tilde{R}$.)

Proof. (3) says that $\tilde{R}$ is a symmetric subspace of R. By T 3.5 and T 3.6, the intersection Δ of all symmetric subspaces is again a symmetric subspace, and hence the smallest symmetric subspace of R. Therefore, since Δ is also the smallest $\sigma(R, B_R)$-closed subspace of R which contains $\mathbf{D}$, this Δ is the $\sigma(R, B_R)$-closed subspace of R generated by $\mathbf{D}$.

If $\tilde{R}$ were properly larger than Δ, this Δ would be a proper, $\sigma(\tilde{R}, B_R)$-closed subspace of $\tilde{R}$, hence a fortiori $\sigma(\tilde{R}, B_{\tilde{R}})$-closed (since $\sigma(\tilde{R}, B_{\tilde{R}})$ is finer than $\sigma(\tilde{R}, B_R)$; see the proof of T 3.5). Therefore in $B_{\tilde{R}}$ an x would exist which is orthogonal to Δ. Then $\mu(x, y) = 0$ would hold for all $y \in \Delta$ and hence $\mu(x, g) = 0$ for all $g \in \mathscr{L}$ in contrast to (4). Therefore, $\tilde{R} = \Delta$ must hold.

It remains yet to show that $\sigma(B_{\tilde{R}}, \mathscr{L})$ with $\tilde{R} = \Delta$ in fact separates: If there were an $x \in B_{\tilde{R}}$ with $\mu(x, g) = 0$ for all $g \in \mathscr{L}$, the $\sigma(\tilde{R}, B_{\tilde{R}})$-closed subset T spanned by $\mathscr{L}$ in $B'_{\tilde{R}} = \tilde{R}$ would not be all of $\tilde{R} = \Delta$. According to T 3.5, then T would also be a $\sigma(\tilde{R}, B_R)$-closed and hence (by T 3.6) symmetric subspace of R. Thus T would be properly smaller than Δ in contrast to the fact that Δ was the smallest symmetric subspace. □

In the sequel we shall denote the uniquely (by T 3.7) determined Banach space $B_{\tilde{R}}$ by $\mathscr{B}$ and correspondingly $\tilde{R}$ by $\mathscr{B}'$. Thus $\mathscr{K}$ is identified with a subset of $\mathscr{B}$ and $\mathscr{L}$ with a subset of $\mathscr{B}'$. For this embedding we find that $\sigma(\mathscr{B}, \mathscr{L})$ separates while the norm in $\mathscr{B}'$ satisfies (3.1).

Thereby $\mathscr{B}$ and $\mathscr{B}'$ are due to T 3.7 determined uniquely (up to isomorphisms, for $\mathscr{B}'$ can always be embedded isomorphically in R). That $\mathscr{B}$ is the Banach space generated by $\mathscr{K}$, already follows from the fact that the norm of $\mathscr{B}'$ satisfies (3.1). In fact, if the Banach space A generated by $\mathscr{K}$ were not all of $\mathscr{B}$, in $\mathscr{B}'$ there would exist a $y \neq 0$ orthogonal to A. Such a y would make $\mu(w, y) = 0$ for all $w \in \mathscr{K}$ and thus $\|y\| = 0$ according to (3.1), contradicting $y \neq 0$. Since $\sigma(\mathscr{B}, \mathscr{L})$ separates, the $\sigma(\mathscr{B}', \mathscr{B})$-closed subspace of $\mathscr{B}'$ spanned by $\mathscr{L}$ must be all of $\mathscr{B}'$ (as shown in the proof of T 3.7). The Banach subspace of $\mathscr{B}'$ spanned by $\mathscr{L}$, denoted above by $\mathscr{D}$, of course need *not* be all of $\mathscr{B}'$! From (3.1) follows $\|g\| \leqq 1$ for all $g \in \mathscr{L}$, while T 3.3 implies $\|w\| = 1$ for all $w \in \mathscr{K}$. We shall prove other important properties of $\mathscr{B}, \mathscr{B}'$ further below. But before this let us once more reflect on the extent to which $\mathscr{B}, \mathscr{B}'$ are also determined "physically uniquely".

How shall the words "physically uniquely" be interpreted? We have indeed proved in T 3.7 that $\mathscr{B}, \mathscr{B}'$ are uniquely determined by $\mathscr{K}, \mathscr{L}$. But this uniqueness would only be "physical" if $\mathscr{B}, \mathscr{B}'$ were already determined by the sets denoted above by $\bar{\mathscr{K}}$ and $\bar{\mathscr{L}}$ and did *not* depend on the choice of the $\mathscr{K}, \mathscr{L}$ as subsets of $\bar{\mathscr{K}}$ and $\bar{\mathscr{L}}$.

In order to investigate this problem of physical uniqueness, it is advantageous to examine the spaces $\mathscr{D}$ and $\mathscr{D}'$ more closely, introducing some abbreviations.

D 3.5 Let the norm closure of $\mathscr{K}$ in $\mathscr{B}$ be denoted by K.

From (d) (end of § 1) follows immediately that K is a convex set.

D 3.6 Let the $\sigma(\mathscr{B}', \mathscr{B})$-closure of $\mathscr{L}$ in $\mathscr{B}'$ be denoted by L.

From (d) follows directly that L is a convex set.

T 3.8 For $w \in K$ we have $\| w \| = 1$ and $\mu(w, \mathbf{1}) = 1$. For $y \in \mathscr{B}'$ we find

$$\| y \| = \sup_{w \in \mathscr{K}} \mu(w, y) = \sup_{w \in \mathscr{K}} | \mu(w, y) | . \tag{3.5}$$

For $g \in L$ we have $\| g \| \leqq 1$, while $w \in K, g \in L$ give $0 \leqq \mu(w, g) \leqq 1$.

Proof. The equality (3.5) follows immediately since K is the norm closure of $\mathscr{K}$. From (3.5) follows $| \mu(w, y) | \leqq \| y \|$ and hence

$$\| w \| = \sup_{y \in \mathscr{B}'} \frac{| \mu(w, y) |}{\| y \|} \leqq 1 .$$

From $\mu(w, \mathbf{1}) = 1$ for $w \in \mathscr{K}$ also follows $\mu(w, \mathbf{1}) = 1$ for $w \in K$. Thus $\| \mathbf{1} \| = 1$ gives $\| w \| \geqq 1$, such that $\| w \| = 1$.

For each $w \in K$ and $\varepsilon > 0$, for $g \in L$ there is a $\tilde{g} \in \mathscr{L}$ with $| \mu(w, g) - \mu(w, \tilde{g}) | < \varepsilon$. Because of $0 \leqq \mu(w, \tilde{g}) \leqq 1$ there results $0 \leqq \mu(w, g) \leqq 1$; hence (3.5) finally gives $\| g \| \leqq 1$.

T 3.9 For $\hat{\mathscr{L}}$ from § 2, we have $\hat{\mathscr{L}} = L \subset \mathscr{B}'_{|1|}$.

Proof. $\hat{\mathscr{L}}$ is the $\sigma(\mathscr{B}', \mathscr{K})$-completion of $\mathscr{L}$. Since $\mathscr{B}'_{|1|}$ is compact in the $\sigma(\mathscr{B}', \mathscr{B})$-topology, the $\sigma(\mathscr{B}', \mathscr{B})$- and $\sigma(\mathscr{B}', \mathscr{K})$-topologies coincide on $\mathscr{B}'_{|1|}$. Hence, the $\sigma(\mathscr{B}', \mathscr{B})$-closure L of $\mathscr{L}$ in $\mathscr{B}'_{|1|}$ equals the $\sigma(\mathscr{B}', \mathscr{K})$-completion $\hat{\mathscr{L}}$ of $\mathscr{L}$. From this follows again the relation $\hat{\mathscr{L}} \subset \mathscr{B}'_{|1|}$ already established in T 3.8. □

Since the elements of $\mathscr{B}$ are norm-continuous functionals over $\mathscr{D} \subset \mathscr{B}'$, we can also identify $\mathscr{B}$ with a subset of $\mathscr{D}'$ (the Banach space dual to $\mathscr{D}$). Thus $\mathscr{B} \subset \mathscr{D}'$ holds and therefore $\mathscr{K} \subset K \subset \mathscr{D}'$. Because of $\mathbf{D} \subset \mathscr{D} \subset R$ we can view $\mathscr{D}$ as a special set $\tilde{R}$ from the theorems T 3.3, T 3.4, T 3.5. Thus we can define the Banach spaces $B_{\mathscr{D}}$ and $B'_{\mathscr{D}}$.

T 3.10 The inclusions $\mathscr{B} \subset B_{\mathscr{D}} \subset \mathscr{D}'$ hold.

Proof. Since $\mathscr{D}'$ is a Banach space relative to the norm

$$\| x \|_{\mathscr{D}} = \sup_{y \in \mathscr{D}} \frac{| \mu(x, y) |}{\| y \|}$$

and $B_{\mathscr{D}}$ the Banach space generated from $\mathbf{B}$ in the norm $\| x \|_{\mathscr{D}}$, this $B_{\mathscr{D}}$ is the closure of $\mathbf{B}$ in $\mathscr{D}'$ in the $\| . \|_{\mathscr{D}}$-norm; hence we get $B_{\mathscr{D}} \subset \mathscr{D}'$. Because of $\mathscr{D} \subset \mathscr{B}'$, we find $\| x \|_{\mathscr{D}} \leqq \| x \|$ with $\| x \|$ as the norm in $\mathscr{B}$ (since $\| x \| = \| x \|_{\mathscr{B}'}$). If $x \in \mathscr{B}$, there is a sequence $x_\nu \in \mathbf{B}$ such that $\| x_\nu - x \| \to 0$. Thus we also have $\| x_\nu - x \|_{\mathscr{D}} \to 0$ and hence $x \in B_{\mathscr{D}}$. □

T 3.11 The inclusions $\mathscr{D} \subset B'_{\mathscr{D}} \subset \mathscr{B}'$ hold. On $\mathscr{D}$, the norm of $B'_{\mathscr{D}}$ equals that of R, hence that of $\mathscr{B}'$. On $B'_{\mathscr{D}} \backslash \mathscr{D}$, the norm of $B'_{\mathscr{D}}$ can at most be larger than that of $\mathscr{B}'$. For $\mathscr{D} \neq \mathscr{B}'$ we have $\mathscr{D} \neq B'_{\mathscr{D}}$.

Proof. From T 3.4 immediately follows $\mathscr{D} \subset B'_{\mathscr{D}}$. Since $\mathscr{B}'$ was the smallest symmetric subspace of R, for $\mathscr{D} \neq \mathscr{B}'$ we must also have $\mathscr{D} \neq B'_{\mathscr{D}}$.

$B'_{\mathscr{D}}$ is the set of linear functionals on $B_{\mathscr{D}}$ continuous in the $\|\cdot\|_{\mathscr{D}}$-norm. Because of $\mathscr{B} \subset B_{\mathscr{D}}$ they are therefore also functionals on $\mathscr{B}$ continuous in the $\|\cdot\|_{\mathscr{D}}$-norm. As elements of R, those of $B'_{\mathscr{D}}$ are already uniquely determined by their values in $\mathscr{B}$. Inasmuch as $\|x\|_{\mathscr{D}} \leqq \|x\|$, they are a fortiori continuous on $\mathscr{B}$ relativ to the norm in $\mathscr{B}$, such that $B'_{\mathscr{D}} \subset \mathscr{B}'$.

From $\|x\|_{\mathscr{D}} \leqq \|x\|$ follows by duality that the norm of $\mathscr{B}'$ is smaller than that of $B'_{\mathscr{D}}$. But by T 3.4 both coincide on $\mathscr{D}$. □

D 3.7 Let $\bar{A}^{\sigma}$ be the $\sigma(\mathscr{D}', \mathscr{D})$-closure of a subset $A \subset \mathscr{D}'$.

From APK immediately follows that $\bar{\mathscr{K}}^{\sigma}$ is a convex set. Since on $\mathscr{B}$ the norm topology is stronger than the $\sigma(\mathscr{D}', \mathscr{D})$-topology, we directly conclude $K \subset \bar{\mathscr{K}}^{\sigma}$ and hence $\bar{K}^{\sigma} = \bar{\mathscr{K}}^{\sigma}$.

T 3.12 The unit ball $\mathscr{D}'_{|1|}$ of $\mathscr{D}'$ gives $\mathscr{D}'_{|1|} = \mathrm{co}(\bar{K}^{\sigma} \cup -\bar{K}^{\sigma})$, hence

$$\mathscr{D}'_{|1|} = \bigcup_{0 \leqq \lambda \leqq 1} (\lambda \bar{K}^{\sigma} - (1-\lambda) \bar{K}^{\sigma}).$$

For $w \in \bar{K}^{\sigma}$ we have $\|w\|_{\mathscr{D}} = 1$. Each element $x \in \mathscr{D}'$ can be written in the form $x = \alpha w - \beta v$, where $\alpha \geqq 0$, $\beta \geqq 0$, $w \in \bar{K}^{\sigma}$ $v \in \bar{K}^{\sigma}$ and $\alpha + \beta = \|x\|_{\mathscr{D}}$. Both $\mathscr{B}_{|1|}$ and $\mathscr{B}$ are $\sigma(\mathscr{D}' \mathscr{D})$-dense in $\mathscr{D}'_{|1|}$ resp. $\mathscr{D}'$.

Proof. The norm of $\mathscr{D}'$ is just

$$\|x\|_{\mathscr{D}} = \sup_{y \in \mathscr{D}} \frac{|\mu(x, y)|}{\|y\|}.$$

Because of $\|x\|_{\mathscr{D}} \leqq \|x\|$ (see the proof of T 3.10) we get $\mathscr{B}_{|1|} \subset \mathscr{D}'_{|1|}$. Since the norm in $\mathscr{D}$ is given by (3.1), the set $\mathscr{D}'_{|1|}$ is bipolar to $\mathscr{K} \cup -\mathscr{K}$ so that [7] IV § 1.5 gives $\mathscr{D}'_{|1|} = \overline{\mathrm{co}}^{\sigma}(\mathscr{K} \cup -\mathscr{K})$. This makes $\bar{K}^{\sigma} \in \mathscr{D}'_{|1|}$ and $(-\bar{K}^{\sigma}) \subset \mathscr{D}'_{|1|}$. Since $\bar{K}^{\sigma}$ and $(-\bar{K}^{\sigma})$ are compact (being closed subsets of the $\sigma(\mathscr{D}', \mathscr{D})$-compact set $\mathscr{D}'_{|1|}$), the set $\mathrm{co}(\bar{K}^{\sigma} \cup -\bar{K}^{\sigma})$ is already compact and hence closed in the $\sigma(\mathscr{D}', \mathscr{D})$-topology, such that $\mathscr{D}'_{|1|} = \mathrm{co}(\bar{K}^{\sigma} \cup -\bar{K}^{\sigma})$. Because of $\bar{K}^{\sigma} \subset \mathscr{D}'_{|1|}$ we get $\|w\|_{\mathscr{D}} \leqq 1$ for $w \in \bar{K}^{\sigma}$. From $\mu(w, \mathbf{1}) = 1$ for $w \in K$ also follows $\mu(w, \mathbf{1}) = 1$ for $w \in \bar{K}^{\sigma}$, therefore $\|w\|_{\mathscr{D}} = 1$. Since $x' = \|x\|_{\mathscr{D}}^{-1} x$ with $x \in \mathscr{D}'$ has the norm $\|x'\|_{\mathscr{D}} = 1$, it equals $x' = \lambda w - (1-\lambda) v$ with $0 \leqq \lambda \leqq 1$ and $w, v \in \bar{K}^{\sigma}$. From this follows $x = \|x\|_{\mathscr{D}} \lambda w - \|x\|_{\mathscr{D}} (1-\lambda) v$ and $\|x\|_{\mathscr{D}} = \|x\|_{\mathscr{D}} \lambda + \|x\|_{\mathscr{D}} (1-\lambda)$.

Hence the set of all $\alpha w - \beta v$ with $w, v \in \mathscr{K}$ is $\sigma(\mathscr{D}', \mathscr{D})$-dense in $\mathscr{D}'$. Therefore $\mathscr{B}$ due to $\mathscr{K} \subset \mathscr{B}$ is $\sigma(\mathscr{D}', \mathscr{D})$-dense in $\mathscr{D}'$. For $\alpha + \beta \leqq 1$ we have $\|\alpha w - \beta v\| \leqq \alpha \|w\| + \beta \|v\| = \alpha + \beta \leqq 1$, and thus $\alpha w - \beta v \in \mathscr{B}_{|1|}$. The $\lambda w - (1-\lambda) v$ with $w, v \in \mathscr{K}$ are $\sigma(\mathscr{D}', \mathscr{D})$-dense in $\mathscr{D}'_{|1|}$; hence $\mathscr{B}_{|1|}$ is a $\sigma(\mathscr{D}', \mathscr{D})$-dense subset of $\mathscr{D}'_{|1|}$.

T 3.13 $\hat{\mathscr{K}} = \bar{\mathscr{K}}^{\sigma} = \bar{K}^{\sigma}$ holds.

Proof. Already above we showed $\bar{K}^{\sigma} = \bar{\mathscr{K}}^{\sigma}$.

As a closed subset of the $\sigma(\mathscr{D}', \mathscr{D})$-compact set $\mathscr{D}'_{|1|}$, also $\bar{K}^{\sigma}$ is $\sigma(\mathscr{D}', \mathscr{D})$-compact. Thus on $\bar{K}^{\sigma}$ the weaker topology $\sigma(\mathscr{D}', \mathscr{L})$ equals $\sigma(\mathscr{D}', \mathscr{D})$. Therefore, the completion $\hat{\mathscr{K}}$ of $\mathscr{K}$ in the $\sigma(\mathscr{D}',\mathscr{L})$-topology is identical with the $\sigma(\mathscr{D}',\mathscr{D})$-closure of $\mathscr{K}$. □

Thus $\bar{\mathscr{K}}$ is the subset of all $w \in \bar{K}^\sigma$ for which $\mu(w, g)$ with $g \in \mathscr{L}$ is $\sigma(\mathscr{D}, \mathscr{K})$-continuous on $\mathscr{L}$.

For that reason we adopt

D 3.8 Let $\tilde{\mathscr{B}}$ be the set of all $y \in \mathscr{D}'$ for which $\mu(y, g)$ is $\sigma(\mathscr{D}, \mathscr{K})$-continuous on $\mathscr{L}$.

This gives $\bar{\mathscr{K}} \subset \tilde{\mathscr{B}} \cap \bar{K}^\sigma = \tilde{\mathscr{B}} \cap \bar{\mathscr{K}}^\sigma$.

For our subsequent considerations, we will observe that $\mathscr{K}$ and $\mathscr{L}$ are countable sets. Then we easily obtain

T 3.14 $\mathscr{B}$ and $\mathscr{D}$ are separable Banach spaces.

By [10] § 21.3 (4) this in turn yields

T 3.15 The topologies $\sigma(\mathscr{B}', \mathscr{B})$ and $\sigma(\mathscr{D}', \mathscr{D})$ are metrizable on each set that is norm bounded in $\mathscr{B}'$ resp. $\mathscr{D}'$.

The following important theorem then holds with the $\tilde{\mathscr{D}}$ defined at the beginning of this section.

T 3.16 $\tilde{\mathscr{D}} = \mathscr{D}$ and hence $\bar{\mathscr{L}} = \hat{\mathscr{L}} \cap \mathscr{D} = L \cap \mathscr{D}$.

Proof. Let $l \in \tilde{\mathscr{D}}$, i.e. l be a linear form on **B** which is $\sigma(\mathbf{B}, \mathscr{L})$-continuous on $\mathscr{K}$. Since $\mathscr{K}$ is a subset of the $\sigma(\mathscr{D}',\mathscr{D})$-compact set $\bar{K}^\sigma$ and $\sigma(\mathscr{D}',\mathscr{L})$ is separating on $\mathscr{D}'$, the topologies $\sigma(\mathscr{D}',\mathscr{L})$ and $\sigma(\mathscr{D}',\mathscr{D})$ are identical on $\mathscr{K}$.

Since $\mathscr{K}$ is rational convex by APK, we can extend l as an affine functional to the $\sigma(\mathscr{D}',\mathscr{D})$-completion of $\mathscr{K}$, i.e. to $\bar{K}^\sigma$. Since $\mathscr{D}'$ is linearly spanned by $\bar{K}^\sigma$, we can extend l as a linear functional to all of $\mathscr{D}'$, and this extended l is $\sigma(\mathscr{D}',\mathscr{D})$-continuous on $\bar{K}^\sigma$. It follows immediately that l is norm-continuous on all of $\mathscr{D}'$, because it must be bounded on the compact set $\bar{K}^\sigma$ so that T 3.8 easily yields

$$\sup_{x \in \mathscr{D}'_{|1|}} |l(x)| = \sup_{w \in \bar{K}^\sigma} |l(w)|.$$

We will show that l is even $\sigma(\mathscr{D}',\mathscr{D})$-continuous on all of $\mathscr{D}'$. The functional l is $\sigma(\mathscr{D}',\mathscr{D})$-continuous on all of $\mathscr{D}'$ if the subspace F of all $x \in \mathscr{D}$ with $l(x) = 0$ is a $\sigma(\mathscr{D}',\mathscr{D})$-closed subspace of $\mathscr{D}'$; and this holds if $F \cap \mathscr{D}'_{|1|}$ is $\sigma(\mathscr{D}',\mathscr{D})$-closed ([7] IV § 6.4). Let $x \in \mathscr{D}'_{|1|}$ be an accumulation point of $F \cap \mathscr{D}'_{|1|}$. Since the $\sigma(\mathscr{D}',\mathscr{D})$-topology on $\mathscr{D}'_{|1|}$ is metrizable by T 3.15, there is a sequence $x_\nu \in F \cap \mathscr{D}'_{|1|}$ with $x_\nu \to x$ in the $\sigma(\mathscr{D}',\mathscr{D})$-topology. From T 3.12 follows $x_\nu = \alpha_\nu w_\nu - \beta_\nu v_\nu$ with $\alpha_\nu + \beta_\nu = \| x_\nu \| \leqq 1$. Hence there is a subsequence of the x_ν (briefly denoted again by x_ν) such that $\alpha_\nu \to \alpha$ and $\beta_\nu \to \beta$ hold with $\alpha + \beta \leqq 1$. From this follows $\| x_\nu - \bar{x}_\nu \| \to 0$ with $\bar{x}_\nu = \alpha w_\nu - \beta v_\nu$, hence also $\bar{x}_\nu \to x$ in the $\sigma(\mathscr{D}',\mathscr{D})$-topology. Since l is norm continuous, from $l(\bar{x}_\nu) - l(x_\nu) = l(\bar{x}_\nu)$ we obtain $l(\bar{x}_\nu) \to 0$.

Since $\bar{K}^\sigma$ is compact, we can select a subsequence of $\bar{x}_\nu$ (again denoted briefly by $\bar{x}_\nu$) so that $w_\nu \to w$ and $v_\nu \to v$ hold in the $\sigma(\mathscr{D}',\mathscr{D})$-topology. With $\bar{x}_\nu \to x$ this implies $x = \alpha w - \beta v$. Since l is $\sigma(\mathscr{D}',\mathscr{D})$-continuous on $\bar{K}^\sigma$, we obtain $l(w_\nu) \to l(w)$ and $l(v_\nu) \to l(v)$ and hence $l(\bar{x}_\nu) = \alpha\, l(w_\nu) - \beta\, l(v_\nu) \to \alpha\, l(w) - \beta\, l(v) = l(x)$. Because of $l(\bar{x}_\nu) \to 0$ follows $l(x) = 0$, hence $x \in F$.

Since $l(x)$ is $\sigma(\mathscr{D}',\mathscr{D})$-continuous on all of $\mathscr{D}'$, we have $l \in \mathscr{D}$. Therefore, $\tilde{\mathscr{D}} \in \mathscr{D}$ holds, and, with (3.2), finally $\tilde{\mathscr{D}} = \mathscr{D}$. □

Therefore, from T 3.9 and T 3.16 follows that the spaces $\mathscr{B}'$ and $\mathscr{D}$ contain only the physical structure of the sets $\bar{\mathscr{L}}$, $\hat{\mathscr{L}} = L$ since $\hat{\mathscr{L}} = L \cap \mathscr{D}$ and $\mathscr{D}$ is the Banach subspace generated by $\bar{\mathscr{L}}$ in $\mathscr{B}'$. Therefore, the choice of $\mathscr{L}$ as a subset of $\bar{\mathscr{L}}$ does not enter into $\mathscr{B}'$ and $\mathscr{D}$. But does perhaps the choice of $\mathscr{K}$ and not only $\bar{\mathscr{K}}$ enter the structure of $\mathscr{B}$ and $\mathscr{B}'$?

T 3.17 The inclusions $\mathscr{B} \subset B_{\mathscr{D}} \subset \tilde{\mathscr{B}} \subset \mathscr{D}'$ and hence $K \subset \bar{\mathscr{K}}$ hold.

Proof. According to T 3.10 and D 3.8, we must only show $B_{\mathscr{D}} \subset \tilde{\mathscr{B}}$. Let $l \in B_{\mathscr{D}}$; then l is a $\sigma(B'_{\mathscr{D}}, B_{\mathscr{D}})$-continuous linear functional on $B'_{\mathscr{D}}$. By T 3.11, the norm of $B'_{\mathscr{D}}$ coincides on $\mathscr{D}$ with that of $\mathscr{B}'$. Therefore $\mathscr{L} \subset \mathscr{D}$ implies $\mathscr{L} \subset B'_{\mathscr{D}|1|}$. Since $B'_{\mathscr{D}|1|}$ is compact in the $\sigma(B'_{\mathscr{D}}, B_{\mathscr{D}})$-topology and $\sigma(B'_{\mathscr{D}}, \mathscr{K})$ is weaker but separating on $B'_{\mathscr{D}}$ (T 3.11 makes $B'_{\mathscr{D}} \subset \mathscr{B}'$), the $\sigma(B'_{\mathscr{D}}, B_{\mathscr{D}})$-topology on $\mathscr{L}$ equals the $\sigma(B'_{\mathscr{D}}, \mathscr{K})$-topology. Therefore, l is $\sigma(B'_{\mathscr{D}}, \mathscr{K})$-continuous on $\mathscr{L}$, so that $B_{\mathscr{D}} \subset \tilde{\mathscr{B}}$.

From $\mathscr{B} \subset B_{\mathscr{D}} \subset \tilde{\mathscr{B}} \subset \mathscr{D}'$ follows $\bar{K} = \tilde{\mathscr{B}} \cap \bar{K}^\sigma \supset \mathscr{B} \cap \bar{K}^\sigma$ and $K \subset \bar{K}^\sigma$ finally gives $K \subset \mathscr{B} \cap \bar{K}^\sigma \subset \bar{\mathscr{K}}$. □

$K = \bar{\mathscr{K}}$ would mean that the vector space $\mathscr{B}$ and the set K are determined "physically uniquely". The description of the probability function $\mathscr{K} \times \mathscr{L} \xrightarrow{\mu} [0, 1]$ within the dual pair $\mathscr{B}, \mathscr{B}'$ is therefore independent of "random" choices (of $\mathscr{K}$ as a subset of $\bar{\mathscr{K}}$ and of $\mathscr{L}$ as a subset of $\bar{\mathscr{L}}$) just then, when we can yet show $K = \bar{\mathscr{K}}$.

If $\tilde{\mathscr{B}} = \mathscr{B}$ and $\mathscr{B} \cap \bar{K}^\sigma = K$, holds, they imply $\bar{\mathscr{K}} = \tilde{\mathscr{B}} \cap \bar{K}^\sigma = K$. By T 3.17 from $\tilde{\mathscr{B}} = \mathscr{B}$ also follows $B_{\mathscr{D}} = \mathscr{B}$ (as sets). Nevertheless, in principle we could have $\| x \|_{\mathscr{D}} \leqq \| x \|$ and hence $B'_{\mathscr{D}} \subsetneqq \mathscr{B}'$.

Therefore, we first ask under what presumptions $\tilde{\mathscr{B}} = \mathscr{B}$ would hold. In the following theorem T 3.18 we shall state an assumption which will later (after the introduction of further axioms, e.g. AV 1.2s in VI), turn out fulfilled.

From $L \subset \mathscr{B}'_{|1|}$ follows with $\mathbf{1} \in \mathscr{L}$, that

$$\mathbf{1} - 2L \subset \mathscr{B}'_{|1|} \tag{3.6}$$

holds due to $\mu(w, \mathbf{1} - 2g) = \mu(w, \mathbf{1}) - 2\mu(w, g) = 1 - 2\mu(w, g)$. From (*c*) (end of § 1) follows

$$L = \mathbf{1} - L \tag{3.7}$$

such that

$$\mathbf{1} - 2L = \mathbf{1} - 2(\mathbf{1} - L) = -(\mathbf{1} - 2L). \tag{3.8}$$

Therefore, $\mathbf{1} - 2L$ is a set symmetric about $\mathbf{0}$.

If the convex set L possesses any interior element g_0 (relative to the norm topology of $\mathscr{B}'$), then $\mathbf{0}$ is an interior element of $\mathbf{1} - 2L$, hence $(1/2)\mathbf{1}$ an interior element of L. This can easily be shown on the basis of (*c*) (end of § 1):

If g_0 is an interior element, there is an $\alpha > 0$ with $g_0 + \alpha\, \mathscr{B}'_{[1]} \subset L$. With (c) follows $\mathbf{1} - g_0 \subset L$. Since L is convex, we obtain

$$\frac{1}{2}(g_0 + \alpha\, \mathscr{B}'_{[1]}) + \frac{1}{2}(\mathbf{1} - g_0) = \frac{1}{2}\mathbf{1} + \frac{\alpha}{2}\mathscr{B}'_{[1]} \subset L$$

hence $\frac{1}{2}\mathbf{1}$ is an interior point of L and thus $\mathbf{0}$ an interior point of $\mathbf{1} - L$.

T 3.18 If L in the norm topology of $\mathscr{B}'$ has an interior point, we obtain $\tilde{\mathscr{B}} = \mathscr{B}$.

Proof. The proof proceeds analogous to that of T 3.16, so we can be brief. An $l \in \tilde{\mathscr{B}}$ can be extended to L as a $\sigma(\mathscr{B}', \mathscr{B})$-continuous affine functional, because on L the $\sigma(\mathscr{B}', \mathscr{B})$-topology coincides with the $\sigma(\mathscr{B}', \mathscr{K})$-topology.

Since $\mathbf{0}$ is an interior point of $\mathbf{1} - 2L$, there is an $\alpha > 0$ with $\alpha\, \mathscr{B}'_{[1]} \subset \mathbf{1} - 2L$. Since $\mathbf{1} \in L$ holds, L therefore spans the whole space $\mathscr{B}'$. Therefore, l can be extended as a linear functional to all of $\mathscr{B}'$. Because of $\alpha\, \mathscr{B}'_{[1]} \subset \mathbf{1} - 2L$, this l is $\sigma(\mathscr{B}', \mathscr{B})$-continuous on $\alpha\, \mathscr{B}'_{[1]}$ and hence ([7] I § 4.2 and IV § 6.4) on all of $\mathscr{B}'$. Thus we get $l \in \mathscr{B}$, hence $\tilde{\mathscr{B}} \subset \mathscr{B}$. This, together with T 3.17, implies $\tilde{\mathscr{B}} = \mathscr{B}$. □

The second condition, $\mathscr{B} \cap \bar{K}^{\sigma} = K$, means that K is closed as a subset of $\mathscr{B}$ in the $\sigma(\mathscr{B}, \mathscr{D})$-topology. Let $x \in \mathscr{B} \cap \bar{K}^{\sigma}$; then $\mathbf{1} \in \mathscr{L}$ implies $\mu(x, \mathbf{1}) = 1$. From $0 \leqq \mu(w, y)$ for all $w \in \bar{K}^{\sigma}$ and $y \in \mathscr{D}_+$, with $\mathscr{D}_+$ from § 4, we also get $0 \leqq \mu(x, y)$ for all $y \in \bar{\mathscr{D}}_+^{\sigma}$, where $\bar{\mathscr{D}}_+^{\sigma}$ is the $\sigma(\mathscr{B}', \mathscr{B})$-closure of $\mathscr{D}_+$ in $\mathscr{B}'$. As shown in § 4, we have $\mathscr{D}_+ = \mathscr{D} \cap \mathscr{B}'_+$ and hence $\bar{\mathscr{D}}_+^{\sigma} \subset \bar{\mathscr{B}}_+'^{\sigma}$. If $\bar{\mathscr{D}}_+^{\sigma} = \mathscr{B}'_+$, by T 4.3 follows $x \in \mathscr{B}_+$ and hence $x \in K$, i.e. $\mathscr{B} \cap \bar{K}^{\sigma} = K$. We have thus proven

T 3.19 If $\mathscr{D} \cap \mathscr{B}'_+$ is $\sigma(\mathscr{B}', \mathscr{B})$-dense in $\mathscr{B}'_+$ and if L has an interior point relative to the norm topology, we obtain $K = \bar{\mathscr{K}}$.

According to VI T 2.2.1, from AV 2 follows that the $\sigma(\mathscr{B}', \mathscr{B})$-closed cone generated by L is *all* of $\mathscr{B}'_+$; hence $\mathscr{D} \cap \mathscr{B}'_+$ is $\sigma(\mathscr{B}', \mathscr{B})$-dense in $\mathscr{B}'_+$. By VI T 2.2.2, the same follows from AV 1.2s. From AV 1.2s also follows that L has an interior point relative to the norm topology. Therefore from AV 1.2s follows $K = \bar{\mathscr{K}}$.

In concluding this section, we again emphasize that the spaces $\mathscr{B}$, $\mathscr{D}$, $\mathscr{B}'$, $\mathscr{D}'$ do not affect the structure of the "physical" sets $\mathscr{K}$, $\mathscr{L}$, $\bar{\mathscr{K}}$, $\bar{\mathscr{L}}$ and $\hat{\mathscr{K}}$, $\hat{\mathscr{L}}$. It is altogether legal that (together with the sets $\mathscr{K}$, $\mathscr{L}$, $\bar{\mathscr{K}}$, $\bar{\mathscr{L}}$ and $\hat{\mathscr{K}}$, $\hat{\mathscr{L}}$ and the uniform structures of physical imprecision) we use other mathematical structures, such as the spaces $\mathscr{B}$, $\mathscr{D}$ and the norm topology in $\mathscr{B}$ (all uniquely determined by $\mathscr{K}$, $\mathscr{L}$), because these only reflect properties of $\mathscr{K}$, $\mathscr{L}$, $\bar{\mathscr{K}}$, $\bar{\mathscr{L}}$ and $\hat{\mathscr{K}}$, $\hat{\mathscr{L}}$ that are already at hand. The spaces $\mathscr{B}$, $\mathscr{D}$, $\mathscr{B}'$, $\mathscr{D}'$ do not depend on the choice of $\mathscr{L}$ as a subset of $\bar{\mathscr{L}}$, but only on $\bar{\mathscr{L}}$. For the case $\bar{\mathscr{K}} = K$ (e.g. if AV 1.2s holds) the spaces $\mathscr{B}$, $\mathscr{D}$, $\mathscr{B}'$, $\mathscr{D}'$ also do not depend on the choice of $\mathscr{K}$ as a subset of $\bar{\mathscr{K}}$, but only on $\bar{\mathscr{K}}$. Just this shows that the structure of the spaces $\mathscr{B}$, $\mathscr{D}$, $\mathscr{B}'$, $\mathscr{D}'$ depicts the true physical structure of the probability μ.

§ 4 $\mathscr{B}, \mathscr{B}', \mathscr{D}, \mathscr{D}'$ Considered as Ordered Vector Spaces

Here we shall sketch some basic structures of the spaces $\mathscr{B}, \mathscr{B}', \mathscr{D}, \mathscr{D}'$, which are necessary and sufficient to characterize the situations described by the sets $\bar{\mathscr{K}}$ and $\bar{\mathscr{L}}$. Together with $\bar{\mathscr{L}} = L \cap \mathscr{D}$ (valid by T 3.16), in this context we have in mind the relation $\bar{\mathscr{K}} = \mathscr{K}$. Due to further axioms, this will later follow from T 3.19. We first introduce several concepts.

A vector space E is called an *ordered* vector space if an ordering relation is given which makes

$$x_1 \leqq x_2 \Rightarrow x_1 + x \leqq x_2 + x \text{ for all } x \in E,$$

$$x_1 \leqq x_2 \Rightarrow \alpha x; \leqq \alpha x_2 \text{ for all } \alpha \geqq 0.$$

As a cone C of E we denote a subset of E which makes $C + C \subset C$ and $\alpha C \subset C$ for all $\alpha \geqq 0$. A cone C is called proper if $C \cap (-C) = \{\mathbf{0}\}$.

The structures of a proper cone C and of an ordering are equivalent via $x \geqq 0 \Leftrightarrow x \in C$.

In the sequel we shall by E_+ denote a proper cone C which defines an ordering in E.

In $\mathscr{B}$ the set $\bigcup_{\lambda \geqq 0} \lambda K$ is a proper cone: It is a cone since K is convex. This cone is proper since $\lambda_1 w_1 = -\lambda_2 w_2$ with $w_1, w_1 \in K$ would imply $\lambda_1 w_1 + \lambda_2 w_2 = 0$ whereas, on the other hand,

$$\| \lambda_1 w_1 + \lambda_2 w_2 \| = (\lambda_1 + \lambda_2) \left\| \frac{\lambda_1}{\lambda_1 + \lambda_2} w_1 + \frac{\lambda_2}{\lambda_1 + \lambda_2} w_2 \right\| = \lambda_1 + \lambda_2$$

holds because K is convex and $\| w \| = 1$ for $w \in K$.

D 4.1 By $\mathscr{B}_+$ we denote the cone defined in $\mathscr{B}$ by $\bigcup_{\lambda \geqq 0} (\lambda K)$. Via this cone, $\mathscr{B}$ becomes an ordered vector space.

T 4.1 $\mathscr{B}_+$ is closed in the norm topology and in the $\sigma(\mathscr{B}, \mathscr{B}')$-topology.

Proof. If $\lambda_\nu w_\nu \to x$ in norm then the norm of $\lambda_\nu w_\nu$ is also convergent: $\| \lambda_\nu w_\nu \| = \lambda_\nu \| w_\nu \| = \lambda_\nu \to \lambda$. Then

$$\| \lambda w_\nu - x \| \leqq \| (\lambda - \lambda_\nu) w_\nu \| + \| \lambda_\nu w_\nu - x \| = | \lambda - \lambda_\nu | + \| \lambda_\nu w_\nu - x \|$$

holds and hence $\lambda w_\nu \to x$. Since K is norm-closed, we have $w_\nu \to w \in K$ and $x = \lambda w \in \mathscr{B}_+$.

But a convex set, norm-closed in $\mathscr{B}$, is also $\sigma(\mathscr{B}, \mathscr{B}')$-closed in $\mathscr{B}$ ([7] II § 9.2). $\square$

For an $x \in \mathscr{B}_+$ we easily see that the representation $x = \lambda w$, with $w \in K$, is unique because $\| x \| = \lambda \| w \| = \lambda$.

A subset M of a cone C is called a *basis* of C, when $C = \bigcup_{\lambda \geqq 0} \lambda M$ and $x = \lambda_1 w_1 = \lambda_2 w_2$ $(w_1, w_2 \in M)$ always imply $\lambda_1 = \lambda_2$ and hence $w_1 = w_2$. Therefore, K is a basis of the cone $\mathscr{B}_+$.

T 4.2 $\mathscr{B}_{|1|} = \overline{\text{co}}\,(K \cup - K)$ (in the sense of norm closure).

Proof. By the definition of the norm in $\mathscr{B}'$, the set $\mathscr{B}_{|1|}$ is bipolar to $K \cup - K$ and hence equal to $\overline{\text{co}}^{\sigma}\,(K \cup (-K))$ (in the sense of the closure in the $\sigma\,(\mathscr{B}, \mathscr{B}')$-topology). Since every norm closed convex subset of $\mathscr{B}$ is also $\sigma\,(\mathscr{B}, \mathscr{B}')$-closed, the assertion follows. □

Theorem T 4.2 says explicitly that $\mathscr{B}_{|1|}$ is the norm closure of the set

$$\bigcup_{0 \leqq \lambda \leqq 1} (\lambda\, K - (1 - \lambda)\, K)\,.$$

D 4.2 An ordered Banach space $\mathscr{B}$ is called basis normed if a basis K of $\mathscr{B}_+$ makes $\mathscr{B}_{|1|} = \overline{\text{co}}\,(K \cup - K)$.

We easily recognize $K = \{x \mid x \in \mathscr{B}_+ \text{ and } \mu\,(x, \mathbf{1}) = 1\}$, i.e. K is the section of $\mathscr{B}_+$ by the hyperplane $\{x \mid \mu\,(x, \mathbf{1}) = 1\}$.

It easily follows that the set of all $y \in \mathscr{B}'$, with $\mu\,(x, y) \geqq 0$ for all $x \in \mathscr{B}_+$, is a cone. This cone is proper: From $\mu\,(x, y_1) \geqq 0$, $\mu\,(x, y_2) \geqq 0$ and $y_1 = - y_2$, i.e. $y_1 + y_2 = 0$ follows $\mu\,(x, y_1) + \mu\,(x, y_2) = 0$ and hence $\mu\,(x, y_1) = 0$ for all $x \in K$. Since K separates the space $\mathscr{B}'$, we get $y_1 = 0$.

D 4.3 We denote the proper cone $\mathscr{B}'_+ = \{y \mid y \in \mathscr{B}' \text{ and } \mu\,(x, y) \geqq 0 \text{ for all } x \in \mathscr{B}_+\}$ as dual to $\mathscr{B}_+$.

We immediately find $\mathscr{B}'_+ = \{y \mid y \in \mathscr{B}' \text{ and } \mu\,(w, y) \geqq 0 \text{ for all } w \in K\}$.

T 4.3 $\mathscr{B}'_+$ is $\sigma\,(\mathscr{B}', \mathscr{B})$-closed and makes

$$\mathscr{B}_+ = \{x \mid x \in \mathscr{B} \text{ with } \mu\,(x, y) \geqq 0 \text{ for all } y \in \mathscr{B}'_+\}\,.$$

Proof. Being polar to $\mathscr{B}_+$, the set $\mathscr{B}'_+$ is $\sigma\,(\mathscr{B}', \mathscr{B})$-closed. Thus, $\{x \mid x \in \mathscr{B}$ with $\mu\,(x, y) \geqq 0$ for all $y \in \mathscr{B}'_+\}$ is the set bipolar to $\mathscr{B}_+$ and hence equal to $\mathscr{B}_+$ (since $\mathscr{B}_+$ is convex and $\sigma\,(\mathscr{B}', \mathscr{B})$-closed). □

T 4.4 $\mathscr{B}'_{|1|} = [-\mathbf{1}, \mathbf{1}]$, where $[y_1, y_2]$ denotes the order interval $\{y \mid y_1 \leqq y \leqq y_2\}$.

Proof. By the definition of the norm of y, the set $\mathscr{B}'_{|1|}$ consists of all y with

$$\sup_{w \in K} |\,\mu\,(w, y)\,| \leqq 1$$

i.e. $-1 \leqq \mu\,(w, y) \leqq 1$. Because of $\mu\,(w, \mathbf{1}) = 1$, this is equivalent to $\mu\,(w, \mathbf{1} - y) \geqq 0$ and $\mu\,(w, y - \mathbf{1}) \geqq 0$, i.e. $-\mathbf{1} \leqq y \leqq \mathbf{1}$. □

T 4.5 The inclusion $L \subset [\mathbf{0}, \mathbf{1}]$ holds.

Proof. This follows immediately from $0 \leqq \mu\,(w, g) \leqq 1$ for $g \in L$. □

D 4.4 An ordered Banach space is called an order-unit space if the unit ball equals the order interval $[-\mathbf{1}, \mathbf{1}]$.

Thus $\mathscr{B}$ is a base normed space with the basis K, and $\mathscr{B}'$ is an order-unit space.

The following theorem states the most important property of the base normed space $\mathscr{B}$ and of its dual order-unit space.

T 4.6 The equalities $\mathscr{B} = \mathscr{B}_+ - \mathscr{B}_+$ and $\mathscr{B}' = \mathscr{B}'_+ - \mathscr{B}'_+$ hold.

Proof. The last relation can be shown very simply: For $y \in \mathscr{B}'$ we have $\| y \|^{-1} y \in \mathscr{B}'_{[1]}$, i.e. $-\mathbf{1} \leqq \| y \|^{-1} y \leqq \mathbf{1}$ and hence $\| y \|^{-1} y + \mathbf{1} \geqq 0$, i.e. $\| y \|^{-1} y + \mathbf{1} = y_1$ with $y_1 \in B'_+$. From this follows $y = \| y \| y_1 - \| y \| \mathbf{1}$, where $\| y \| y_1 \in B'_+$ and $\| y \| \mathbf{1} \in \mathscr{B}'_+$.

The relation $\mathscr{B} = \mathscr{B}_+ - \mathscr{B}_+$ is more profound since K need not be compact. From [7] V § 3.1 (e) follows first that the cone $\mathscr{B}_+$ is normal because $y = \sup_{w \in K} | \mu(w, y) |$ for $y_1, y_2 \in \mathscr{B}'_+$ implies $\| y_1 \| \leqq \| y_1 + y_2 \|$. Since $\mathscr{B}_+$ is normal and closed (by T 4.1), by [7] V § 3.5 follows $\mathscr{B} = \mathscr{B}_+ - \mathscr{B}_+$. □

According to [7] V § 5.5, for $\mathscr{B}'_+$ we find

T 4.7 The set of *all* positive linear forms on $\mathscr{B}_+$ equals $\mathscr{B}'_+$, so that

$$\begin{aligned} \mathscr{B}'_+ &= \{ y \mid y \in \mathscr{B}^* \text{ and } \mu(x, y) \geqq 0 \text{ for all } x \in \mathscr{B}_+ \} \\ &= \{ y \mid y \in \mathscr{B}^* \text{ and } \mu(w, y) \geqq 0 \text{ for all } w \in K \}; \end{aligned}$$

hence $\mathscr{B}'_+$ is also the set of all positive affine functionals on K.

This T 4.7 implies a relation stronger than T 4.3: $\mathscr{B}'_+$ is $\sigma(\mathscr{B}', \mathscr{B})$-complete. According to T 4.7, $\mathscr{B}'_+$ in fact is the cone polar to $\mathscr{B}_+$ in $\mathscr{B}^*$ and hence $\sigma(\mathscr{B}^*, \mathscr{B})$-closed in $\mathscr{B}^*$. Since $\mathscr{B}^*$ is $\sigma(\mathscr{B}^*, \mathscr{B})$-complete, $\mathscr{B}'_+$ is so.

T 4.8 The norm in $\mathscr{B}$ makes $\| x \|$ the gauge functional $p(x)$ of $\operatorname{co}(K \cup -K)$, i.e.

$$\begin{aligned} \| x \| &= \inf \{ \lambda \mid \lambda \geqq 0, x \in \lambda \operatorname{co}(K \cup -K) \} \\ &= \inf \{ \alpha + \beta \mid a, \beta \geqq 0, x = \alpha w_1 - \beta w_2 \text{ and } w_1, w_2 \in K \}. \end{aligned}$$

Proof. By T 4.6 we have $\mathscr{B} = \mathscr{B}_+ - \mathscr{B}_+$ and hence $\operatorname{co}(K \cup -K)$ is radial; therefore, $p(x)$ is a seminorm. Due to $\| \alpha w_1 - \beta w_2 \| \leqq \alpha \| w_1 \| + \beta \| w_2 \| = \alpha + \beta$, from $p(x) = 0$ follows $\| x \| = 0$, i.e. $x = 0$; therefore, $p(x)$ is a norm.

Then the sets $U_n = \{ x \mid \| x \| \leqq 1/2^n \}$ are closed in the $\| \cdot \|$-topology and obey $U_{n+1} + U_{n+1} \subset U_n$.

According to [7] V § 3.4, the sets

$$V_n = U_n \cap \mathscr{B}_+ - U_n \cap \mathscr{B}_+$$

then form a neighborhood basis of a topology $\mathscr{T}$ in which $\mathscr{B} = \mathscr{B}_+ - \mathscr{B}_+$ is complete. With $\mathscr{B}_+ = \bigcup_{\lambda \geqq 0} (\lambda K)$ and $\| \lambda w \| = \lambda$ for $w \in K$, we obtain

$$V_n = \bigcup_{\lambda_1 \leqq \frac{1}{2^n}} \lambda_1 K - \bigcup_{\lambda_2 \leqq \frac{1}{2^n}} \lambda_2 K .$$

Therefore, the inclusions

$$\left\{ x \mid p(x) \leqq \frac{1}{2^n} \right\} \subset V_n \subset \left\{ x \mid p(x) \leqq \frac{1}{2^{n-1}} \right\}$$

hold and hence $\mathscr{T}$ is the topology of the norm $p(x)$ in $\mathscr{B}$. Therefore, $\mathscr{B}$ is also a Banach space relative to $p(x)$.

From T 4.2 follows $p(x) \geqq \| x \|$. Then, according to [7] III § 2.1, the topologies of $p(x)$ and $\| x \|$ are the same and hence the set $\mathscr{B}'$ of the continuous linear functionals is the same for $p(x)$ and $\| x \|$. Since the norm in $\mathscr{B}$ also makes

$$\| x \| = \sup_{y \in \mathscr{B}'} \frac{|\mu(x, y)|}{\| y \|},$$

the norms $p(x)$ and $\| x \|$ are equal if the norms

$$\| y \| = \sup_{w \in K} |\mu(w, y)| = \sup \{ |\mu(x, y)| \, | \, \| x \| \leqq 1 \}$$

and

$$\| y \|_1 = \sup \{ |\mu(x, y)| \, | \, p(x) \leqq 1 \}$$

coincide. Because of $p(x) \geqq \| x \|$ we have $\| y \|_1 \leqq \| y \|$. On the other hand, because $p(w) \leqq 1$ for $w \in K$ we have

$$\| y \|_1 \geqq \sup_{w \in K} |\mu(w, y)| = \| y \| . \quad \square$$

From T 4.8 immediately follows

T 4.9 For each $x \in \mathscr{B}$ and $\varepsilon > 0$, there is a representation $x = \alpha w_1 - \beta w_2$ with $\alpha, \beta \geqq 0$ and $w_1, w_2 \in K$, such that $\alpha + \beta \geqq \| x \| > \alpha + \beta - \varepsilon$.

For $\| x \| < 1$ we therefore have $x \in \operatorname{co}(K \cup -K)$, i.e. $\mathscr{B}_{|1|}$ can differ from $\operatorname{co}(K \cup -K)$ only by elements x with $\| x \| = 1$.

We say that $\mathscr{B}$ possesses the minimal decomposition property if each x has a decomposition $x = \alpha w_1 - \beta w_2$ with $\| x \| = \alpha + \beta$. We easily see that the minimal decomposition property is equivalent to $\operatorname{co}(K \cup -K)$ being norm closed, since $p(x) = \| x \|$. There are base normed spaces that do not posses the minimal decomposition property ([11] page 30).

But T 4.9 can be sharpened to

T 4.10 For each $x \in B$ and $\varepsilon > 0$ there is a decomposition $x = \alpha w_1 - \beta w_2$ with $\alpha, \beta \geqq 0$ and $w_1, w_2 \in K$ such that there is a $y_0 \in [-\mathbf{1}, \mathbf{1}]$ which makes $\mu(w_2, y_0) = -1$ and $\mu(w_1, y_0) \geqq 1 - \varepsilon$, and also a $y_0' \in [-\mathbf{1}, \mathbf{1}]$ with $\mu(w_1, y_0') = 1$ and $\mu(w_2, y_0') \leqq -1 + \varepsilon$. If $x = \alpha w_1 - \beta w_2$ is a minimal decomposition so that $\| x \| = \alpha + \beta$, then there exists a $y_0 \in [-\mathbf{1}, \mathbf{1}]$ with $\mu(w_1, y_0) = 1$, $\mu(w_2, y_0) = -1$.

Proof. If $x = \alpha w_1 - \beta w_2$ is a minimal decomposition, we have

$$\alpha + \beta = \sup \{ \mu(x, y) \mid y \in [-\mathbf{1}, \mathbf{1}] \} .$$

Since $[-\mathbf{1}, \mathbf{1}]$ is $\sigma(\mathscr{B}', \mathscr{B})$-compact, $\mu(x, y)$ reaches its supremum on $[-\mathbf{1}, \mathbf{1}]$, i.e. there is a y_0 with $\mu(x, y_0) = \alpha + \beta = \alpha \mu(w_1, y_0) - \beta \mu(w_2, y_0)$. Since $|\mu(w_1, y_0)| \leqq 1$ and $|\mu(w_2, y_0)| \leqq 1$, we must therefore have $\mu(w_1, y_0) = 1$ and $\mu(w_2, y_0) = -1$.

If $\mathscr{B}$ lacks the minimal decomposition property, then (by A IV, T 6 and T 7) for each $x \in \mathscr{B}$ and $\varepsilon > 0$ there is an element $y_e \in \mathscr{B}'$ with $y_e \in [\mathbf{0}, \mathbf{1}]$ and $\| \mathbf{1} - y_e \| < \varepsilon/2$, so that x has a minimal decomposition relative to the basis $K_e = \mathscr{B}_+^\cdot \cap \{x \mid \mu(x, y_e) = 1\}$. Hence, there is a decomposition $x = \tilde{\alpha} x_1 - \tilde{\beta} x_2$ with $x_1, x_2 \in K_e$ and

$$\sup\{\mu(x, y) \mid y \in [-y_e, y_e]\} = \tilde{\alpha} + \tilde{\beta}.$$

Likewise follows that there is a $\tilde{y}_0$ with $\tilde{y}_0 \in [-y_e, y_e]$ and $\mu(x_1, \tilde{y}_0) = 1$, $\mu(x_2, \tilde{y}_0) = -1$. Because of $x_1, x_2 \in \mathscr{B}_+$, we obtain $x_1 = \lambda_1 w_1$ and $x_2 = \lambda_2 w_2$ with $w_1, w_2 \in K$ and $\lambda_1 = \| x_1 \| = \mu(x_1, \mathbf{1})$, $\lambda_2 = \| x_2 \| = \mu(x_2, \mathbf{1})$. Therefore, $x = \alpha w_1 - \beta w_2$ holds with $\alpha = \tilde{\alpha} \lambda_1$, $\beta = \tilde{\beta} \lambda_2$.

From $\mu(x_1, y_e) = 1$ and $\| \mathbf{1} - y_e \| < \varepsilon/2$ we conclude

$$\big| \| x_1 \| - 1 \big| = | \mu(x_1, \mathbf{1} - y_e) | < \| x_1 \| \frac{\varepsilon}{2}$$

hence

$$\frac{1}{1 + \frac{\varepsilon}{2}} < \| x_1 \| < \frac{1}{1 - \frac{\varepsilon}{2}}.$$

The same follows for $\| x_2 \|$, such that

$$\frac{1}{1 + \frac{\varepsilon}{2}} < \lambda_1, \quad \lambda_2 < \frac{1}{1 - \frac{\varepsilon}{2}} \tag{4.1}$$

holds. Because of $\tilde{y}_0 \in [-y_e, y_e]$ and $y_e \in [\mathbf{0}, \mathbf{1}]$, we have $y_0 = \tilde{y}_0 + y_e - \mathbf{1} \in [-\mathbf{1}, \mathbf{1}]$ and $y_0' = -y_e + \tilde{y}_0 + \mathbf{1} \in [-\mathbf{1}, \mathbf{1}]$.

From $\mu(x_1, \tilde{y}_0) = 1$, $\mu(x_2, \tilde{y}_0) = -1$, with $\mu(x_1, y_e) = \mu(x_2, y_e) = 1$ we get

$$\mu(x_1, y_0) = 2 - \lambda_1, \qquad \mu(x_2, y_0) = -\lambda_2,$$
$$\mu(x_1, y_0') = \lambda_1, \qquad \mu(x_2, y_0') = -2 + \lambda_2$$

and hence

$$\mu(w_1, y_0) = \frac{2 - \lambda_1}{\lambda_1}, \qquad \mu(w_2, y_0) = -1,$$
$$\mu(w_1, y_0') = 1, \qquad \mu(w_2, y_0') = \frac{-2 + \lambda_2}{\lambda_2}.$$

This together with (4.1) implies

$$\mu(w_1, y_0) > 1 - \varepsilon, \ \mu(w_2, y_0') < -1 + \varepsilon. \quad \square$$

The situation for $\mathscr{D}$ and $\mathscr{D}'$ is simpler. From T 3.12 immediately follows that $\mathscr{D}'$, with the cone $\mathscr{D}_+' = \bigcup_{\lambda \geqq 0} \lambda \bar{K}^\sigma$, is a base-normed space (basis $\bar{K}^\sigma$) and that $\mathscr{D}'$ possesses the minimal decomposition property.

The cone $\mathscr{D}_+$ dual to $\mathscr{D}_+'$ in $\mathscr{D}$ is given by

$$\mathscr{D}_+ = \{y \mid y \in \mathscr{D} \text{ with } \mu(x, y) \geqq 0 \text{ for } x \in \mathscr{D}_+'\}.$$

From $\mu(w, y) \geqq 0$ for $w \in K$ and $y \in \mathscr{D}$ also follows $\mu(w, y) \geqq 0$ for $w \in \bar{K}^\sigma$. Therefore we get

$$\mathscr{D}_+ = \mathscr{B}_+' \cap \mathscr{D}.$$

This $\mathscr{D}_+$ is norm-closed in $\mathscr{D}$ and $\sigma(\mathscr{D}, \mathscr{D}')$-closed.

Since $\mathscr{D}'_+$ is $\sigma(\mathscr{D}', \mathscr{D})$-closed, we also have

$$\mathscr{D}'_+ = \{x \mid x \in \mathscr{D}' \text{ and } \mu(x, y) \geqq 0 \text{ for } y \in \mathscr{D}_+\}.$$

Since the norm of $\mathscr{D}$ coincides with that of $\mathscr{B}'$ and $\mathbf{1} \in \mathscr{S} \subset \mathscr{D}$ holds, we have $\mathscr{D}_{|1|} = \mathscr{B}'_{|1|} \cap \mathscr{D}$ and hence $\mathscr{D}_{|1|} = [-\mathbf{1}, \mathbf{1}]_{\mathscr{D}}$ where $[-\mathbf{1}, \mathbf{1}]_{\mathscr{D}}$ is the order interval in $\mathscr{D}$. Therefore, $\mathscr{D}$ is an order unit Banach space.

Whenever $L = [\mathbf{0}, \mathbf{1}]$ (as further axioms will imply), from $\mathscr{S} \subset \mathscr{D}$ and $\mathscr{S} \subset \mathscr{B}'_{|1|}$ follows that $\mathscr{D}_{|1|}$ is $\sigma(\mathscr{B}', \mathscr{B})$-dense in $\mathscr{B}'_{|1|}$. By T 3.16 then follows $\bar{\mathscr{S}} = [\mathbf{0}, \mathbf{1}]_{\mathscr{D}}$ and (by T 3.19) the relation $K = \bar{\mathscr{K}}$.

So far no satisfying way is known to physically transparent axioms which characterize $\mathscr{D}$ more precisely as a subspace of $\mathscr{B}'$. Hence only the following properties remain for the choice of $\mathscr{D}$:

$\mathscr{D}$ is a separable Banach subspace of $\mathscr{B}'$; $\mathbf{1} \in \mathscr{D}$; $\mathscr{D}_{|1|} = \mathscr{D} \cap \mathscr{B}'_{|1|}$ is $\sigma(\mathscr{B}', \mathscr{B})$-dense in $\mathscr{B}'_{|1|}$. Then $\mathscr{D}$ is also $\sigma(\mathscr{B}', \mathscr{B})$-dense in $\mathscr{B}'$, and $\mathscr{D} \cap [\mathbf{0}, \mathbf{1}] = [\mathbf{0}, \mathbf{1}]_{\mathscr{D}}$ is $\sigma(\mathscr{B}', \mathscr{B})$-dense in $[\mathbf{0}, \mathbf{1}]$, and $\mathscr{D} \cap \mathscr{B}'_+$ is $\sigma(\mathscr{B}', \mathscr{B})$-dense in $\mathscr{B}'_+$ so that T 3.19 gives $K = \bar{\mathscr{K}}$.

Some further considerations regarding the physical problem of the space $\mathscr{D}$ can be read in [2] and [45]. In this book, we need not further occupy ourselves with the space $\mathscr{D}$.

§ 5 The Faces of K and L

Here we shall introduce yet several concepts important for later considerations.

If C is a closed convex set in a topological vector space E, we define a closed face of C as a subset $F \subset C$ which is closed and obeys:

(i) F is convex.

(ii) From $x \in F$ and $x = \lambda x_1 + (1 - \lambda) x_2$, with $0 < \lambda < 1$ and $x_1, x_2 \in C$, follows $x_1, x_2 \in F$.

We see immediately that the closed faces, ordered by set-theoretic inclusion, form a complete lattice (with C as unit element and $\emptyset$ as zero element), in which the lattice-theoretic intersection equals the set-theoretic one.

If a face consists of only one point, this point is called an extreme point of C. The set of extreme points of C is often denoted by $\partial_e C$.

Let A be the set of affine, continuous functionals on C. Let A_+ denote the subset of positive functionals. For $y \in A_+$ and $x \in C$, one easily sees that $\{x \mid y(x) = 0\}$ is a closed face of C. A face F of C for which there is a $y \in A_+$ with $F = \{x \mid y(x) = 0\}$, is called an *exposed* face.

If an exposed face consists of only one point, this point is called an *exposed* point of C. Therefore, the set of exposed points is a subset of $\partial_e C$. But not every extreme point need be an exposed point!

Because of T 4.6 and T 4.7, for the basis K of B the set of positive affine functionals on K can be identified with B'_+. An exposed face of K is therefore given by a set

$$K_0(y) = \{w \mid w \in K \text{ and } \mu(w, y) = 0 \text{ for a } y \in B'_+\}. \tag{5.1}$$

For the convex sets C: $[-\mathbf{1}, \mathbf{1}]$, $[\mathbf{0}, \mathbf{1}]$, L and $2L-\mathbf{1}$, one obtains the set of $\sigma(\mathscr{B}', \mathscr{B})$-continuous affine functionals from the elements of $\mathscr{B}$ by adding arbitrary constants. One thus obtains all the exposed faces of these convex sets C as the sets $\{y \mid y \in C \text{ and } \mu(x, y) = \inf_{y' \in C} \mu(x, y')\}$. Since the given sets C are $\sigma(\mathscr{B}', \mathscr{B})$-compact, for each $x \in \mathscr{B}$ the functional $\mu(x, y)$ attains its infimum on C, i.e. each $x \in \mathscr{B}$ in this way determines an exposed face of C. Since C is $\sigma(\mathscr{B}', \mathscr{B})$-closed, it equals the intersection of all half-spaces of the form

$$\{y \mid \mu(x, y) \geqq \inf_{y' \in C} \mu(x, y')\},$$

which are generated by the hyperplanes

$$\{y \mid \mu(x, y) = \inf_{y' \in C} \mu(x, y')\}.$$

Since C is compact, its intersection with *each* of these hyperplanes is a (nonempty) exposed face of C.

According to the Hahn-Banach-theorem ([7] III § 9.1), the same holds for a convex set that has an interior point:

If C is a convex set in a real topological vector space (e.g. in $\mathscr{B}$) and C has an interior point (e.g. C is the unit ball $\mathscr{B}_{|1|}$ of $\mathscr{B}$), for each x on the boundary of C there is a continuous linear functional l such that $l(x) = \inf_{x' \in C} l(x')$.

Then the set $\{x \mid l(x) = \inf_{x' \in C} l(x')\}$ is an exposed face of C. If C is closed, it equals the intersection of all half-spaces of the form $\{x \mid l(x) \geqq \inf_{x' \in C} l(x')\}$ for those l which attain their infimum on C. The same holds for K, even though it need not be compact or have an interior point:

T 5.1 With $J = \{y \mid y \in [\mathbf{0},\mathbf{1}] \text{ and } K_0(y) \neq \emptyset\}$, we have $K = \{x \mid \mu(x, \mathbf{1}) = 1$ and $\mu(x, y) \geqq 0$ for all $y \in J\}$.

Proof. Let $x \notin K$ and $\mu(x, \mathbf{1}) = 1$. By a decomposition $x = \alpha w_1 - \beta w_2$, with $\alpha, \beta \geqq 0$ and $w_1, w_2 \in K$ from $\mu(x, \mathbf{1}) = 1$ follows $\alpha - \beta = 1$, i.e. $x = \alpha w_1 + (1-\alpha) w_2$, where α must be greater than 1 in order that $x \notin K$ holds.

By T 4.10, the decomposition can be chosen so that, for given $\varepsilon > 0$ there is a $y_0' \in [-\mathbf{1}, \mathbf{1}]$ with $\mu(w_1, y_0') = 1$, $\mu(w_2, y_0') < -1 + \varepsilon$. For $y = \frac{1}{2}(\mathbf{1} - y_0)$ we then have $y \in [\mathbf{0}, \mathbf{1}]$ and $\mu(w_1, y) = 0$, $\mu(w_2, y) \geqq 1 - \frac{\varepsilon}{2}$; that is, $y \in J$ and $\mu(x, y) = (1-\alpha)\,\mu(w_2, y_0) < 0$. □

We could also have proved T 5.1 directly by means of the Bishop-Phelps theorem (A IV, T 5):

If K is bounded closed convex set of the Banach space $\mathscr{B}$, then the set of y for which $\mu(x, y)$ attains the value $\inf_{x' \in C} \mu(x', y)$ on K, is norm-dense in $\mathscr{B}'$.

According to this theorem, the set of y with $y \in [\mathbf{0}, \mathbf{1}]$ for which (x, y) attains its $\inf_{w \in K} \mu(w, y)$ on K, is norm-dense in $[\mathbf{0}, \mathbf{1}]$. For $y \in [\mathbf{0}, \mathbf{1}]$ we have $y' = y - \inf_{w \in K} (\mu(w, y))\, \mathbf{1} \in [\mathbf{0}, \mathbf{1}]$ and $\inf_{w \in K} \mu(w, y') = 0$. The set J in T 5.1 is therefore norm-dense in the set

$$\{y \mid y \in [\mathbf{0}, \mathbf{1}] \text{ and } \inf_{w \in K} \mu(w, y) = 0\}$$

which again provides T 5.1.

Somewhat more generally than in (5.1), for a subset $N \subset \mathscr{B}'_+$ we define

$$K_0(N) = \{w \mid w \in K \text{ and } \mu(w, y) = 0 \text{ for } y \in N \subset \mathscr{B}'_+\}. \tag{5.2}$$

T 5.2 $K_0(N)$ is an exposed face of K.

Proof. Since $\mathscr{B}'$ is separable in the $\sigma(\mathscr{B}', \mathscr{B})$-topology, there is a countable subset $\{y_\nu\}$ of N which is $\sigma(\mathscr{B}', \mathscr{B})$-dense in N. From $\mu(w, y_\nu) = 0$ for all y_ν then follows $\mu(w, y) = 0$ for all $y \in N$. Since $\mathscr{B}'_+$ is convex, $y_0 = \sum_\nu \lambda_\nu y_\nu$ with $\lambda_\nu > 0$ and $\sum_\nu \lambda_\nu = 1$ is an element of $\mathscr{B}'_+$. From $\mu(w, y_0) = 0$ follows $\mu(w, y_\nu) = 0$ for all y_ν. Therefore, $K_0(N) = K_0(y_0)$. □

D 5.1 An exposed face F_1 of K is called *orthogonal* to another exposed face F_2 if there is a $y \in [\mathbf{0}, \mathbf{1}]$ such that $F_1 \subset K_0(y)$ and $F_2 \subset K_0(\mathbf{1} - y)$. For this we briefly write $F_1 \perp F_2$.

We see immediately that the relation $F_1 \perp F_2$ is symmetric, since $\mathbf{1} - (\mathbf{1} - y) = y$.

The following sets of faces of K will carry special interest to us. Since norm-closed convex sets in $\mathscr{B}$ are always $\sigma(\mathscr{B}, \mathscr{B}')$-closed, we simply call a convex set in $\mathscr{B}$ closed.

We shall use the notations

$$\mathscr{W} = \{F \mid F \text{ is a closed face of } K\}, \tag{5.3}$$

$$\mathscr{W}_{\mathscr{B}'} = \{K_0(N) \mid \text{with } N \subset \mathscr{B}'_+\}, \tag{5.4}$$

$$\mathscr{W}_L = \{K_0(N) \mid \text{with } N \subset L\}, \tag{5.5}$$

with $K_0(N)$ defined in (5.2).

Because $L \subset \mathscr{B}'_+$, we immediately see

$$\mathscr{W} \supset \mathscr{W}_{\mathscr{B}'} \supset \mathscr{W}_L. \tag{5.6}$$

Since $y \in \mathscr{B}'_+$ also makes $\|y\|^{-1} y \in \mathscr{B}'_+$, (5.4) becomes

$$\mathscr{W}_{\mathscr{B}'} = \{K_0(N) \mid \text{ with } N \subset [\mathbf{0}, \mathbf{1}]\}. \tag{5.7}$$

According to T 5.2, this $\mathscr{W}_{\mathscr{B}'}$ is also the set of all exposed faces of K! By (5.2), we can regard K_0 as the surjective mapping

$$\mathscr{P}(\mathscr{B}'_+) \xrightarrow{K_0} \mathscr{W}_{\mathscr{B}'}. \tag{5.8}$$

As is easy to see, here we have

$$K_0\left(\bigcup_\lambda N_\lambda\right) = \bigcap_\lambda K_0(N_\lambda). \tag{5.9}$$

Since the intersection of faces yields faces, for each set $k \subset K$ there is a smallest closed face that contains k. We call it the face generated by k and write it $C(k)$. If

k consists of only one element w, simply $C(w)$. Therefore, C is a surjective mapping

$$\mathscr{P}(K) \xrightarrow{C} \mathscr{W}, \tag{5.10}$$

where $\mathscr{W} \xrightarrow{C} \mathscr{W}$ is the identity mapping.

In an quite analogous way, we can define sets of faces for L and the order interval $[\mathbf{0}, \mathbf{1}]$: With

$$L_0(k) = \{y \mid y \in L \text{ and } \mu(w, y) = 0 \text{ for } w \in k \subset K\}, \tag{5.11}$$

$$\tilde{L}_0(k) = \{y \mid y \in [\mathbf{0}, \mathbf{1}] \text{ and } \mu(w, y) = 0 \text{ for } w \in k \subset K\}, \tag{5.12}$$

we introduce

$$\mathscr{U} = \{L_0(k) \mid k \subset K\}, \tag{5.13}$$

$$\mathscr{V} = \{\tilde{L}_0(k) \mid k \subset K\}. \tag{5.14}$$

One can regard L_0 and $\tilde{L}_0$ as the surjective mappings

$$\mathscr{P}(K) \xrightarrow{L_0} \mathscr{U}, \tag{5.15}$$

$$\mathscr{P}(K) \xrightarrow{\tilde{L}_0} \mathscr{V}, \tag{5.16}$$

where

$$L_0\Big(\bigcup_\lambda k_\lambda\Big) = \bigcap_\lambda L_0(k_\lambda)$$

and

$$\left.\tilde{L}_0\Big(\bigcup_\lambda k_\lambda\Big) = \bigcap_\lambda \tilde{L}_0(k_\lambda).\right\} \tag{5.17}$$

T 5.3 The sets $\mathscr{W}, \mathscr{W}_{\mathscr{B}'}, \mathscr{W}_L, \mathscr{U}, \mathscr{V}$ are complete set-lattices, where the set-theoretic and lattice-theoretic intersections coincide.

Proof. The proof for $\mathscr{W}$ has been indicated above. For $\mathscr{W}_{\mathscr{B}'}$ and in precisely the same way for $\mathscr{W}_L$, the proof follows from $K = K_0(\emptyset) \in \mathscr{W}_{\mathscr{B}'}$ and (5.9). The latter implies that the intersection of elements of a subset of $\mathscr{W}_{\mathscr{B}'}$ is again an element of $\mathscr{W}_{\mathscr{B}'}$. The proof for $\mathscr{U}, \mathscr{V}$ follows with (5.17). □

T 5.4 The mappings $\mathscr{V} \xrightarrow{K_0} \mathscr{W}_{\mathscr{B}'}$, $\mathscr{U} \xrightarrow{K_0} \mathscr{W}_L$, $\mathscr{W}_{\mathscr{B}'} \xrightarrow{\tilde{L}_0} \mathscr{V}$, $\mathscr{W}_L \xrightarrow{L_0} \mathscr{U}$ are dual isomorphisms of the lattices ("dual" as they reverse the ordering). Here L_0, K_0 resp. $\tilde{L}_0, K_0$ are mutually inverse mappings, e.g. $K_0 L_0$ is the identity mapping of $\mathscr{W}_L$ onto itself. The set of those $k \subset K$ for which the $L_0(k)$ are equal, has a largest element, namely $K_0 L_0(k)$; something analogous holds for the other mappings.

Proof. As an example, we perform the proof for the mappings $\mathscr{U} \xrightarrow{K_0} \mathscr{W}_L$ and $\mathscr{W}_L \xrightarrow{L_0} \mathscr{U}$. It is directly clear that these mappings reverse the ordering.

Just as easily follows $k \subset K_0 L_0(k)$ and $N \subset L_0 K_0(N)$. Since K_0 reverses the ordering, the last relation implies $K_0(N) \supset K_0 L_0 K_0(N)$. From $k \subset K_0 L_0(k)$ with $k = K_0(N)$ follows $K_0(N) \subset K_0 L_0 K_0(N)$ and hence $K_0(N) = K_0 L_0 K_0(N)$, hence $K_0 L_0$ is the identity mapping on $\mathscr{W}_L$. In the same way, one shows that $L_0 K_0$ is the identity mapping on $\mathscr{U}$; and so forth.

Because of $L_0(k) = L_0 K_0 L_0(k)$ and $k \subset K_0 L_0(k)$, we find $K_0 L_0(k)$ as the largest of all the subsets k for which the sets $L_0(k)$ are equal. □

T 5.5 The inclusions $C(k) \subset K_0 L_0(k)$ and $L_0(k) = L_0 C(k)$ hold.

Proof. From $k \subset K_0 L_0(k)$ follows that $K_0 L_0(k)$ is a closed face which comprises k so that $C(k) \subset K_0 L_0(k)$. Using L_0 we conclude $L_0 C(k) \supset L_0 K_0 L_0(k) = L_0(k)$. From $k \subset C(k)$ follows $L_0(k) \supset L_0 C(k)$; therefore, $L_0(k) = L_0 C(k)$. □

T 5.6 For each $C \in \mathscr{W}$, there is *one* element $w \in K$ with $C = C(w)$. For each element in $\mathscr{U}$ (resp. $\mathscr{V}$) there is *one* $w \in K$ such that this element can be written $L_0(k) = L_0(w)$ (resp. $\tilde{L}_0(k) = L_0(w)$).

Proof. Since $\mathscr{B}$ is norm-separable, in $C \in \mathscr{W}$ there is a countable set $\{w_\nu\}$ with $w_\nu \in C$, which is norm-dense in C. With numbers $\lambda_\nu > 0$ and $\sum_\nu \lambda_\nu = 1$, we have $\sum_\nu \lambda_\nu w_\nu \in C$. One can easily show that $\sum_{\nu=1}^{N} \lambda_\nu w_\nu$ converges in norm as $N \to \infty$ and that the

$$\left(\sum_{\nu=1}^{N} \lambda_\nu\right)^{-1} \left(\sum_{\nu=1}^{N} \lambda_\nu w_\nu\right) \in C \tag{5.18}$$

have the same limit in the norm as the $\sum_{\nu=1}^{N} \lambda_\nu w_\nu$ have. Since C is closed, we find $\sum_\nu \lambda_\nu w_\nu \in C$.

From

$$\sum_\nu \lambda_\nu w_\nu = \lambda_{\nu_0} w_{\nu_0} + (1 - \lambda_{\nu_0}) \sum_{\nu \neq \nu_0} \frac{\lambda_\nu}{1 - \lambda_\nu} w_\nu$$

and $\lambda_{\nu_0} \neq 0$ follows $w_{\nu_0} \in C$, since C is a face. Therefore, $C = C(w)$.

From $C(k) = C(w)$ and $L_0(k) = L_0 C(k)$, by T 5.3 follows $L_0(k) = L_0 C(w)$ and $L_0(w) = L_0 C(w)$, i.e. $L_0(k) = L_0(w)$. □

T 5.6 means that the elements of $\mathscr{U}$ and $\mathscr{V}$ are exposed faces of L resp. of $[\mathbf{0}, \mathbf{1}]$.

A $w \in K$ is called *effective* if $\mu(w, g) = 0$ for a $g \in L$ implies $g = 0$. If $\{w_\nu\}$ is a countable norm-dense subset in K, then $w_0 = \sum_\nu \lambda_\nu w_\nu$ (with $\lambda_\nu > 0$, $\sum_\nu \lambda_\nu = 1$) is an effective element of K. For, from $\mu(w_0, g) = 0$ follows $\mu(w_\nu, g) = 0$ for all w_ν (because $\lambda_\nu \neq 0$ and $\mu(w_\nu, g) \geqq 0$). From this follows $\mu(w, g) = 0$ for all $w \in K$ and hence $g = 0$.

The above w_0 makes $C(w_0) = K$ since $\omega_\nu \in C(w_0)$ holds, the w_ν are dense in K, and $C(w_0)$ is norm-closed. Conversely, if $C(w) = K$ holds for some $w \in K$ then w is effective:

From $\mu(w, g) = 0$ for a $g \in L$ follows $g \in L_0(w)$. By T 5.3 we find $L_0(w) = L_0 C(w) = L_0 K$ and hence $g = 0$, since $L_0 K$ contains only the elements $g = 0$.

§ 6 Some Convergence Theorems

The following two theorems will be of importance later.

T 6.1 An increasing bounded sequence (resp. a decreasing sequence) $x_\nu \in \mathscr{B}_+$ is a Cauchy sequence in the norm. Then $x_\nu \to x \in \mathscr{B}_+$ holds.

Proof. If x_ν is increasing, for $\nu > \mu$ follows $x_\nu - x_\mu \in \mathscr{B}_+$ and hence $x_\nu = x_\mu + \lambda w$ with $\lambda \geqq 0$, $w \in K$. With $x_\mu = \lambda_\mu w_\mu$ follows

$$x_\nu = (\lambda_\mu + \lambda)\left[\frac{\lambda_\mu}{\lambda_\mu + \lambda} w_\mu + \frac{\lambda}{\lambda_\mu + \lambda} w\right].$$

Since the brackets contain an element of K, we get $\|x_\nu\| = \lambda_\mu + \lambda = \|x_\mu\| + \|x_\nu - x_\mu\|$. Since the x_ν are bounded, the increasing sequence $\|x_\nu\|$ is convergent and thus (because $\|x_\nu - x_\mu\| = \|x_\nu\| - \|x_\mu\|$) the sequence of the x_ν is a Cauchy sequence. From T 4.1 follows $x \in \mathscr{B}'_+$.

If x_ν is decreasing, for $\nu > \mu$ follows $\|x_\nu - x_\mu\| = \|x_\mu\| - \|x_\nu\|$ and thus (because $\|x_\nu\| \geqq 0$) the x_ν's form a Cauchy sequence. □

For $\mathscr{D}$ a theorem corresponding to T 6.1 does not hold in the norm topology! One must certainly go over to $\mathscr{B}'$ and the $\sigma(\mathscr{B}', \mathscr{B})$-topology.

T 6.2 A norm-bounded upward directed (resp. downward directed) subset R of $\mathscr{B}'_+$ has a supremum (resp. infimum) $y \in \mathscr{B}'_+$, which is an accumulation point of R in the $\sigma(\mathscr{B}', \mathscr{B})$-topology and to which the section filter of R converges.

Proof. Because of T 4.3 the set $\mathscr{B}'_+ \cap (\alpha\, \mathscr{B}'_{\|1\|})$ with $\alpha > 0$ is a $\sigma(\mathscr{B}', \mathscr{B})$-compact set. Hence $\bar{R}^\sigma$ (the $\sigma(\mathscr{B}', \mathscr{B})$-closure of R) is compact.

We will show that $\bar{R}^\sigma$ has a greatest element if R is directed upward (resp. a smallest element if R is directed downward). On $\bar{R}^\sigma$ the $\sigma(\mathscr{B}', \mathscr{B})$-topology equals the $\sigma(\mathscr{B}', K)$-topology. Therefore (and because $y_1 \leqq y_2 \Leftrightarrow \mu(w, y_1) \leqq \mu(w, y_2)$ for all $w \in K$!) we find a $\sigma(\mathscr{B}', \mathscr{B})$-accumulation point l of R determined by $l(w) = \sup_{y \in R} \mu(w, y)$. Hence an element l of $\bar{R}^\sigma$ is given with $l \geqq y$ for all $y \in R$. Something analogous holds for a set directed downward. □

Of course, T 6.2 holds in particular for increasing bounded (resp. decreasing) sequences from $\mathscr{B}'_+$.

The two theorems T 6.1 and T 6.2 evidence the significant nonsymmetry between $\mathscr{B}$ and $\mathscr{D}$, based on the fact that $\mathscr{B}$ is a base-normed space whereas $\mathscr{D}$ is an order unit space. Whereas $\mathscr{B}$ suffices to mathematically include the limit elements of increasing (resp. decreasing) sequences, $\mathscr{D}$ does not: We obtain limit elements only when we proceed from $\mathscr{D}$ to $\mathscr{B}'$. This is one of the mathematical reasons why one intensively deals with the dual pair $\mathscr{B}$, $\mathscr{B}'$ and scarcely with $\mathscr{D}$ (and $\mathscr{D}'$).

We shall later see that "physically interpretable" axioms can be expressed especially "simply" as relations in $\mathscr{B}$, $\mathscr{B}'$ K, L.

T 6.3 Let $F (\neq K)$ be an exposed face of K. Then there is in $\tilde{L}_0(F)$ a maximal g with $F = K_0(g)$, which makes $\sup_{w \in K} \mu(w, g) = 1$.

Proof. Since F is exposed there is an $y \in \mathbf{[0, 1]}$ with $K_0(y) = F$. Zorn's Lemma and T 6.2 imply that there is a maximal $g \in \tilde{L}_0(F)$ with $g \geqq y$. Therefore $K_0(g) = F$. If $\sup \mu(w, g) = \alpha \neq 1$ then $\alpha^{-1} g \gneqq g$. □

D 6.1 A $y \in [\mathbf{0}, \mathbf{1}]$ may be called *extremal*, if $K_0(y) \vee K_0(\mathbf{1} - y) = K$. For an extremal y, a complement of $K_0(y)$ in the lattice $\mathscr{W}_{\mathscr{B}'}$ is $K_0(\mathbf{1} - y)$.

T 6.4 An extremal y is maximal in $L_0 K_0(y)$.

Proof. If y were not maximal, there would be a $z \in [\mathbf{0}, \mathbf{1}]$ with $y + z \in L_0 K_0(y)$. It follows $K_0(z) \subset K_0(y)$ and $K_0(z) \subset K_0(\mathbf{1} - y)$ and therefore $K_0(z) \subset K_0(y) \vee K_0(\mathbf{1} - y) = K$, so that $z = 0$. □

It has not been proven that every maximal y is also extremal. There can be different maximal $y_i (i = 1, 2)$ with $K_0(y_1) = K_0(y_2)$ and $K_0(\mathbf{1} - y_1) \neq K_0(\mathbf{1} - y_2)$ (this is impossible if AV 1.1 and AVid are presumed; IV § 3).

T 6.5 The $\sigma(\mathscr{B}', \mathscr{B})$-closed convex set generated by the maximal elements is $[\mathbf{0}, \mathbf{1}]$.

Proof. See the proof of VI T 3.1 and replace e_0 by a maximal element $\bar{g} \geqq g_0$. □

If $L = [\mathbf{0}, \mathbf{1}]$, then we can by analogy to [2] IV (8.2.15) introduce "generalized properties" in the following way. Using the sets

$$\mathscr{D}_F = \{a \mid a \in \mathscr{D}' \text{ and } \varphi(a) \in F\},$$

$$\mathscr{R}_F = \{b \mid b \in \mathscr{R} \text{ and there is a } b_0 \in \mathscr{R}_0 \text{ with } \psi(b_0, b) \leqq \mathbf{1} - y \text{ where } y \text{ is maximal and } K_0(y) \subset F\},$$

we define

$$p_F = \Big[\bigcup_{a \in \mathscr{D}_F} a\Big] \cup \Big[\bigcup_{b \in \mathscr{R}_F}\Big] \quad \text{where} \quad F \in W_{\mathscr{B}'}$$

and call it a generalized property. We shall not pursue this possibility. We wanted only to demonstrate that propositions of the form "x has the (generalized) property p_F" are not restricted to the case of an orthomodular lattice $\mathscr{W}_{\mathscr{B}'}$ treated in [2] IV § 8.3.

In conclusion, let us yet note an important property of L and $[\mathbf{0}, \mathbf{1}]$: Since L and $[\mathbf{0}, \mathbf{1}]$ are $\sigma(\mathscr{B}', \mathscr{B})$-compact, by the Krein-Milman theorem ([10] § 25.1) follows

T 6.6 $\overline{\text{co}}^\sigma\, \partial_e L = L$; $\overline{\text{co}}^\sigma\, \partial_e [\mathbf{0}, \mathbf{1}] = [\mathbf{0}, \mathbf{1}]$.

V Observables and Preparators

If we forget all about the originally introduced structures, up to the derived sets K, L with the structure $K \times L \xrightarrow{\mu} [0, 1]$, which we investigated in IV, then we have forgotten "too much" to describe physical systems as effect carriers. Let us try to amend this by introducing the concepts "observable" and "preparator". These concepts represent abstract and idealized residues from the structure of preparation and registration procedures. Without falling back on these physically interpreted procedures, we must use many descriptive words in order to give a physical sense to the concepts observable and preparator. That this is difficult is shown not only by the uncertainty if introducing the concept observable via the "correspondence principle" (see [1] XI § 1.7), but in particular by the fact that one has not yet used the concept of preparator and for this reason encountered great difficulties with the Einstein-Podolski-Rosen paradox.

In our treatment, the idealization of the abstract concepts observable and preparator and their physical interpretation will be especially clear from the axioms AOb of § 5 and Apr of § 8.

This chapter is really only a brief summary of the considerations from [2] IV. There one finds a detailed presentation, where more special sets K and L are chosen than the sets K, L from IV which still are little structured. In [2] IV, many proofs were knowingly so carried out that the special properties of the sets K, L used there did *not* enter. Therefore we can limit ourselves in the next sections to describe the most important structures and leave the proof to the reader (referring to [2] IV).

We called the elements of $\mathscr{K}$ ensembles, those of $\mathscr{L}$ effects. Since K and L are certain topological completions of $\mathscr{K}$ resp. $\mathscr{L}$ let us call all the elements of K ensembles and all the elements of L effects.

§ 1 Coexistent Effects and Observables

The very practical concept of observable often used in quantum mechanics turns out to be *derivable* from $\mathscr{R}_0, \mathscr{R}$ after idealizations.

§ 1.1 Coexistent Registrations

So far we have only considered the image set $\mathscr{L}$ of the mapping $\mathscr{F} \xrightarrow{\psi} \mathscr{L}$, i.e. in $\mathscr{L}$ the $f=(b_0, b)$ occur only as elements of a set $\mathscr{F}$. Nothing has entered from the fact that these elements f are pairs (b_0, b) of a registration method $b_0 \in \mathscr{R}_0$ and a registration procedure $b \in \mathscr{R}$ with $b \subset b_0$. The registration procedures $b \in \mathscr{R}(b_0)$ represent an important physical structure, namely as "indications" on one and the same "device" characterized by b_0. This structure is lost if one forgets $\mathscr{R}_0$, $\mathscr{R}$ completely, considering only the set $\mathscr{L} \subset L$ as basis set of a structure.

We shall now attempt to introduce in idealized form something from the structure of the sets $\mathscr{R}(b_0)$ for various $b_0 \in \mathscr{R}_0$ in the theory characterized by K, L and $K \times L \xrightarrow{\mu} [0, 1]$. The idealizations enable us to define a concept really generally, first disregarding concrete, experimental realizations.

In order to specify briefly the additional structure to be considered in $\mathscr{F}$, we introduce

D 1.1.1 The registration procedures $b \in \mathscr{R}(b_0)$ are called *coexistent* relative to the registration method b_0. Several $(b_0, b) \in \mathscr{F}$ relative to the same b_0 are called *coexistent* effects procedures.

Therefore, a subset $A \subset \mathscr{F}$ just then is a set of coexistent effect procedures, when all elements in A have the same first component b_0. How is this structure of coexistent effect procedures reflected *after* the mapping ψ of $\mathscr{F}$ into L?

§ 1.2 Coexistent Effects

For fixed b_0, the mapping $\mathscr{F} \xrightarrow{\psi} L$ defines a mapping $\mathscr{R}(b_0) \xrightarrow{\psi_0} L$ by $\psi_0(b) = \psi(b_0, b)$. We know that $\mathscr{R}(b_0)$ is a Boolean ring. What property has the mapping ψ_0 relative to the Boolean ring structure of $\mathscr{R}(b_0)$?

The following definition is customary:

D 1.2.1 A mapping F of a Boolean ring Σ into the order interval $[0, u]$ of an ordered vector space, which has $F(\varepsilon) = u$ (where ε the unit element of Σ) and

$$F(\sigma_1 \vee \sigma_2) = F(\sigma_1) + F(\sigma_2) \quad \text{for} \quad \sigma_1 \wedge \sigma_2 = 0,$$

is called an (additive) *measure* on Σ.

In particular, if $[0, u]$ is the interval $[0, 1]$ of real numbers, we say that F is a real (normed) measure. From III T 4.2 (vi) follows:

The mapping ψ_0 from $\mathscr{R}(b_0)$ into $[\mathbf{0}, \mathbf{1}] \subset \mathscr{B}'$ defined by $\psi(b_0, b)$ (for fixed b_0) is an additive measure on $\mathscr{R}(b_0)$. Of course, $\psi_0\, \mathscr{R}(b_0) \subset \mathscr{L} \subset L$ holds (see [2] IV Th. 1.2.1).

This suggests the following definition (generalized in idealized form):

D 1.2.2 A set $A \subset L$ is called set of coexistent effects if there is a Boolean ring Σ with an additive measure $\Sigma \xrightarrow{F} L$ such that $A \subset F\,\Sigma$.

The following definition will be useful.

D 1.2.3 An additive measure F over Σ is called *effective* if $F(\sigma) = \mathbf{0}$ implies $\sigma = 0$.

If $\Sigma \xrightarrow{F} L$ is an additive measure, an additive effective measure $\tilde{F}$ on the Boolean ring Σ/Σ_0 with $\Sigma_0 = \{\sigma \mid \sigma \in \Sigma \text{ and } F(\sigma) = 0\}$ is defined by

$$\tilde{F}(\tilde{\sigma}) = F(\sigma) \quad \text{for a } \sigma \in \tilde{\sigma}$$

(see [2] IV Th 1.2.2).

To investigate coexistent effects we therefore need only consider Boolean rings with effective measures. We see immediately that ψ maps a set of coexistent effect procedures into a set of coexistent effects. It also follows easily that the mapping $\mathscr{R}(b_0) \xrightarrow{\psi_0} L$ is an effective measure on $\mathscr{R}(b_0)$. For, from $\psi_0(b) = 0$, i.e. $\psi(b_0, b) = 0$, follows $\lambda_{\mathscr{S}}(a \cap b_0, a \cap b) = 0$ for all $a \in \mathscr{Q}'$ combinable with b_0. Therefore, $a \cap b = \emptyset$ holds for all $a \in \mathscr{Q}'$ combinable with b_0. According to APS 5.2, this implies $b = \emptyset$.

The definition D 1.2.2 has (in comparison with the situation $\mathscr{R}(b_0) \xrightarrow{\psi_0} L$) the essential advantage that initially one need not worry, whether for each set A of coexistent effects there also exists a registration method $b_0 \in \mathscr{R}_0$ with $A \subset \psi_0 \mathscr{R}(b_0)$. Indeed we shall in § 5 require that there is "approximately" a "realization" by an $\mathscr{R}(b_0)$ for each Boolean ring Σ with an effective, additive measure $\Sigma \xrightarrow{F} L$. But just from the mathematical complication of the "approximate realization" we are relieved by considering a general additive, effective measure over a Boolean ring.

Now we have the interesting theorem that for $g_1, g_2 \in L$ the following two conditions are equivalent:

(i) $\{g_1, g_2\}$ are coexistent;
(ii) There exist three elements $g_1', g_2', g_{12} \in L$ such that $g_1 = g_1' + g_{12}$, $g_2 = g_2' + g_{12}$ and $g_1' + g_2' + g_{12} = g_1 + g_2' = g_2 + g_2' \in L$.

(See [2] IV Th 1.2.4; here one must notice that also the proof in [2] IV Th 1.2.3 persists since $g \in L$ always implies $1 - g \in L$.)

§ 1.3 Observables

The concept of an observable is an idealized generalization of the structure $\mathscr{R}(b_0) \xrightarrow{\psi_0} L$, already considered in § 1.1 and § 1.2. But not the general structure $\Sigma \xrightarrow{F} L$, with Σ a Boolean ring and F an additive measure, shall be called an observable. Rather let us impose several additional requirements on the concept of an observable. Starting from $\Sigma \xrightarrow{F} L$, we therefore introduce a uniform structure in Σ.

T 1.3.1 The sets

$$N_{w,\varepsilon} = \{(\sigma_1, \sigma_2) \mid \sigma_1, \sigma_2 \in \Sigma,\ \mu(w, F(\sigma_1 \dot{+} \sigma_2)) < \varepsilon\}$$

with $w \in K$ and $\sigma_1 \dot{+} \sigma_2 = (\sigma_1 \vee \sigma_2) \wedge (\sigma_1 \wedge \sigma_2)^*$ form a fundamental system for a uniform structure $\mathscr{U}_g$ of the Boolean ring Σ with the additive measure $\Sigma \xrightarrow{F} L$. This $\mathscr{U}_g$ separates Σ if F is effective. $\mathscr{U}_g$ is metrizable; one can use

$$d(\sigma_1, \sigma_2) = \mu(w_0, F(\sigma_1 \dot{+} \sigma_2))$$

as a metric, where any effective $w \in K$ can be chosen for w_0 (w is called effective if $g \in L$ and $\mu(w, g) = 0$ imply $g = 0$); see [2] IV Th 1.7 and [2] IV Th 2.1.11.

Let Σ with the uniform structure $\mathscr{U}_g$ be briefly denoted by Σ_g.

T 1.3.2 The uniform completion $\hat{\Sigma}_g$ of Σ_g is a (lattice-theoretically) complete ring, and the additive measure $\Sigma_g \xrightarrow{F} L$ is extended to an additive measure $\hat{\Sigma}_g \xrightarrow{F} L$. If F is effective on Σ_g, then also on $\hat{\Sigma}_g$ (see [2] IV Th 1.4.2 and [2] IV Th 1.4.3).

The extension of F to $\hat{\Sigma}_g$ would not be possible if we had not completed $\mathscr{L}$ to L in the $\sigma(\mathscr{B}', \mathscr{B})$-topology. In the case of preparators, it will suffice to have K norm-complete (see § 6)!

The point of departure for our discussion was the structure $\Sigma \xrightarrow{F} L$ as an idealization of $\mathscr{R}(b_0) \xrightarrow{\psi_0} L$. In particular, if $\Sigma = \mathscr{R}(b_0)$ and $F = \psi_0$, then one can extend the measure ψ_0 (by the methods of measure theory, see e.g. [34]), to a larger set of "measurable" subsets of b_0 than the $b \in \mathscr{R}(b_0)$. Let the set of "measurable" subsets be $\bar{\mathscr{R}}(b_0)$ such that $\mathscr{R}(b_0) \subset \bar{\mathscr{R}}(b_0)$.

Then, using

$$\mathscr{J}(b_0) = \{b \mid b \in \bar{\mathscr{R}}(b_0) \text{ and } \psi_0(b) = 0\},$$

we find $\hat{\Sigma}_g$ isomorphic with $\bar{\mathscr{R}}(b_0)/\mathscr{J}(b_0)$. Moreover, $F = \psi_0$ extended to $\hat{\Sigma}_g$ is connected with ψ_0 extended to $\bar{\mathscr{R}}(b_0)$ by $F(\eta) = \psi_0(b)$ with $\eta \in \bar{\mathscr{R}}(b_0)/\mathscr{J}(b_0)$ and $b \in \eta$. Here let us not pursue this extension $\bar{\mathscr{R}}(b_0)$ further.

In [2] IV § 1.4 we discussed a uniform structure of "physical imprecision" on Σ, which we shall not repeat here. We only emphasize that $\mathscr{U}_g$ need not be the uniform structure of physical imprecision because in general $\hat{\Sigma}_g$ is not compact.

Since $\Sigma \xrightarrow{F} L$ represents an idealization and abstraction of the structures $\mathscr{R}(b_0) \xrightarrow{\psi_0} L$, and $\mathscr{R}(b_0)$ is countable, it is natural to introduce the concept of an observable by

D 1.3.1 A Boolean ring Σ with an additive, effective measure $\Sigma \xrightarrow{F} L$ such that Σ is complete and *separable*, relative to the uniform structure $\mathscr{U}_g$ (determined by F) shall be called an *observable*.

The following theorem holds:

T 1.3.3 $\Sigma \xrightarrow{F} L$ is an observable if and only if Σ is lattice-theoretically complete, F is σ-additive (relative to the $\sigma(\mathscr{B}', \mathscr{B})$-topology in L) and there is a countable Boolean sublattice Σ_a of Σ whose lattice-theoretic completion in Σ equals Σ (see [2] IV Th 1.4.6).

§ 2 Mixture Morphisms

In this section, let us only define and investigate some mathematical structures needed later, which play a large role in many parts of quantum mechanics.

We start from two convex sets K_1 and K_2 which are bases of the base-normed Banach spaces $\mathscr{B}_1$ and $\mathscr{B}_2$.

D 2.1 An affine mapping S of K_1 into K_2 is called a *mixture morphism*.

T 2.1 A mixture morphism $K_1 \xrightarrow{S} K_2$ can be extended uniquely as a linear mapping of $\mathscr{B}_1$ into $\mathscr{B}_2$ with $\| S \| = 1$.

Proof. Since $\mathscr{B}_1$ is spanned linearly by K_1, we can extend S to $\mathscr{B}_1$. Since $w \in K_1$ implies $S w \in K_2$, all $w \in K_1$ make $\| S w \| = 1$ so that $\| S \| \geqq 1$. With $x = \alpha w_1 - \beta w_2$ and $\| x \| \geqq \alpha + \beta - \varepsilon$ ($\varepsilon > 0$ arbitrary), we find $\| S x \| \leqq \alpha + \beta \leqq \| x \| + \varepsilon$ and thus $\| S x \| \leqq \| x \|$, i.e. $\| S \| \leqq 1$. □

T 2.2 To each mixture morphism S corresponds a dual mapping $\mathscr{B}_2' \xrightarrow{S'} \mathscr{B}_1'$ with $[\mathbf{0}, \mathbf{1}]_2 \xrightarrow{S'} [\mathbf{0}, \mathbf{1}]_1$ and $S' \mathbf{1} = \mathbf{1}$. Here, S' is $\sigma(\mathscr{B}_2', \mathscr{B}_2) - \sigma(\mathscr{B}_1', \mathscr{B}_1)$-continuous and has $\| S' \| = 1$.

Proof. That S' exists and is continuous in the σ-topology, follows directly from the fact that (by T 2.1) S is norm-continuous. From $K_1 \xrightarrow{S} K_2$ follows $\mu_2(S w, \mathbf{1}) = 1 = \mu_1(w, S' \mathbf{1})$ for all $w \in K_1$ and hence $S' \mathbf{1} = \mathbf{1}$. □

It follows easily that a bijective mixture-morphism $K_1 \xrightarrow{S'} K_2$ is a mixture-isomorphism, for, $v = \lambda v_1 + (1 - \lambda) v_2$ with $v, v_1, v_2 \in K$ and $w = \lambda w_1 + (1 - \lambda) w_2$ with $w_1 = S^{-1} v_1$, $w_2 = S^{-1} v_2$, imply $S w = \lambda v_1 + (1 - \lambda) v_2 = v$ and hence

$$S^{-1} v = w = \lambda S^{-1} v_1 + (1 - \lambda) S^{-1} v_2 .$$

T 2.3 For a mixture-morphism S, the following conditions are equivalent:

(i) S is a mixture-isomorphism,
(ii) $K_1 \xrightarrow{S} K_2$ is injective and $S K_1$ is norm-dense in K_2,
(iii) $[\mathbf{0}, \mathbf{1}]_2 \xrightarrow{S'} [\mathbf{0}, \mathbf{1}]_1$ is bijective,
(iv) S' is an isomorphic mapping of Banach spaces,
(v) S is an isomorphic mapping of Banach spaces.

Proof.
(i) ⇒ (ii) is trivial.
(ii) ⇒ (iii): Since $S K_1$ is norm-dense in K_2,

$$\| S' y \| = \sup_{w \in K_1} | \mu_1(w, S' y) | = \sup_{w \in K_1} | \mu_2(S w, y) | = \sup_{\tilde{w} \in K_2} | \mu_2(\tilde{w}, y) | = \| y \|$$

holds, i.e. S' is norm-preserving and hence injective. From this follows that $S' \mathscr{B}_2'$ is a norm-closed subspace of $\mathscr{B}_1'$.

From $K_1 \xrightarrow{S} K_2$ injective follows that $\mathscr{B}_1 \xrightarrow{S} \mathscr{B}_2$ is also injective, since each $x \in \mathscr{B}_1$ can be written $x = \alpha w_1 - \beta w_2$ with $\alpha, \beta \geqq 0$ and $w_1, w_2 \in K_1$. Because $S w_1$, $S w_2 \in K_2$ then follows from $0 = S x = \alpha S w_1 - \beta S w_2$, $\alpha = \beta$; hence we must have $S w_1 = S w_2$. Since $K_1 \xrightarrow{S} K_2$ is injective, $w_1 = w_2$ follows and thus finally $x = 0$.

Since $\mathscr{B}_1 \xrightarrow{S} \mathscr{B}_2$ is injective, $S' \mathscr{B}_2'$ must be $\sigma(\mathscr{B}_1', \mathscr{B}_1)$-dense in $\mathscr{B}_1'$. Since the unit ball $[-\mathbf{1}, \mathbf{1}]_2$ is $\sigma(\mathscr{B}_2', \mathscr{B}_2)$-compact, the set $A = S'[-\mathbf{1}, \mathbf{1}]_2$ is $\sigma(\mathscr{B}_1', \mathscr{B}_1)$-compact. Since S' preserves the norm, we get $A = S' \mathscr{B}_2' \cap [-\mathbf{1}, \mathbf{1}]_1$. Thus $S' \mathscr{B}_2'$ is also $\sigma(\mathscr{B}_1', \mathscr{B}_1)$-closed ([7] IV § 6.1) and hence $S \mathscr{B}_2' = \mathscr{B}_1'$. Therefore, $S'[-\mathbf{1}, \mathbf{1}]_2 = [-\mathbf{1}, \mathbf{1}]_1$ holds and thus (iii) is proven.

(iii) ⇒ (iv) follows easily.

(iv) ⇒ (v): Being the inverse mapping of the σ-continuous bijective mapping of the σ-compact unit ball, $(S')^{-1}$ is also σ-continuous on the unit ball and hence σ-continuous throughout. From this follows that S^{-1} exists with $(S^{-1})' = (S')^{-1}$. Since S' maps the unit ball bijectively, we have $\| S x \| = \| x \|$, whence (v) follows.

(v) ⇒ (i): The existence of $(S')^{-1}$ and that $(S')^{-1} = (S^{-1})'$, follow straightaway. Since T 2.2 gives $S' \mathbf{1} = \mathbf{1}$, we also have $(S')^{-1} \mathbf{1} = \mathbf{1}$. From this follows $\mu_1 (S^{-1} w, \mathbf{1}) = \mu_2 (w, (S^{-1})' \mathbf{1}) = \mu_2 (w, \mathbf{1}) = 1$ for $w \in K_2$. Since S^{-1} preserves the norm, $w \in K_2$ also makes $\| S^{-1} w \| = 1$. With $S^{-1} w = \alpha w_1 - \beta w_2$ $(\alpha, \beta \geqq 0;\ w_1, w_2 \in K_1)$ and $\| S^{-1} w \| \supseteq \alpha + \beta - \varepsilon$ ($\varepsilon > 0$ arbitrary), we obtain $\alpha - \beta = 1$ and $1 \leqq \alpha + \beta \leqq \alpha + \beta - \varepsilon$, i.e. $1 \leqq \alpha \leqq 1 - \varepsilon/2$ and hence $\alpha = 1$ and $\beta = 0$, whence $S^{-1} K_2 \subset K_1$ follows. Since S must be a mixture-morphism, we also have $S K_1 \subset K_2$. Therefore $K_1 \xrightarrow{S} K_2$ is bijective, whence (i) follows. □

§ 3 Structures in the Class of Observables

In this section we shall describe structures of an observable $\Sigma \xrightarrow{F} L$ that were investigated in [2] IV. We shall use them later, especially in XI.

§ 3.1 The Spaces $\mathscr{B}(\Sigma)$ and $\mathscr{B}'(\Sigma)$ Assigned to a Boolean Ring Σ

We generally start from a Boolean ring Σ with an additive and effective measure $\Sigma \xrightarrow{m_0} [0, 1] \subset \mathbf{R}$ relative to which Σ_g (i.e. Σ with the uniform structure $\mathscr{U}_g$ defined by the metric $d(\sigma_1, \sigma_2) = m_0(\sigma_1 \dot{+} \sigma_2))$ is complete. According to [2] IV Th 1.4.4, a lattice-theoretically complete Boolean ring Σ, with a σ-additive, effective measure, is also $\mathscr{U}_g$-complete. The effective measure can be of the form $m_0(\sigma) = \mu(w_0, F(\sigma))$, with w_0 an effective ensemble from K.

D 3.1.1 A function $\Sigma \xrightarrow{x} \mathbf{R}$ is called a *signed* σ-additive measure over Σ, if there is a number c with $|x(\sigma)| < c$ for all $\sigma \in \Sigma$, and if $\sigma = \bigvee_\nu \sigma_\nu$ (with $\sigma_\nu \wedge \sigma_\mu = 0$ for $\nu \neq \mu$) implies

$$x(\sigma) = \sum_\nu x(\sigma_\nu)$$

for countably many σ_ν.

We can use all the considerations from [2] IV § 2.1. Let us only mention

D 3.1.2 Let the set of all signed σ-additive measures over Σ be denoted by $\mathscr{B}(\Sigma)$ and the set of all σ-additive (normed) measures by $K(\Sigma)$.

Then the following fundamental theorems are known to hold:

$\mathscr{B}(\Sigma)$ is a base-normed Banach space with the basis $K(\Sigma)$ (see Th 2.1.8 through Th 2.1.10 in [2] IV).

$\mathscr{B}(\Sigma)$ is separable if Σ is (see [2] IV Th 2.1.12).

Since $\mathscr{B}(\Sigma)$ is a base-normed Banach space, $\mathscr{B}'(\Sigma)$ is an order-unit space. The "unit" (briefly $\mathbf{1}$) is that linear functional on $\mathscr{B}(\Sigma)$ such that $\mathbf{1}(m) = 1$ for all $m \in K(\Sigma)$.

The unit ball of $\mathcal{B}'(\Sigma)$ is the order interval $[-\mathbf{1}, \mathbf{1}]$ and is $\sigma(\mathcal{B}', \mathcal{B})$-compact. Therefore, by the Krein-Milman theorem, $[-\mathbf{1}, \mathbf{1}] = \overline{\text{co}}^{\sigma}\, \partial_e [-\mathbf{1}, \mathbf{1}]$.

D 3.1.3 $L(\Sigma) \overset{\text{def}}{=} [\mathbf{0}, \mathbf{1}] \subset \mathcal{B}'(\Sigma)$.

Therefore, $L(\Sigma)$ is topologically isomorphic to the unit ball since $L(\Sigma) = \frac{1}{2}\{[-\mathbf{1}, \mathbf{1}] + \mathbf{1}\}$.

By $l_\sigma(m) = m(\sigma)$ for fixed $\sigma \in \Sigma$, a bounded, linear functional l_σ is defined with $l_\sigma \in L(\Sigma)$. The set $\partial_e L(\Sigma)$ equals $\{l_\sigma \mid \sigma \in \Sigma\}$ (see [2] IV Th 2.1.16).

Therefore, one can identify Σ with $\partial_e L(\Sigma)$ by $\sigma \leftrightarrow l_\sigma$.

§ 3.2 The Mixture Morphism Corresponding to an Observable

By an observable $\Sigma \xrightarrow{F} L$, σ-additive measures $m(\sigma) = \mu(w, F(\sigma))$ for all $w \in K$ are defined ([2] IV Th 1.4.3). Obviously,

$$w \to \mu(w, F(\sigma)) \tag{3.2.1}$$

defines an affine mapping S by

$$K \xrightarrow{S} K(\Sigma)\,.$$

According to D 2.1, S is a mixture morphism; by T 2.1 it can be extended as a norm-continuous mapping

$$\mathcal{B} \xrightarrow{S} \mathcal{B}(\Sigma) \quad \text{with} \quad \| S \| = 1\,.$$

T 3.2.1 The mixture-morphism S defined by (3.2.1) maps effective ensembles $w \in K$ into effective measures $m \in K(\Sigma)$. The adjoint mapping S' of $\mathcal{B}'(\Sigma)$ into $\mathcal{B}'$ maps $L(\Sigma)$ into $[\mathbf{0}, \mathbf{1}]$; its restriction to $\partial_e L(\Sigma)$ equals the measure $\Sigma \xrightarrow{F} L$, if one identifies Σ with $\partial_e L(\Sigma)$ according to § 3.1. This restriction of S' to $\partial_e L(\Sigma)$ uniquely determines S (see [2] IV Th 2.2.1).

T 3.2.2 If S is a mixture morphism $K \xrightarrow{S} K(\Sigma)$ which maps effective ensembles into effective measures, then the adjoint mapping S', restricted to $\partial_e L(\Sigma)$, defines a σ-additive, effective measure $\Sigma \xrightarrow{F} [\mathbf{0}, \mathbf{1}] \subset \mathcal{B}'$ by $F(\sigma) = S'\sigma$. Here $F(\varepsilon) = \mathbf{1}$ holds. With $\mathcal{U}_g$ as the uniform structure determined by F, Σ_g is complete. If $F\Sigma \subset L$ holds and Σ_g is separable, then $\Sigma \xrightarrow{F} L$ is an observable. (See [2] IV Th 2.2.2, where one must begin the proof by saying that S' due to T 2.2 maps the set $L(\Sigma)$ into $[\mathbf{0}, \mathbf{1}] \subset \mathcal{B}'$.)

In particular, if $L = [\mathbf{0}, \mathbf{1}]$ (true by subsequent axioms), then the observables $\Sigma \xrightarrow{F} L$ and mixture-morphisms S by (3.2.1) are in one-to-one correspondence if $\mathcal{B}(\Sigma)$ is separable.

§ 3.3 The Kernel of an Observable

If $\Sigma \xrightarrow{F} L$ is an observable, its elements $\sigma \in \Sigma$ in abstract form symbolize the various $b \in \mathcal{R}(b_0)$. If $\sigma, \sigma_1, \sigma_2$ thus are three "indications" in Σ, which in particular obey

$$F(\sigma) = \alpha F(\sigma_1) + (1 - \alpha) F(\sigma_2), \tag{3.3.1}$$

then it is experimentally not necessary to read the indication σ; reading σ_1 and σ_2 suffices. Here, σ can be interpreted as a statistical mixture of the two σ_1 and σ_2 with the weights α resp. $1-\alpha$. Thus the question arises whether or not a Boolean subring Σ_a of Σ suffices so that the $\sigma \in \Sigma \backslash \Sigma_a$ can be regarded as statistical mixtures of the $\sigma \in \Sigma_a$.

We begin our discussion with a definition:

D 3.3.1 Let $\Sigma \xrightarrow{F} L$ be an observable. Then $\overline{\mathrm{co}}^\sigma(F\Sigma)$ is called the *convex image set* of the observable F.

Since L is $\sigma(\mathscr{B}', \mathscr{B})$-compact, its closed subset $\overline{\mathrm{co}}^\sigma(F\Sigma)$ is also $\sigma(\mathscr{B}', \mathscr{B})$-compact; hence the Krein-Milman theorem gives $\overline{\mathrm{co}}^\sigma(F\Sigma) = \overline{\mathrm{co}}^\sigma \partial_e(\overline{\mathrm{co}}^\sigma(F\Sigma))$. Thus $\partial_e \overline{\mathrm{co}}^\sigma(F\Sigma)$ is the really essential set of effects in the measurement of the observable $\Sigma \xrightarrow{F} L$.

Therefore let us adopt

D 3.3.2 $\partial_e \overline{\mathrm{co}}^\sigma(F\Sigma)$ is called the *extremal kernel* of the observable $\Sigma \xrightarrow{F} L$.

D 3.3.3 Two observables, $\Sigma_1 \xrightarrow{F_1} L$ and $\Sigma_2 \xrightarrow{F_2} L$, are called *convex equivalent*, if $\overline{\mathrm{co}}^\sigma(F_1\Sigma_1) = \overline{\mathrm{co}}^\sigma(F_2\Sigma_2)$.

Thus the following two assertions are equivalent by the Krein-Milman theorem:

(i) $\Sigma_1 \xrightarrow{F} L$ and $\Sigma_2 \xrightarrow{F} L$ are convex equivalent.
(ii) $\Sigma_1 \xrightarrow{F} L$ and $\Sigma_2 \xrightarrow{F} L$ have the same extremal kernel.

For physics this suggests to seek a smallest possible subring Σ_a of Σ so that $\Sigma_a \xrightarrow{F} L$ and $\Sigma \xrightarrow{F} L$ still are convex equivalent. We can solve this problem on the basis of

T 3.3.1 Let $\Sigma \xrightarrow{F} L$ be an effective, additive measure and $\Sigma = \hat{\Sigma}_g$ ($\hat{\Sigma}_g$ need not be separable). Let S be the mixture-morphism which due to (2.3.1) corresponds to F, and S' the mapping adjoint to S. Then we find

(i) $S' L(\Sigma) = \overline{\mathrm{co}}^\sigma(F\Sigma)$.
(ii) For each $g_e \in \partial_e \overline{\mathrm{co}}^\sigma(F\Sigma) = \partial_e S' L(\Sigma)$, there is one and only one $\sigma \in \Sigma$ with $g_e = F(\sigma)$.
(iii) $\partial_e L \cap \overline{\mathrm{co}}^\sigma(F\Sigma) = \partial_e L \cap \partial_e \overline{\mathrm{co}}^\sigma(F\Sigma) = \partial_e L \cap (F\Sigma)$.

Proof. For (i) and (ii), see [2] V Th 2.3.4. From (ii) follows $\partial_e \overline{\mathrm{co}}^\sigma(F\Sigma) \subset F\Sigma$.

Because $\overline{\mathrm{co}}^\sigma(F\Sigma) \subset L$, we must have $\partial_e L \cap \overline{\mathrm{co}}^\sigma(F\Sigma) = \partial_e L \cap \partial_e \overline{\mathrm{co}}^\sigma(F\Sigma)$, hence $\partial_e L \cap \overline{\mathrm{co}}^\sigma(F\Sigma) \subset \partial_e L \cap (F\Sigma)$. Because $F\Sigma \subset \overline{\mathrm{co}}^\sigma(F\Sigma)$, we also have $\partial_e L \cap \overline{\mathrm{co}}^\sigma(F\Sigma) \supset \partial_e L \cap (F\Sigma)$. □

D 3.3.4 For an observable $\Sigma \xrightarrow{F} L$, the subset

$$N = \{\sigma \mid \sigma \in \Sigma \text{ and } F(\sigma) \in \partial_e \overline{\mathrm{co}}^\sigma(F\Sigma)\}$$

is called the *kernel* of the observable.

According to T 3.3.1, $\sigma \to F(\sigma)$ defines a bijection of N onto $\partial_e \overline{\mathrm{co}}^\sigma (F\Sigma)$.

The kernel N of an observable is the part essential for measuring practice: Only for the "indications" $\sigma \in N$ the frequency $\mu(w, F(\sigma))$ need be "measured".

T 3.3.2 The kernel N of a measure $\Sigma \xrightarrow{F} L$ (with $\Sigma = \hat{\Sigma}_g$ but $\hat{\Sigma}_g$ not necessarily separable; see [39]) is $\mathscr{U}_g$-separable. There is an observable $\Sigma_1 \xrightarrow{F_1} L$ with $\Sigma_1 \subset \Sigma$ and $F_1 = F|_{\Sigma_1}$, whose kernel is $N_1 = N$ with N the kernel of $\Sigma \xrightarrow{F} L$. The complete Boolean ring $\Sigma_2 \subset \Sigma$ generated by the kernel N is separable; hence $\Sigma_2 \xrightarrow{F_2} L$ with $F_2 = F|_{\Sigma_2}$ is an observable (see [2] IV Th 2.3.5).

The complete Boolean ring Σ_2 generated by the kernel N (which exists by T 3.3.2) therefore yields the smallest possible observable convex equivalent to $\Sigma \xrightarrow{F} L$.

D 3.3.5 An observable $\Sigma \xrightarrow{F} L$ for which the complete Boolean ring generated by the kernel N equals Σ, is briefly called a "kernel observable".

§ 3.4 De-mixing of Observables

We have introduced the concept of an observable (as an abstraction and idealization of the structure $\mathscr{R}(b_0) \xrightarrow{\psi_0} L$ for fixed b_0) in order to compensate the "loss" in forgetting the sets $\mathscr{Q}$, $\mathscr{R}_0$, $\mathscr{R}$. In the preceding sections, we have only considered *one* observable $\Sigma \xrightarrow{F} L$ that corresponds to *one* $b_0 \in \mathscr{R}_0$. But the structure $\mathscr{R}_0$, $\mathscr{R}$ also establishes a relation among the "real observables", i.e. among the various $b_0 \in \mathscr{R}_0$ and hence among the $\mathscr{R}(b_0)$. Such relations can at least partly be retrieved in the abstract form of observables. Here, we shall be *mainly* interested in a relation corresponding to a de-mixing of registration *methods* (see III § 5.3). Let us begin with a natural definition (see [2] IV § 2.4):

D 3.4.1 If $\Sigma_1 \xrightarrow{F_1} L$ and $\Sigma_2 \xrightarrow{F_2} L$ are two observables, we write $(\Sigma_1 \xrightarrow{F_1} L) \prec (\Sigma_2 \xrightarrow{F_2} L)$ if there is a homomorphism h of the Boolean ring Σ_1 into the Boolean ring Σ_2 such that we get the diagram

$$\begin{array}{ccc} \Sigma_1 & \xrightarrow{h} & \Sigma_2 \\ & {\scriptstyle F_1} \searrow \quad \swarrow {\scriptstyle F_2} & \\ & L & \end{array} \quad . \tag{3.4.1}$$

We then say that $\Sigma_2 \xrightarrow{F_2} L$ is *more comprehensive* than $\Sigma_1 \xrightarrow{F_1} L$.

It follows (see [2] IV § 2.4) that h is an isomorphism $\Sigma_1 \xrightarrow{h} h\,\Sigma_1$ of the Boolean rings Σ_1 and $h\,\Sigma_1$ (as a subring of Σ_2), also relative to the uniform structures $\mathscr{U}_g$ of the Boolean rings induced by F_1 resp. F_2. Therefore, $h\,\Sigma_1$ is a complete Boolean sublattice of Σ_2.

D 3.4.1 expresses formally what only would formulate intuitively as follows: $\Sigma_2 \xrightarrow{F_2} L$ "measures more" than $\Sigma_1 \xrightarrow{F_1} L$ since the measurement of $\Sigma_1 \xrightarrow{F_1} L$ is "contained" in that of $\Sigma_2 \xrightarrow{F_2} L$.

One can show (see [2] IV § 2.4) that the relation $\prec$ represents a sort of pre-ordering, so that one can define an equivalence:

D 3.4.2 Two observables $\Sigma_1 \xrightarrow{F_1} L$ and $\Sigma_2 \xrightarrow{F_2} L$ are said to be *equivalent* if $(\Sigma_1 \xrightarrow{F_1} L) \prec (\Sigma_2 \xrightarrow{F_2} L)$ and $(\Sigma_2 \xrightarrow{F_2} L) \prec (\Sigma_1 \xrightarrow{F_1} L)$ hold.

By a de-mixing of a registration method $b_0 \in \mathscr{R}_0$ we understand a decomposition

$$b_0 = \bigcup_{i=1}^{n} b_{0i} \text{ with } b_{0i} \in \mathscr{R}_0 \text{ and } b_{0i} \cap b_{0k} = \emptyset \text{ for } i \neq k . \tag{3.4.2}$$

Then III T 4.2 (viii) provides

$$\psi(b_0, b) = \sum_{i=1}^{n} \lambda_i \psi(b_{0i}, b_{0i} \cap b) \tag{3.4.3}$$

with

$$\lambda_i = \lambda_{\mathscr{R}_0}(b_0, b_{0i}) . \tag{3.4.4}$$

Because of $\lambda_{\mathscr{R}_0}(b_0, b_{0i}) = \lambda_{\mathscr{S}}(a \cap b_0, a \cap b_{0i})$, we have

$$\lambda_{\mathscr{R}_0}(b_0, b_{0i}) = \mu(w, \psi(b_0, b_{0i}))$$

for all $w \in K$. Since this implies

$$\psi(b_0, b_{0i}) = \lambda_i \mathbf{1} , \tag{3.4.5}$$

the λ_i are determined by the mapping ψ.

In order to imitate this de-mixing in abstract form for observables, we note that the mappings

$$\mathscr{R}(b_0) \xrightarrow{\psi_0} L \quad \text{and} \quad \mathscr{R}(b_0) \xrightarrow{\psi_{0i}} L$$

with

$$\psi_0(b) = \psi(b_0, b) \quad \text{and} \quad \psi_{0i}(b) = \psi(b_{0i}, b_{0i} \cap b)$$

are additive measures on the Boolean ring $\mathscr{R}(b_0)$ while (3.4.3) implies

$$\psi_0 = \sum_{i=1}^{n} \lambda_i \psi_{0i} . \tag{3.4.6}$$

This can easily be imitated for abstract observables:

D 3.4.3 An observable $\Sigma \xrightarrow{F} L$ is called a *mixture* of the observables $\Sigma \xrightarrow{F_i} L$ if

$$F = \sum_{i=1}^{n} \lambda_i F_i \quad \text{holds with} \quad \lambda_i > 0, \sum_{i=1}^{n} \lambda_i = 1 .$$

The λ_i are called the *weights* of the individual "mixture components" F_i. If $F = \lambda F_1 + (1 - \lambda) F_2$ with $0 < \lambda < 1$ always implies $F_1 = F$ (hence also $F_2 = F$), then F is called an *irreducible* observable. For the F_i in D 3.4.3 one need *only* presume that they are additive measures. Then it follows that they must also be σ-additive (as F is).

Since mixtures of observables cannot yield more information about the physical systems than the mixture components do, it appears worth striving for the irreducible observables. For fixed Σ, the observables $\Sigma \xrightarrow{F} L$ form a convex set, briefly to be called $K(\Sigma, L)$. Hence the irreducible observables are the extreme points of $K(\Sigma, L)$. But in general there are *not* enough extreme points of $K(\Sigma, L)$ so that $K(\Sigma, L) = \overline{\text{co}}\, \partial_e K(\Sigma, L)$ holds in a suitable topology of the physical imprecision on $K(\Sigma, L)$. (Example: The theory of classical point mechanics.)

To be sure, one thing follows immediately:

For $F\Sigma \subset \partial_e L$, $\Sigma \xrightarrow{F} L$ is kernel observable and irreducible.

Proof. This follows immediately from T 3.3.1 and the definition of an extreme point. That $F\Sigma \subset \partial_e L$ holds for each irreducible kernel observable is in general (also for quantum mechanics) not correct (see [2] IV § 2.4).

§ 3.5 Measurement Scales of Observables and Totally Ordered Subsets of *L*

A most remarkable prejudice is that a quantitative measurement characterized by a scale is essential for physics. In fact quantitative measurement is *not* of fundamental significance for physics, though in many cases eminently *practical.* Since we have here begun an axiomatic foundation *of* quantum mechanics, one should not wonder that the measurement scales of observables so far did not appear. Only additional structures would allow us to distinguish "practical" measurement scales and "interesting" observables. We shall not treat such structures in this book because they are in detail presented in [2] V through VIII. In spite of this, we can introduce the general concept of a measurement scale, without distinguishing "definite" scales.

A measurement scale is simply a practical device to provide a view over the complete Boolean ring Σ. Hence the scale is an "ordering" principle for the elements of Σ and does not concern the mapping $\Sigma \xrightarrow{F} L$. Let us adopt

D 3.5.1 An element $y \in \mathscr{B}'(\Sigma)$ is called a *measurement scale* for the complete Boolean ring Σ.

Instead of "measurement scale" the phrase "random variable" is also in use; we do not use this phrase in order to avoid prejudices which the adjective "random" might evoke.

The investigations in [2] IV § 2.1 and § 2.5 can easily be transferred so that we need not describe them here. Let us only introduce several concepts of later importance.

For each $y \in \mathscr{B}'(\Sigma)$ and real α, one can define $\sigma(y \leqq \alpha) \in \Sigma$ as the largest element of the set $\{\sigma \mid \sigma \in \Sigma \text{ and } s(y \mid \sigma) \leqq \alpha\}$. Here we have

$$s(y, \sigma) = \sup \{\langle m, y\rangle \mid m \in K(\Sigma) \text{ and } m(\sigma) = 1\}$$

with $\langle \ldots, \ldots \rangle$ the canonical bilinear form of $\mathscr{B}(\Sigma)$, $\mathscr{B}'(\Sigma)$. For $-\infty < \alpha < +\infty$, $\sigma(y \leqq \alpha)$ is called the *spectral family* of y.

D 3.5.2 The complete Boolean subring of Σ generated by the spectral family $\sigma(y \leqq \alpha)$ shall be called $\Sigma(y)$.

If Σ is separable, there is a scale $y \in \mathscr{B}'(\Sigma)$ with $\Sigma(y) = \Sigma$ (see [2] IV Th 2.5.6). Therefore, if $\Sigma \xrightarrow{F} L$ is an observable, a y exists with $\Sigma(y) = \Sigma$.

If y is an arbitrary element of $\mathscr{B}'(\Sigma)$ and $\sigma(y \leqq \alpha)$ its spectral family, then $F(\lambda) = F(\sigma(y \leqq \lambda))$ defines a family of effects that increase with λ. With the mapping S' assigned to the observables by § 3.2, we then get

$$S' y = \int \lambda \, dF(\lambda) . \tag{3.5.1}$$

But S' is not always injective, so that in general one *cannot* retrieve the scale $y \in \mathscr{B}'(\Sigma)$ uniquely from

$$y' = \int \lambda \, dF(\lambda) \in \mathscr{B}' .$$

§ 4 Coexistent and Complementary Observables

The concepts to be introduced here have only for quantum mechanics become of interest since for "classical theories" all observables coexist and there exist no complementary observables.

D 4.1 Two observables $\Sigma_1 \xrightarrow{F_1} L$ and $\Sigma_2 \xrightarrow{F_2} L$ are said to *coexist* if there exist an observable $\Sigma \xrightarrow{F} L$ and two homomorphisms h_1, h_2 which give the diagram

$$\begin{array}{ccccc} \Sigma_1 & \xrightarrow{h_1} & \Sigma & \xleftarrow{h_2} & \Sigma_2 \\ & {\scriptstyle F_1}\searrow & \downarrow{\scriptstyle F} & \swarrow{\scriptstyle F_2} & \\ & & L & & \end{array} .$$

Therefore $(F_1 \Sigma_1) \cup (F_2 \Sigma_2)$ is a set of coexistent effects.

For two observables let us define a relation to describe that somehow the two observables are "extremely" non-coexistent. To this end we first adopt

D 4.2 $\Xi = \{g \mid g \in L$ and g coexist with *each* $g' \in L\}$.

D 4.3 Two observables $\Sigma_1 \xrightarrow{F_1} L$, $\Sigma_2 \xrightarrow{F_2} L$ are called *mutually complementary* if $F_1 \Sigma_1 \nsubseteq \Xi$, $F_2 \Sigma_2 \nsubseteq \Xi$ and each observable $\Sigma \xrightarrow{F} L$ yields either $(F_1 \Sigma_1) \cap (F \Sigma) \subset \Xi$ or $(F_2 \Sigma_2) \cap (F \Sigma) \subset \Xi$.

§ 5 Realization of Observables

The structure analysis of observables was made mathematically transparent by defining observables in a form $\Sigma \xrightarrow{F} L$, more idealized than $\mathscr{R}(b_0) \xrightarrow{\psi_0} L$. But now we must ask whether one can actually realize an observable $\Sigma \xrightarrow{F} L$ in measuring-practice, i.e. whether there exists a $b_0 \in \mathscr{R}_0$ and an isomorphism h such that we obtain

$$\begin{array}{ccc} \Sigma & \xrightarrow{h} & \mathscr{R}(b_0) \\ & {\scriptstyle F}\searrow \swarrow{\scriptstyle \psi_0} & \\ & L & \end{array} .$$

Demanding such an $\mathscr{R}(b_0)$ to exist for each observable $\Sigma \xrightarrow{F} L$ would certainly be too severe, especially because we assumed $\mathscr{Q}$, $\mathscr{R}$ countable.

The following "approximate" realization can be joined to the axioms postulated so far.

AOb: For each observable $\Sigma \xrightarrow{F} L$, and each *finite* Boolean subring $\tilde{\Sigma}$ of Σ, and each $\sigma(\mathscr{B}', \mathscr{B})$-neighborhood U of $\mathbf{0} \in \mathscr{B}'$, there exist an $\mathscr{R}(b_0) \xrightarrow{\psi_0} L$ and a homomorphism $\tilde{\Sigma} \xrightarrow{h} \mathscr{R}(b_0)$ such that $\psi_0 h(\sigma) - F(\sigma) \in U$ for all $\sigma \in \tilde{\Sigma}$.

If one foregoes this as an axiom, one can in any case regard it as a "certain hypothesis" (in the sense of [3] § 10.1) though we shall omit the proof. Proving this (that AOb is a certain hypothesis) we would see that the adjunction of AOb as an axiom leads to no contradictory theory. As explained in [3] § 10.4, it is then only a matter of taste whether one adjoins AOb as an axiom, unless experience strongly indicates that nature presents serious obstacles to constructing all the devices "possible" under AOb (see [3] § 10).

One often expresses AOb briefly by saying: Every observable can be measured "approximately" (in this connection, one pays attention to the fact that $\sigma(\mathscr{B}', \mathscr{B})$ determines the uniform structure of physical imprecision in L; see IV § 2 and § 3).

The coexistence of two observables (defined in D 4.1) has as a consequence, on the basis of AOb, that there is a measurement method $b_0 \in \mathscr{R}_0$ by which the two observables are measured (approximately) together. One can easily deduce this from the diagram in D 4.1: Let $\tilde{\Sigma}_1$ and $\tilde{\Sigma}_2$ be finite Boolean subrings of Σ_1 resp. of Σ_2. Then $h_1 \tilde{\Sigma}_1$ and $h_2 \tilde{\Sigma}_2$ generate in Σ a finite Boolean subring $\tilde{\Sigma}$ for which due to AOb an $\mathscr{R}(b_0)$ exists with $h\tilde{\Sigma} \subset \mathscr{R}(b_0)$ and $\psi_0 h(\sigma) - F(\sigma) \in U$ whenever a neighborhood U is given. But from this follows

$$\tilde{\Sigma}_i \xrightarrow{h h_i} \mathscr{R}(b_0)$$

for $i = 1, 2$ and likewise

$$\psi_0 h h_i(\sigma) - F_i(\sigma) \in U \text{ for all } \sigma \text{ in } \tilde{\Sigma}_i .$$

Therefore, b_0 is an approximate measurement of $\Sigma_1 \xrightarrow{F_1} L$ as well as of $\Sigma_2 \xrightarrow{F_2} L$, i.e. these coexistent observables can *jointly* be measured approximately, i.e. they can be measured by a *single* method b_0.

Axiom AOb indeed means that it should be "physically possible" (see [3] § 10.4) to measure every observable approximately. But it does not state *how* to find the measurement method b_0 in a concrete case. The theory presented here clearly cannot accomplish this since it contains no mathematical mapping for the technical construction of the "devices" $b_0 \in \mathscr{R}_0$. We shall return to this problem in XI.

The objectivating way of describing experiments in trajectory spaces (as explained in III § 6) enables us to relate the observables closely to the trajectory spaces and hence to obtain a certain overview of all realizable observables.

According to III (6.4.12), for fixed a_2 there is a correspondence

$$\varphi\pi(a_1 \cap b_1) \rightarrow \frac{R_2 \psi'_{1S}(b_{10}, b_1)\, \varphi_{12}(a_1 \times a_2 \cap M)}{\langle \varphi_1(a_1), \psi_{1S}(b_{10}, b_1) \rangle_1} \tag{5.1}$$

which satisfies III (6.4.11). It can at once be extended to a mapping of $\varphi(Q')$ into $K(\hat{S}_2)$ since $\varphi_{12}\pi(a_1 \times a_2)$ and $\psi'_{1s}(b_{10}, b_1)$ are additive measures relative to a_1 resp.

b_1. The mapping by (5.1) is also norm-continuous, since for calculating norms, the supremum on the right side of III (6.4.11) must for fixed a_2 be taken *only* over the different $\psi_{2S}(b_{20}, b_2)$, i.e. over $L(\hat{S}_2)$ (since la $\psi_2(\Phi_2)$ is dense in $L(\hat{Y}_2)$). Being norm-continuous that mapping can be extended to all of K:

$$K \xrightarrow{\tilde{S}_{a_2}} K(\hat{S}_2) . \tag{5.2}$$

The index on $\tilde{S}_{a_2}$ means that it depends on a_2.

Let $\hat{S}_2(a_2)$ be the support of $\tilde{S}_{a_2}K$. Then, for fixed a_2, one can instead of $\hat{S}_2$ use the subset $\hat{S}_2(a_2)$ and regard $\tilde{S}_{a_2}K$ as a subset of $K(\hat{S}_2(a_2))$, so that (5.2) can be written

$$K \xrightarrow{\tilde{S}_{a_2}} K(\hat{S}_2(a_2)) . \tag{5.3}$$

With $\mathscr{B}(\hat{S}_2(a_2))$ as the Borel field belonging to $\hat{S}_2(a_2)$, and $\mathscr{J}(\hat{S}_2(a_2))$ as the sets of $(\tilde{S}_{a_2}K)$-measure zero (see II § 3.3), the elements of $\tilde{S}_{a_2}K$ are σ-additive measures on $\Sigma_{a_2} = \mathscr{B}(\hat{S}_2(a_2))/\mathscr{J}(\hat{S}_2(a_2))$. The set $\hat{S}_2(a_2)$ consists of those physically possible trajectories on the systems 2 which were prepared according to a_2 and on which the various system 1 can act. The open subsets of $\hat{S}_2(a_2)$ are not elements of $\mathscr{J}(\hat{S}_2(a_2))$.

Since one can also regard $\tilde{S}_{a_2}K$ as a subset of $K(\Sigma_{a_2})$, for (5.3) one can also write

$$K \xrightarrow{\tilde{S}_{a_2}} K(\Sigma_{a_2}) . \tag{5.4}$$

$\tilde{S}_{a_2}$ maps effective measures into effective measures since an element of Σ_{a_2} cannot have measure zero for all elements $\tilde{S}_{a_2} w$ with $w \in K$.

Therefore, S_{a_2} due to § 3.2 defines a σ-additive measure $\Sigma_{a_2} \xrightarrow{F_{a_2}} [\mathbf{0}, \mathbf{1}]$. The set $L(\hat{S}_2(a_2))$ can be identified with a subset of $\mathscr{B}'(\Sigma_{a_2})$ and $L(\hat{S}_2(a_2))$ is $\sigma(\mathscr{B}'(\Sigma_{a_2}), \mathscr{B}(\Sigma_{a_2}))$-dense in $L(\Sigma_{a_2})$. Hence, $F_{a_2}\Sigma_{a_2} \subset L$ holds because of $\tilde{S}'_{a_2}L(\hat{S}_2(a_2)) \subset L$. Therefore, $\Sigma_{a_2} \xrightarrow{F_{a_2}} L$ is an observable.

The observable $\Sigma_{a_2} \xrightarrow{F_{a_2}} L$ is called the ideal observable *fixed* by a_2. The derivation immediately implies $\psi\varrho(a_2 \cap b_{20}, a_2 \cap b_2) \in \overline{\mathrm{co}}^\sigma(F_{a_2}\Sigma_{a_2})$; hence one can view the measurement of trajectories by method b_{02} as a more or less accurate measurement of the ideal observable. For fixed a_2, the range of every other observables determined by measurements on the systems 2 lies in $\overline{\mathrm{co}}^\sigma(F_{a_2}\Sigma_{a_2})$. Therefore, for *fixed* a_2, all observables resting on various methods of measuring the trajectories are coexistent!

Hence we say that the observable $\Sigma_{a_2} \overset{F_{a_1}}{} L$ is measured "objectively" on the action carriers by the coupling with the registration devices which have been prepared due to a_2. It does not matter how precisely one measures the processes on the registration devices, even being immaterial whether or not one measures these processes at all. A measurement of an observable $\Sigma_{a_2} \xrightarrow{F_{a_1}} L$ is perfect as soon as the registration devices prepared due to a_2 are used.

Neither the "reading" of the results on the registration devices, nor the acceptance of these results into a consciousness is decisive for the final validity of the measurement. This fact is often erroneously interpreted. It rests on the objectivating description of the measuring devices by trajectories of a trajectory space. The ideal observable is fixed by the set Σ_{a_2} of the "objective properties" of the measuring devices.

We have eliminated the more or less exact measurement of the measuring devices $x_2 \in a_2$ by considering the observable $\Sigma_{a_2} \xrightarrow{F_{a_1}} L$. This $\Sigma_{a_2} \xrightarrow{F_{a_2}} L$ represents the ideal limiting case of the "arbitrarily accurate" measurement of the measuring devices $x_2 \in a_2$. Of course this by no means says that $\Sigma_{a_2} \xrightarrow{F_{a_1}} L$ is an especially interesting observable, e.g. an irreducible observable.

For example, a de-mixing of a_2, as an element of $\mathcal{Q}_2'$, yields a de-mixing of the observable $\Sigma_{a_2} \xrightarrow{F_{a_2}} L$ and thus an illustrative example for the discussion from § 3.4. Hence the experimental physicist strives for the "smallest possible" elements of $\mathcal{Q}_2'$, i.e. for elements of $\mathcal{Q}_2'$ that in practice are no longer properly de-mixable. An a_2 is no longer properly de-mixable if and only if in practice there do not appear fluctuations in the quality of the devices $x_2 \in a_2$, i.e. if for all experiments all the elements $x_2 \in a_2$ are practically (macroscopically) "equal".

Thus, we see how the basis structure from III precisely reflects the problems of experimental work. One problem certainly remains completely open: Which "choice" a_2 of measuring devices should one in practice adopt, even when one only considers a_2 which are no longer de-mixable? Which choice a_2 leads to "interesting" experiments?

The experimental physicist tries to find irreducible kernel observables. But which "sort a_2" of measuring devices does approximately yield such observables? Behind a_2 there is hidden the whole problem of constructing measurement devices. Without further theoretical considerations, one cannot attack this problem (see XI § 6 and 7).

§ 6 Coexistent De-mixing of Ensembles

The concept of coexistent effects and observables resulted in a natural way by idealizing a substructure $\mathcal{R}(b_0) \xrightarrow{\psi_0} L$. Equally important is the closer investigation of a substructure $\mathcal{Q}'(a) \xrightarrow{\varphi} K$ of preparing (described in III § 5.3). According to III T 5.3, in the Banach space $\mathcal{B}$ follows: For a de-mixing $a = \bigcup_{i=1}^{n} a_i$ of a preparation procedure a,

$$\varphi(a) = \sum_{i=1}^{n} \lambda_i \varphi(a_i) \tag{6.1}$$

holds with $\lambda_i = \lambda_{\mathcal{Q}}(a, a_i)$, $0 < \lambda_i \leqq 1$ and $\sum_{i=1}^{n} \lambda_i = 1$.

Therefore, two such de-mixings $a = \bigcup_{i=1}^{n} a_i = \bigcup_{k=1}^{m} \tilde{a}_m$ of the *same* $a \in \mathcal{Q}'$ yield

$$\varphi(a) = \sum_{i=1}^{n} \lambda_i \varphi(a_i) = \sum_{k=1}^{m} \tilde{\lambda}_k \varphi(\tilde{a}_k) .$$

But then one can combine both de-mixings into the de-mixing

$$a = \bigcup_{i,k}{}' (a_i \cap \tilde{a}_k) , \tag{6.2}$$

where $\bigcup'$ requires to sum only over the pairs i, k with $a_i \cap \tilde{a}_k \neq \emptyset$. From (6.2) then follows

$$\varphi(a) = \sum_{i,k}{}' \lambda_{ik} \varphi(a_i \cap \tilde{a}_k) \tag{6.3}$$

with $\lambda_{ik} = \lambda_{\mathscr{D}}(a, a_i \cap \tilde{a}_k)$. Moreover we get

$$\varphi(a_i) = \sum_k{}' \lambda_k^i \, \varphi(a_i \cap \tilde{a}_k) ,$$

$$\varphi(\tilde{a}_k) = \sum_i{}' \tilde{\lambda}_i^k \, \varphi(a_i \cap \tilde{a}_k) ,$$

where $\lambda_k^i = \lambda_{ik} \left(\sum_k{}' \lambda_{ik} \right)^{-1}$, $\tilde{\lambda}_i^k = \lambda_{ik} \left(\sum_i{}' \lambda_{ik} \right)^{-1}$.

These relations can be formulated especially simply if one further defines the following set and mapping.

D 6.1 By $\check{K}$ we denote the truncated cone

$$\check{K} = \bigcup_{0 \leqq \lambda \leqq 1} \lambda K .$$

For fixed $a \in \mathscr{D}'$, we introduce the mapping

$$\varphi_a(\tilde{a}) = \lambda_{\mathscr{D}}(a, \tilde{a}) \, \varphi(\tilde{a})$$

of $\mathscr{D}(a)$ into $\check{K}$, where $\mathscr{D}(a) = \{\tilde{a} \mid \tilde{a} \in \mathscr{D}, \tilde{a} \subset a\}$.

With this definition, from (6.1) follows

T 6.1 The mapping φ_a of $\mathscr{D}(a)$ into $\check{K}$ is an additive measure over the Boolean ring $\mathscr{D}(a)$, with $\varphi_a(a) = \varphi(a) \in K$ (see [2] IV Th 5.1).

Two de-mixings of one and the same preparation procedure $a = \bigcup_{i=1}^{n} a_i = \bigcup_{k=1}^{m} \tilde{a}_k$ thus lead to two de-mixings

$$\varphi(a) = \sum_{i=1}^{n} \varphi_a(a_i) = \sum_{k=1}^{m} \varphi_a(\tilde{a}_k)$$

of the ensemble $\varphi(a)$, the components $\varphi_a(a_i)$, $\varphi_a(\tilde{a}_k)$ of which lie in the range of the additive measure φ_a over the Boolean ring $\mathscr{D}(a)$.

This suggests

D 6.2 Two de-mixings

$$w = \sum_{i=1}^{n} w_i = \sum_{k=1}^{m} \tilde{w}_k \quad \text{with} \quad w_i, \tilde{w}_k \in \check{K}$$

of an ensemble are called *coexistent* if there is a Boolean ring Σ with an additive measure $\Sigma \overset{W}{\rightarrow} \check{K}$, such that $W(\varepsilon) = w$ and $w_i, \tilde{w}_k \in W\Sigma$.

Two de-mixings of one and the same preparation procedure a thus always yield coexistent de-mixings of $\varphi(a)$.

D 6.3 A set $A \subset \check{K}$ is called a set of *coexistent components* of w, if there is a Boolean ring Σ with an additive measure $\Sigma \overset{W}{\rightarrow} \check{K}$, where $W(\varepsilon) = w$ and $A \subset W\Sigma$.

Similarly to the case of coexistent effects, one can restrict one-self to effective measures; and just as there it is natural to idealize the situation $\Sigma \overset{W}{\rightarrow} \check{K}$ mathematically by completion.

T 6.2 If W is an effective, additive measure over a Boolean ring $\Sigma \xrightarrow{W} \check{K}$, with $W(\varepsilon) = w \in K$, then $m_0(\sigma) = \| W(\sigma) \| = \mu(W(\sigma), \mathbf{1})$ is an effective, additive, real measure with $m_0(\varepsilon) = 1$. Here, $d(\sigma_1, \sigma_2) = m_0(\sigma_1 \dotplus \sigma_2)$ is a metric in Σ for which W is uniformly continuous as a mapping in the Banach space $\mathscr{B}$. (See [2] IV Th 5.2.)

As in § 1.3, from T 6.2 follows immediately that Σ can be completed and that W can be extended to the completion. Then W becomes a σ-additive measure on the completion. If Σ is a (lattice theoretically) complete Boolean ring and W a σ-additive measure, then Σ is also complete with respect to the metric $d(\sigma_1, \sigma_2) = \mu(W(\sigma_1 \dotplus \sigma_2), \mathbf{1})$. As for an observable, we therefore adopt

D 6.4 A Boolean ring Σ with additive, effective measure $\Sigma \xrightarrow{W} K$, with $W(\varepsilon) = w \in K$, which is complete and separable in the metric determined by W, is called a *preparator* of w.

The concept of a preparator is not customary because one is not accustomed to distinguish between "measuring" and "trans-preparing". Historically, this goes back to the too restrictive consideration of measuring by J. v. Neumann. According to him, a system should *after* the measurement have the measured value of an observable (see XI § 5 and [2] XVII), although in reality this is rarely the case.

Speaking of preparing in everyday language (to avoid the concept of a preparator) can easily lead to alleged contradictions as in the so-called Einstein-Podolski-Rosen paradox (XII § 2).

T 6.3 If $\Sigma \xrightarrow{W} K$ is a preparator, then there is a mixture-morphism $\mathscr{B}(\Sigma) \xrightarrow{T} \mathscr{B}$ with

$$T m_{0\sigma} = W(\sigma)\,, \tag{6.4}$$

where $m_{0\sigma}$ is determined by the measure m_0 from T 6.2 according to

$$m_{0\sigma}(\tilde{\sigma}) = m_0(\sigma \wedge \tilde{\sigma})\,. \tag{6.5}$$

T is uniquely determined by (6.4).

Proof. The elements of the subspace $\mathscr{T}$ that is linearly spanned by Σ in $\mathscr{B}'(\Sigma)$ are finite linear combinations $\sum_\nu \alpha_\nu \sigma_\nu$. Then $V(\sum_\nu \alpha_\nu \sigma_\nu) = \sum_\nu \alpha_\nu m_{0\sigma_\nu}$ defines a linear mapping $\mathscr{T} \xrightarrow{V} \mathscr{B}(\Sigma)$. For, from $\sum_\nu \alpha_\nu \sigma_\nu = \sum_\mu \beta_\mu \sigma'_\mu$ follows $\sum_\nu \alpha_\nu \sigma_\nu = \sum_{\nu\mu} \gamma_{\nu\mu}(\sigma_\nu \wedge \sigma'_\mu)$ $= \sum_\mu \beta_\mu \sigma'_\mu$; hence the additivity of $m_{0\sigma}$ (as a function of σ!) implies $\sum_\nu \alpha_\nu m_{0\sigma_\nu}$ $= \sum_\mu \beta_\mu m_{0\sigma'_\mu}$.

The mapping V is injective since each element in $\mathscr{T}$ can be written $\sum_\nu \alpha_\nu \sigma_\nu$ with $\sigma_\nu \wedge \sigma_\mu = 0$ for $\nu \neq \mu$, so that $\sum_\nu \alpha_\nu m_{0\sigma_\nu} = 0$ gives $\alpha_\nu = 0$ and thus $\sum_\nu \alpha_\nu \sigma_\nu = 0$.

From the additivity of W likewise follows that $\sum_\nu \alpha_\nu \sigma_\nu \to \sum_\nu \alpha_\nu W(\sigma_\nu)$ defines an extension of W as a mapping $\mathscr{T} \xrightarrow{W} \mathscr{B}$. Then $V\mathscr{T} \xrightarrow{V^{-1}} \mathscr{T} \xrightarrow{W} \mathscr{B}$ defines a linear mapping $V\mathscr{T} \xrightarrow{T} \mathscr{B}$, which is norm-continuous:

Because $\sigma_\nu \wedge \sigma_\mu = 0$ for $\nu \neq \mu$, with the m_0 from T 6.2 follows

$$\Big\| \sum_\nu \alpha_\nu m_{0\sigma_\nu} \Big\| = \sum_\nu |\alpha_\nu|\; m_0(\sigma_\nu) = \sum_\nu |\alpha_\nu|\; \| W(\sigma_\nu) \|\,.$$

On the other hand,

$$\left\| \sum_\nu \alpha_\nu W(\sigma_\nu) \right\| \leqq \sum_\nu |\alpha_\nu| \, \| W(\sigma_\nu) \| = \left\| \sum_\nu \alpha_\nu m_{0\sigma_\nu} \right\| .$$

According to [2] IV Th 2.1.11, $V\mathcal{T}$ is norm-dense in $\mathcal{B}(\Sigma)$. Therefore, T can be extended as a linear norm-continuous mapping $\mathcal{B}(\Sigma) \xrightarrow{T} \mathcal{B}$.

From $\sum_\nu \alpha_\nu m_{0\sigma_\nu} \in K(\Sigma)$ follows $\alpha_\nu \geqq 0$ and $1 = \left\| \sum_\nu \alpha_\nu m_{0\sigma_\nu} \right\| = \sum_\nu \alpha_\nu m_0(\sigma_\nu)$ $= \sum_\nu \alpha_\nu \| W(\sigma_\nu) \| = \left\| \sum_\nu \alpha_\nu W(\sigma_\nu) \right\|$ and hence $\sum_\nu \alpha_\nu W(\sigma_\nu) \in K$. The mapping T is a mixture-morphism.

The proof shows that T is uniquely determined by (6.4). □

T 6.4 If T is a mixture-morphism $\mathcal{B}(\Sigma) \xrightarrow{T} \mathcal{B}$, each effective measure $m_0 \in K(\Sigma)$ by means of $W(\sigma) = T m_{0\sigma}$ (with $m_{0\sigma}$ of (6.5)) defines a preparator $\Sigma \xrightarrow{W} \check{K}$ of $w_0 = T m_0$.

Proof. The additivity of W follows immediately from the linearity of T. But W is in fact σ-additive, since m_0 is σ-additive and T is norm-continuous. □

§ 7 Complementary De-mixings of Ensembles

The classical theories are distinguished by the fact that all de-mixings are coexistent. Indeed, one can prove (see VII Th 5.3.3) that the coexistence of all de-mixings makes $\mathcal{B}$ isomorphic to the space $\mathcal{B}(\Sigma)$ of a Boolean ring Σ.

But it is just important for quantum mechanics that there also exist de-mixings of an ensemble w which are not coexistent. In order to describe this possibility still better, we introduce the following concepts:

D 7.1 A preparator $\Sigma \xrightarrow{W} K$ of the ensemble w is called *more comprehensive* than another $\Sigma_1 \xrightarrow{W_1} \check{K}$ of the same ensemble w, if there is a homomorphism h to give the diagram

$$\begin{array}{ccc} \Sigma_1 & \xrightarrow{h} & \Sigma \\ & {\scriptstyle W_1}\searrow \quad \swarrow{\scriptstyle W} & \\ & \check{K} & \end{array} .$$

D 7.2 Two preparators $\Sigma_1 \xrightarrow{W_1} \check{K}$ and $\Sigma_2 \xrightarrow{W_2} \check{K}$ of the same ensemble w are said to *coexist* if there is a preparator $\Sigma \xrightarrow{W} \check{K}$ which is more comprehensive than either of $\Sigma_1 \xrightarrow{W_1} \check{K}$ and $\Sigma_2 \xrightarrow{W_2} \check{K}$.

For two preparators, the situation opposite to coexistence somewhat differs from that for observables. We are led to diametrically distinct situations by first considering two de-mixings of an ensemble w which arise from de-mixings of preparation procedures.

Thus, let $a = \bigcup_{i=1}^{n} a_i$ be a de-mixing of a preparation procedure a and let $\tilde{a} = \bigcup_{k=1}^{m} \tilde{a}_k$ be a de-mixing of $\tilde{a}$. Moreover, in particular choose $\varphi(a) = \varphi(\tilde{a}) = w$. Then $\varphi_a(a_i) = w_i$ and $\varphi_{\tilde{a}}(\tilde{a}_k) = \tilde{w}_k$ yield $w = \sum_{i=1}^{n} w_i = \sum_{k=1}^{m} \tilde{w}_k$ with $w_i, \tilde{w}_k \in \check{K}$.

Therefore, if these assumptions imply $a \cap \tilde{a} = \emptyset$, the preparation procedures a and $\tilde{a}$ can have nothing in common. Then one must physically exclude (see [3] § 10.4) that a single microsystem can be viewed as being prepared according to a as well as $\tilde{a}$. The preparation procedures a and $\tilde{a}$ are mutually exclusive, although they prepare the *same* (!) ensemble $w = \varphi(a) = \varphi(\tilde{a})$.

Let us formulate this possibility $a \cap \tilde{a} = \emptyset$ in an abstract, idealized form, with Σ_1, Σ_2 instead of $\mathscr{Q}(a)$, $\mathscr{Q}(\tilde{a})$. Initial difficulties are caused by the fact that for $\sigma_1 \in \Sigma_1$, $\sigma_2 \in \Sigma_2$ with two distinct Boolean rings Σ_1, Σ_2, no intersection $\sigma_1 \wedge \sigma_2$ is defined. Hence we proceed as follows:

From a preparator $\Sigma \xrightarrow{W} L$ of an ensemble w, one can easily obtain new preparators as follows: Let $[0, \eta]$ be a section from Σ. Then $[0, \eta] \xrightarrow{\lambda^{-1} W} \check{K}$ with $\lambda = \mu(W(\eta), \mathbf{1})$ is a preparator of the ensemble $w_0 = \lambda^{-1} W(\eta)$. We call such a preparator briefly the preparator determined canonically by $[0, \eta]$.

D 7.3 Two preparations $\Sigma_1 \xrightarrow{W_1} \check{K}$ and $\Sigma_2 \xrightarrow{W_2} \check{K}$ of the same ensemble w are called *mutually exclusive* if there do not exist two sections $[0, \eta_1] \subset \Sigma_1$ and $[0, \eta_2] \subset \Sigma_2$ such that $\lambda_1^{-1} W_1(\eta_1) = \lambda_2^{-1} W_2(\eta_2)$ holds with $\lambda_1 = \mu(W_1(\eta_1), \mathbf{1})$, $\lambda_2 = \mu(W_2(\eta_2), \mathbf{1})$, and the preparators determined canonically by $[0, \eta_1]$ and $[0, \eta_2]$ coexist.

D 7.4 Two preparations $\Sigma_1 \xrightarrow{W_1} \check{K}$ and $\Sigma_2 \xrightarrow{W_2} \check{K}$ of the same ensemble w are called *complementary*, if two preparators are mutually exclusive whenever they are more comprehensive than $\Sigma_1 \xrightarrow{W_1} \check{K}$ resp. $\Sigma_2 \xrightarrow{W_2} \check{K}$.

D 7.5 Two de-mixings

$$w = \sum_{i=1}^{n} w_i = \sum_{k=1}^{m} \tilde{w}_k$$

of an ensemble are called *complementary* whenever two preparators $\Sigma_1 \xrightarrow{W_1} \check{K}$, $\Sigma_2 \xrightarrow{W_2} \check{K}$ with $w_1 \in W_1 \Sigma_1$, $\tilde{w}_k \in W_2 \Sigma_2$ are complementary.

How can one recognize that two de-mixings are complementary? The following theorem gives the answer.

T 7.1 Two de-mixings

$$w = \sum_{i=1}^{n} w_i = \sum_{k=1}^{m} \tilde{w}_k$$

of an ensemble are complementary if and only if each pair w_i, $\tilde{w}_k$ makes

$$w_0 \in \check{K}, \quad w_0 \leqq w_i, \quad w_0 \leqq \tilde{w}_k \Rightarrow w_0 = 0.$$

(see [2] IV Th 6.1).

Complementary de-mixings play a large role in quantum mechanics (see VII § 5.3 and XII § 2). The Einstein-Podolski-Rosen "paradox" presents nothing than examples for preparation devices a, $\tilde{a}$ with $\varphi(a) = \varphi(\tilde{a})$, for which the preparators $\mathscr{Q}(a) \xrightarrow{\varphi_a} \check{K}$ and $\mathscr{Q}(\tilde{a}) \xrightarrow{\varphi_{\tilde{a}}} \check{K}$ are complementary.

§ 8 Realizations of De-mixings

If, for a preparator $\Sigma \xrightarrow{W} K$, there exist a preparation procedure $a \in \mathscr{Q}'$ and a homomorphism h of Σ into $\mathscr{Q}(a)$ such that we get the diagram

$$\begin{array}{ccc} \Sigma & \xrightarrow{h} & \mathscr{Q}(a) \\ & {\scriptstyle W}\searrow \quad \swarrow{\scriptstyle \varphi_a} & \\ & \breve{K} & \end{array},$$

one can identify Σ with the Boolean subring $h\,\Sigma$ of $\mathscr{Q}(a)$. Then the preparation procedure a (with its $\mathscr{Q}(a)$ and φ_a) can be called a realization of the preparator $\Sigma \xrightarrow{W} \breve{K}$. But too strong would be the demand that a realization be physically possible for each preparator (also see § 5). In analogy to § 5 let us therefore impose

Apr: For each preparator $\Sigma \xrightarrow{W} \breve{K}$, and each finite Boolean subring $\tilde{\Sigma}$ of Σ, and each $\sigma(\mathscr{B}', \mathscr{B})$-neighborhood U of $\mathbf{0}$, there exist a $\mathscr{Q}(a) \xrightarrow{\varphi_a} \breve{K}$ and a homomorphism $\tilde{\Sigma} \xrightarrow{h} \mathscr{Q}(a)$ such that $\varphi_a\, h(\sigma) - W(\sigma) \in U$ holds for all $\sigma \in \tilde{\Sigma}$.

In the sense of this axiom one can realize each preparator and hence each de-mixing of $w \in K$ "in a physical approximation".

Similarly to what we did for observables in § 5, let us point out connections between preparators and the trajectory registration procedures on the preparation devices. To this end, we use III (6.4.13). We now keep a_1 fixed and let a_2, b_{20}, b_2 run over all possible values; this is equivalent to testing $a \in \mathscr{Q}_1$ for all possible values of $(b_0, b) \in \mathscr{F}$ because $b_0 = M_1 \times (a_2 \cap b_{20}) \cap M$ and $b = M_1 \times (a_2 \cap b_2) \cap M$ generate all of $\mathscr{R}_0$ and $\mathscr{R}$. Thus III (6.4.13) gives (for fixed a_1!) a correspondence

$$\psi_{1\,S}(b_{01}, b_1) \to \lambda_{\hat{\mathscr{Q}}}(a_0, a)\, \varphi(a) \tag{8.1}$$

with $a_0 = a_1 \cap b_{01} \times M_2 \cap M$, $a = a_1 \cap b_1 \times M_2 \cap M$. Due to D 6.1 this can also be written

$$\psi_{1\,S}(b_{01}, b_1) \xrightarrow{k} \varphi_{a_0}(a) = \varphi_{a_0}\, \pi(a_1 \cap b_1)\,. \tag{8.2}$$

The mapping k is norm-continuous if $L = [\mathbf{0}, \mathbf{1}]$ (or L contains at least one interior point; see IV Th 3.18).

In fact, with III (6.4.13) we step by step obtain

$$\begin{aligned} \tfrac{1}{2}\, \| \varphi_{a_0}(a) - \varphi_{a_0}(\tilde{a}) \| &= \sup_{g \in L} |\mu(\varphi_{a_0}(a) - \varphi_{a_0}(\tilde{a}), g)| \\ &= \sup_{(b_0, b) \in \mathscr{F}} |\mu(\varphi_0(a) - \varphi_{a_0}(\tilde{a}), \psi(b_0, b))| \\ &= \sup_{\substack{(b_{20}, b_2) \in \mathscr{F}_2 \\ a_2 \in \mathscr{Q}_2}} |\langle R_1\, \psi'_{2S}(b_{20}, b_2)\, \varphi_{12}(a_1 \times a_2 \cap M), \psi_{1S}(b_{10}, b_1) - \psi_{1S}(\tilde{b}_{10}, \tilde{b}_1)\rangle| \\ &\leqq \| \psi_{1S}(b_{10}, b_1) - \psi_{1S}(\tilde{b}_{10}, \tilde{b}_1) \|\,. \end{aligned}$$

Therefore, k can be extended as a mapping

$$L(\hat{S}_1) \xrightarrow{k} [\mathbf{0}, \varphi(a_0)] \subset \breve{K}\,.$$

Since $\psi_{1S}(b_{01}, b)$ is an additive measure on $\mathscr{R}_1(b_{01})$, this k is linear, so that it can be extended to all of $C(\hat{S}_1)$.

The dual mapping k' then makes

$$\mathscr{B}' \xrightarrow{k'} C'(\hat{S}_1) \quad \text{with} \quad L \xrightarrow{k'} \check{K}(\hat{S}_1),$$

where $\check{K}(\hat{S}_1)$ is the truncated cone on the basis $K(\hat{S}_1)$.

Let $\hat{S}_1(a_1)$ be the support of $k' L$. Then we can go over from $\hat{S}_1$ to $\hat{S}_1(a_1)$. To each $f \in C(\hat{S}_1)$ there corresponds an $f \in C(\hat{S}_1(a_1))$ by restriction to $\hat{S}_1(a_1)$. Then one can regard k as the mapping

$$L(\hat{S}_1(a_1)) \xrightarrow{k} [\mathbf{0}, \varphi(a_0)] \subset \check{K}, \tag{8.3}$$

because $\langle k' g, f \rangle_1 = \langle k' g, f|_{\hat{S}_1(a_1)} \rangle_{1a_1}$ holds with $\langle \ldots, \ldots \rangle_{1a_1}$ the canonical dual form of $C'(\hat{S}_1(a_1))$, $C(\hat{S}_1(a_1))$. The mapping k in (8.3) is injective.

With $\Sigma_{a_1} = \mathscr{B}(\hat{S}_1(a_1))/\mathscr{J}(\hat{S}_1(a_1))$, where $\mathscr{J}(\hat{S}_1(a_1))$ are the sets of measure zero relative to $k' L$, one can regard $L(\hat{S}_1(a_1))$ as a subset of $L(\Sigma_{a_1})$. Then one can extend k to the elements of Σ_{a_1} as a σ-additive measure (Σ_{a_1} regarded as a subset of $L(\Sigma_{a_1})$; see § 3.1).

This follows because for a decreasing (or increasing) sequence f_ν from $L(\hat{S}_1(a_1))$ the sequence $k f_\nu$ is norm-convergent (IV T 6.1), while for a decreasing sequence $\sigma_\mu \in \Sigma_{a_1}$ with $\bigwedge_\mu \sigma_\mu = 0$ the sequence $k\,\sigma_\mu$ is norm-convergent to a $w \in [\mathbf{0}, \varphi(a_0)]$. Here we have

$$\langle \sigma_\mu, k' g \rangle_{1a_1} = \mu(k\,\sigma_\mu, g) \to \mu(w, g)$$

and $\langle \sigma_\mu, k' g \rangle_{1a_1} \to 0$, hence $w = 0$. Since the k in (8.3) is injective,

$$\Sigma_{a_1} \xrightarrow{k} [\mathbf{0}, \varphi(a_0)] \tag{8.4}$$

is an effective, σ-additive measure with $k(\varepsilon) = \varphi(a_0)$. Therefore, (8.4) is a preparator of $\varphi(a_0)$. We call it the ideal preparator belonging to $a_1 \in \mathscr{Q}_1'$. The mixture components $k\,\sigma$ (with $\sigma \in \Sigma_{a_1}$) of $\varphi(a_0)$ are given by the objective properties $\sigma \in \Sigma_{a_1}$ of the systems 1 prepared by a_1. Every $k f$ with $f \in \hat{S}_1(a_1)$ is a mixture of the $k\,\sigma$. For fixed a_1, all possible de-mixings coexist with the ideal preparator (8.4).

The considerations from § 5 about de-mixings of a_2 can at once be transferred to a_1: The experimenter will if possible choose such a_1, i.e. preparing devices which possibly no longer fluctuate in their norms.

At this point, we can not yet ask for those a_1 which are especially interesting for experiments with microsystems.

§ 9 Preparators and Faces of *K*

For a preparator $\Sigma \xrightarrow{W} \check{K}$ of the ensemble $w = W(\varepsilon)$, all $\sigma \in \Sigma$ make $W(\sigma) \leqq w$. The range $W\Sigma$ of a preparator of w is therefore a subset of the order interval $[\mathbf{0}, w]$. Conversely, for $w_1 \in [\mathbf{0}, w]$ we can construct a preparator of the ensemble w, in whose range w_1 lies. Consider the Boolean ring of the subsets of the two-element set $\{\alpha, \beta\}$. With $\sigma_1 = \{\alpha\}$, $\sigma_2 = \{\beta\}$, and $\varepsilon = \{\alpha, \beta\}$ we set $W(\varepsilon) = w$, $W(\sigma_1) = w_1$, $W(\sigma_2) = w - w_1$.

The order intervals $[\mathbf{0}, w]$ are intimately connected with the closed faces $C(w)$ of K generated by w:

T 9.1 With $\check{C}(w) = \bigcup_{0 \leqq \lambda \leqq 1} \lambda C(w)$ we get $[\mathbf{0}, w] = \check{C}(w)$. There is a $w_0 \in C(w)$ such that $C(w)$ is the norm-closure of

$$A(w_0) = \{\tilde{w} \mid \tilde{w} = w' \mu(w', \mathbf{1})^{-1} \quad \text{with} \quad w' \neq 0,\ w' \in [\mathbf{0}, w_0]\}.$$

For arbitrary $w_1 \in K$, the set $A(w_1)$ is a face of K (in general not closed).

Proof. For $w_1 \neq w$ and $w_1 \neq 0$, from $w_1 \leqq w$ follows

$$w = \lambda \tilde{w}_1 + (1 - \lambda) \tilde{w}_2,$$

where $\tilde{w}_1 = w_1 \mu(w_1, \mathbf{1})^{-1}$, $\tilde{w}_2 = (w - w_1)(1 - \mu(w_1, \mathbf{1}))^{-1}$ and $\lambda = \mu(w_1, \mathbf{1})$. This gives $\tilde{w}_1, \tilde{w}_2 \in C(w)$, hence $w_1 = \lambda \tilde{w}_1 \in \check{C}(w)$, which proves $[\mathbf{0}, w] \subset \check{C}(w)$.

For a set $\{w_\nu\}$ dense in $C(w)$ and $\lambda_\nu > 0$, $\sum_\nu \lambda_\nu = 1$, we have $w_0 = \sum_\nu \lambda_\nu w_\nu \in C(w)$ (see IV T 5.6). From this follows $w_\nu \in A(w_0)$, whence $A(w_0)$ is norm-dense in $C(w)$.

$[\mathbf{0}, w_1]$ convex implies $A(w_1)$ convex. $\tilde{w} \in A(w_1)$ implies $\tilde{w} = w' \mu(w', \mathbf{1})^{-1}$ with $w' \in [\mathbf{0}, w_1]$, hence $\lambda [\mathbf{0}, \tilde{w}] \subset [\mathbf{0}, w_1]$ with $\lambda = \mu(w', \mathbf{1})$. Thus $A(\tilde{w}) \subset A(w_1)$ so that $A(w_1)$ is a face. □

In general, no $w_0 \in C(w)$ exists with $A(w_0) = C(w_0)$; but we find

T 9.2 If $C(w)$ is finite-dimensional, then $A(w) = C(w)$.

Proof. From T 9.1 follows that $A(w) \subset C(w)$ holds and $A(w)$ is a face. Thus, we need only show that a finite-dimensional $A(w)$ is closed.

If $A(w)$ is finite-dimensional, w is interior to the convex set $A(w)$. For, w is interior to a convex subset of the same dimension as $A(w)$, generated by *finitely* many w_i (with $\lambda_i w_i \leqq w$ if $0 < \lambda_i < 1$). Since w is interior to $A(w)$, each boundary point w_r of $A(w)$ gives $w = \alpha w_r + (1 - \alpha) \tilde{w}_r$ with $0 < \alpha < 1$ and $\tilde{w}_r \in A(w)$. □

Hence for any $w \in K$ the range of *all* its possible preparators is just the order interval $[0, w]$. This interval and $C(w)$ correspond one-to-one. If w_0 from $C(w)$ is suitably chosen, $A(w_0)$ is norm-dense in $C(w)$ so that we can regard $C(w)$ as the set of all possible mixture components of w_0. In fact, $A(w_0)$ and $C(w)$ are physically indistinguishable, since $A(w_0)$ is norm-dense in $C(w)$. Thus the closed faces of K become physically important, describing all possibilities of acquiring ensembles by preparators of w_0. This result will lead us intuitively to axioms for preparation and registration possibilities (see VI § 2.1).

After this discussion of the physical meaning of the closed faces of K, let us emphasize again: That the sets $\mathscr{W}$, $\mathscr{W}_{\mathscr{B}'}$, $\mathscr{W}_L$ of faces are complete lattices (IV T 5.3), is a fundamental structure solely based on the description of directed actions by physical systems as action carriers. Herein lies the "general validity" of this structure for a wide area in physics. In contrast, microsystems as action carriers are distinguished by more special structures of the set K.

§ 10 Physical Objects as Action Carriers

In VII § 5.3, we shall define when in a fundamental domain to say that the action carriers are "classical systems". There some axioms from VI (laws for the fundamental domain) will be presumed. Here, however, we shall seek to give a definite meaning to the basic concept of "objective properties" of the action carriers. Since this problem has also been discussed in [2] III § 4, we here proceed faster in order to exhibit the essential viewpoint.

It must be emphasized that the approach from Chapter II § 5 cannot be transferred to action carriers because there exist no *pretheories* by which objective properties of the action carriers could be introduced. Therefore, there remains only to seek a way to the objective properties from the preparation and registration procedures. We shall not consider purely "imagined" properties.

As the first step let us complete the registration and preparation procedures by "idealized" elements*). We adopt

D 10.1 A set $c \subset M$ be an *idealized* registration procedure, if a $b_0 \in \mathscr{R}_0$ exists with $c \subset b_0$ while

$$c = \bigcup_{\substack{b \in \mathscr{R} \\ b \subset c}} b \quad \text{and} \quad b_0 \backslash c = \bigcup_{\substack{b \in \mathscr{R} \\ b \subset b_0 \backslash c}} b .$$

Via the following considerations, one recognizes that the idealized registration procedures $c \subset b_0$ can be identified with some elements of the Boolean ring $\hat{\Sigma}$, which in the sense of T 1.3.3 is the abstract (!) completion of $\Sigma = \mathscr{R}(b_0)$.

In general we cannot uniquely identify all of $\hat{\Sigma}$ with subsets of b_0 (one can only do so up to sets of "measure zero").

Therefore, ideal registration procedures can be approximated arbitrarily well "from above and below" by real registration procedures.

According to V T 6.2, the mapping $\mathscr{F} \xrightarrow{\psi} L$ can be extended to the ideal registration procedures $c \subset b_0$ by

$$\psi(b_0, c) = \sup_{\substack{b \in \mathscr{R} \\ b \subset c}} \psi(b_0, b) = \inf_{\substack{b \in \mathscr{R} \\ b_0 \supset b \supset c}} \psi(b_0, b);$$

and likewise we extend the function $\lambda_{\mathscr{F}}$. Thus, one can extend $\mathscr{R}$ by the idealized registration procedures to a system $\hat{\mathscr{R}}$ of selection procedures (see [3] § 12.3).

D 10.2 A set $p \subset M$ is called ideally *registrable* if, for every $b_0 \subset \mathscr{R}_0$, $b_0 \cap p$ is an idealized registration procedure.

Then an ideally registrable set p (see [3] § 12.3) obeys

$$p = \bigcup_{\substack{b \in \mathscr{R} \\ b \subset p}} b \quad \text{and} \quad M \backslash p = \bigcup_{\substack{b \in \mathscr{R} \\ b \in M \backslash p}} b . \tag{10.1}$$

If this holds for p, then p is also ideally registrable.

*) A different extension can be performed by the methods from [41].

Thus, an ideally registrable p has one structure to be expected of an objective property. But p is still too arbitrary as long as one does not fix how p is connected with the preparation procedures. The set of those systems from an $a \in \mathscr{Q}'$ which also have the property p, should be $a \cap p$. Hence this set should also be interpretable as a selection procedure, namely as the "procedure" which selects according to $a \in \mathscr{Q}'$ *and* according to the property p. This suggests a definition analogous to D 10.2:

D 10.3 A $p \subset M$ is called *ideally preparable* if

$$p = \bigcup_{\substack{a \in \mathscr{Q} \\ a \subset p}} a \quad \text{and} \quad M \setminus p = \bigcup_{\substack{a \in \mathscr{Q} \\ a \in M \setminus p}} a .$$

Again, one can extend the mapping $\mathscr{Q}' \xrightarrow{\varphi} K$ uniquely to $\hat{\mathscr{Q}}' \xrightarrow{\varphi} K$ with $\hat{\mathscr{Q}}'$ the system generated by all $\{(a \cap p) \mid a \in \mathscr{Q}', p \text{ is ideally preparable}\}$.

Then we are led to

D 10.4 The set $\mathscr{E}_m$ of all ideally registrable *and* ideally preparable subsets $p \subset M$ is called the set of objective properties of the action carriers.

This $\mathscr{E}_m$ is a Boolean set ring, as follows immediately from (10.1) and D 10.3.

In [3] § 12.3 it is shown, that

$$\chi(p) = \psi(b_0, b_0 \cap p)$$

does not depend on b_0, so that $\mathscr{E}_m \xrightarrow{\chi} L$ is an additive measure. Therefore, with $\hat{\mathscr{E}}_m$ as the completion of $\mathscr{E}_m$ in the sense of T 1.3.3, $\hat{\mathscr{E}}_m \xrightarrow{\chi} L$ is an observable.

By § 3.2, to this observable corresponds a mixture-morphism S, with $\mathscr{B} \xrightarrow{S} \mathscr{B}(\hat{\mathscr{E}}_m)$. The dual S' maps $\hat{\mathscr{E}}_m$ (regarded as a subset $\hat{\mathscr{E}} = \partial_e L(\hat{\mathscr{E}}_m)$ of $L(\hat{\mathscr{E}}_m)$) into L, and $S' p = \chi(p)$ holds.

In [3] § 12.3, we have by means of [2] IV 2.1.11 shown that SK is norm-dense in $K(\hat{\mathscr{E}}_m)$.

Till now we have only introduced some definitions and theorems. The sets $\mathscr{E}_m$ and $\hat{\mathscr{E}}_m$ could in principle be trivial, e.g. consist only of M and $\emptyset$. We shall denote action carriers as physical objects if $\mathscr{E}_m$ is sufficiently comprehensive, i.e. if "it suffices" to perform only the $b_0 \cap p$ with $p \in \mathscr{E}_m$ instead of all registration procedures. Here "it suffices" means that $\chi(\mathscr{E}_m)$ separates the $w \in K$. Hence we adopt

D 10.5 The action carriers are called *physical objects* if $\chi(\mathscr{E}_m)$ separates the set K (better, if "$\chi(\mathscr{E}_m)$ separates K" is a certain hypothesis in the sense of [3] § 10.1).

Therefore, if $\chi(\mathscr{E}_m)$ separates the set K, then $K \xrightarrow{S} K(\hat{\mathscr{E}}_m)$ is injective and SK norm-dense in $K(\hat{\mathscr{E}}_m)$. Due to T 2.3, this conclusion makes S an isomorphism between K and $K(\hat{\mathscr{E}}_m)$ while S' becomes an isomorphism between $L = [\mathbf{0}, \mathbf{1}]$ and $L(\hat{\mathscr{E}}_m)$.

For physical objects, we thus can identify $\mathscr{B}, \mathscr{B}'$ with $\mathscr{B}(\hat{\mathscr{E}}_m)$, $\mathscr{B}'(\hat{\mathscr{E}}_m)$ and K, L with $K(\hat{\mathscr{E}}_m)$, $L(\hat{\mathscr{E}}_m)$.

Therefore, a set of objective properties that is "sufficient for objects" (a set $\mathscr{E}_m$ for which $\chi(\mathscr{E}_m)$ separates K), does only exist if $\mathscr{B} = \mathscr{B}(\hat{\mathscr{E}}_m)$. Then $\hat{\mathscr{E}}_m$ is uniquely determined by $\mathscr{B}$; hence (by [3] § 10.5) $\hat{\mathscr{E}}_m$ is a set of real situations. Thus, the properties p can be ascribed "objectively" to the action carriers in the sense of $x \in p$, i.e. independently of all preparation and registration procedures. The objective properties are approximately registrable in the above sense, i.e. approximately measurable. Other "imagined" properties would not be uniquely defined and hence not recognizable as "real".

Conversely, if $\mathscr{B} = \mathscr{B}(\Sigma)$, with the given axioms it is a certain hypothesis that $\Sigma = \hat{\mathscr{E}}_m$ holds for a suitable $\mathscr{E}_m$.

Therefore, action carriers are physical objects if and only if $\mathscr{B}$ and $\mathscr{B}'$ have the form $\mathscr{B}(\Sigma)$ resp. $\mathscr{B}'(\Sigma)$.

§ 11 Operations and Transpreparators

For later, let us here introduce concepts that are already meaningful in the structure $\mathscr{B}$, $\mathscr{B}'$ with $\mathscr{B}$ a base-normed Banach space.

As in § 2 we start from two base-normed Banach spaces $\mathscr{B}_1$, $\mathscr{B}_2$ with the bases K_1, K_2, and somewhat more generally than in D 2.1 adopt

D 11.1 An affine mapping S of $\check{K}_1$ into $\check{K}_2$ is called an *operation* (for $\check{K}$ see D 6.1).

By T 2.1, a mixture-morphism is a special operation, which maps the subset K_1 of $\check{K}_1$ into K_2. For an operation, K_1 can thus be mapped "only" into $\check{K}_2$.

Just as T 2.1, one can easily prove

T 11.1 An affine mapping $K_1 \xrightarrow{S} \check{K}_2$ can be uniquely extended as a linear mapping $\mathscr{B}_1 \to \mathscr{B}_2$ with $\| S \| \leqq 1$.

But we also obtain

T 11.2 Every positive, linear, norm-continuous mapping $\mathscr{B}_1 \xrightarrow{S} \mathscr{B}_2$ with $\| S \| \leqq 1$ is (restricted to $\check{K}_1$) an operation. Every positive linear mapping $\mathscr{B}_1 \xrightarrow{S} \mathscr{B}_2$ is norm-continuous and $\| S \|^{-1} S$ is an operation.

Proof. See [2] V Th 4.1.2.

T 11.3 To each operation S there corresponds a dual mapping S' of $\mathscr{B}_2'$ into $\mathscr{B}_1'$ with $[\mathbf{0}, \mathbf{1}]_2 \xrightarrow{S'} [\mathbf{0}, \mathbf{1}]_1$, while S is a mixture-morphism if and only if $S' \mathbf{1} = \mathbf{1}$.

Proof. See [2] V Th 4.1.3.

As is well known, the norm-continuous mappings of $\mathscr{B}_1$ into $\mathscr{B}_2$ form a Banach algebra $\mathscr{A}(\mathscr{B}_1, \mathscr{B}_2)$ and hence a Banach space with the norm

$$\| S \| = \sup \{ \| S x \| \mid x \in \mathscr{B}_1, \| x \| \leqq 1 \} = \sup \{ \| S w \| \mid w \in K_1 \}.$$

A positive cone $\mathscr{A}_+(\mathscr{B}_1, \mathscr{B}_2)$ is in $\mathscr{A}(\mathscr{B}_1, \mathscr{B}_2)$ defined by

$$S \geqq 0 \quad \text{is equivalent to} \quad \{S x \geqq 0 \text{ for all } x \geqq 0\}.$$

This $\mathscr{A}_+(\mathscr{B}_1, \mathscr{B}_2)$ is the set of "positive" mappings.

As is well known, $\mathscr{A}(\mathscr{B}_1, \mathscr{B}_2)$ is complete not only in the norm topology but also in the topology of simple convergence. Hence for each sequence of S_n with $S_n w$ norm-convergent in $\mathscr{B}_2$ for all $w \in K_1$, there is an $S \in \mathscr{A}(\mathscr{B}_1, \mathscr{B}_2)$ which makes $\| S_n w - S w \| \to 0$ for all $w \in K_1$ and also $\| S_n x - S x \| \to 0$ for all $x \in \mathscr{B}_1$. According to T 11.2, this $S \in \mathscr{A}(\mathscr{B}_1, \mathscr{B}_2)$ is an operation if and only if $S \in \mathscr{A}_+(\mathscr{B}_1, \mathscr{B}_2)$ and $\| S \| \leqq 1$ hold.

The set of operations, which therefore is the intersection of $\mathscr{A}_+(\mathscr{B}_1, \mathscr{B}_2)$ with the unit ball, shall be denoted by Π.

D 11.2 An additive mapping of a Boolean ring $\Sigma \xrightarrow{\chi} \Pi$ for which $\chi(\varepsilon)$ (with ε the unit element) is a mixture-morphism, will be called an *operation measure* over Σ.

Given an effective ensemble w_0, one can in Σ introduce a uniform structure by

$$d(\sigma_1, \sigma_2) = \mu(\chi(\sigma_1 \dot{+} \sigma_2) w_0, \mathbf{1}).$$

This equals the metric in T 6.2, if there one puts $W(\sigma) = \chi(\sigma) w_0$.

Then easily follows

T 11.4 The mapping $\Sigma \xrightarrow{\chi} \Pi$ is uniformly continuous relative to the metric in Σ and to the uniform structure of simple convergence in Π.

Proof. See [2] V Th 4.3.2.

This theorem and the fact that Π is closed in the topology of simple convergence, as in § 6 imply that we can complete Σ and extend χ to the completion.

With D 6.4 follows:

By $\Sigma \xrightarrow{\chi} \Pi$ with a complete and separable Σ for each $w \in K_1$, the mapping $\Sigma \xrightarrow{\chi w} K_2$ defines a preparator of the ensemble $\chi(\varepsilon) w$.

Therefore we adopt

D 11.3 An additive measure $\Sigma \xrightarrow{\chi} \Pi$ over a separable Boolean ring Σ that is complete relative to $d(\sigma_1, \sigma_2)$ is called a *transpreparator*.

D 11.4 A set $\tilde{\mathscr{A}}$ of operations is said to *coexist* if there exist a Boolean ring Σ and an additive measure $\Sigma \xrightarrow{\chi} \Pi$, with $\tilde{\mathscr{A}} \subset \chi \Sigma$.

VI Main Laws of Preparation and Registration

So far the theory of physical systems as action carriers is far from a g. G.-closed theory (in the sense of [3] § 10.3), i.e. too little structured to characterize the directed actions in nature. Hence, we must add axioms (laws in the sense of [3] § 7.3) and thus try to approach (by standard extension in the sense of [3] § 8) a g. G.-closed theory of "microsystems". Here, the word "microsystem" is to characterize (initially inexactly) the fundamental domain $\mathcal{G}$ of the intended theory (for $\mathcal{G}$ see [3] §§ 2, 3 and 5).

In this chapter let us introduce basic axioms for preparation and registration, which therefore shall be called the main natural laws (briefly, main laws) of preparation and registration. They do not yet characterize the individual structures of the various microsystems (electrons, atoms, etc.), but are universal, like the main laws of (phenomenological) thermodynamics are universal, not specifying the individual properties of thermodynamic systems.

We shall proceed by discussing the physical meaning of the individual main laws for the sets $\mathcal{Q}$, $\mathcal{R}_0$, $\mathcal{R}$, in order then finally to give them a concise mathematical form. The two sets K, L and the mapping $K \times L \xrightarrow{\mu} [0, 1]$ will turn out especially suitable for such a concise form. Expressed in yet another way:

We shall prove mathematical theorems between relations in $\mathcal{Q}$, $\mathcal{R}_0$, $\mathcal{R}$ and relations in K, L, the physical meaning of those in $\mathcal{Q}$, $\mathcal{R}_0$, $\mathcal{R}$ being more transparent. Due to these theorems, as axioms one could choose certain relations in $\mathcal{Q}$, $\mathcal{R}_0$, $\mathcal{R}$, or others in K, L. But we shall formulate the axioms as relations in K, L where they are mathematically more concise, just because K and L have arisen from $\mathcal{K}$ resp. $\mathcal{L}$ by "completions".

In deriving further theorems from the axioms we shall presume the properties of $K \times L \xrightarrow{\mu} [0, 1]$ described in IV, but indicate which further axioms are used.

All the axioms introduced in this chapter are ultimately just structure statements for the convex set K. As we saw in IV, the set $[\mathbf{0}, \mathbf{1}] \subset \mathcal{B}'$ is uniquely determined by K as the set of the affine functionals l on K with $0 \leqq l(w) \leqq 1$. When we formulate axioms for $[\mathbf{0}, \mathbf{1}]$, they are thus indirect axioms for K. We shall formulate axioms for $L \subset [\mathbf{0}, \mathbf{1}]$ instead of $[\mathbf{0}, \mathbf{1}]$. Then they are "not only" axioms for K, since they also express something about the magnitude of L as a subset of $[\mathbf{0}, \mathbf{1}]$. Often one uses to fix $L = [\mathbf{0}, \mathbf{1}]$ in advance (see AV 1.2s in § 1.4). Then the axioms for L are "only" axioms on K. However, from the axioms AV 1.2 and AV 2 formulated only for L, the relation $L = [\mathbf{0}, \mathbf{1}]$ will follow as a theorem.

Together with this validity of $L = [\mathbf{0}, \mathbf{1}]$ as a theorem, the further axioms will in particular determine the lattice $\mathcal{W}_{\mathcal{B}'}$ of exposed faces in K. The first decisive restriction by axioms AV 1 through AV 2 will make $\mathcal{W}_{\mathcal{B}'}$ an orthocomplemented, orthomodular lattice. How this is achieved by physically interpretable axioms will be described in §§ 1 and 2. Therefore recall from IV § 5 that the lattice structure of $\mathcal{W}_{\mathcal{B}'}$ needs *no* foundation by further axioms (see IV T 5.3). Difficult is it only to determine by axioms (natural laws) the *further* structure of the lattice $\mathcal{W}_{\mathcal{B}'}$.

§ 1 Main Laws for the Increase in Sensitivity of Registrations

The laws to be introduced here make assertions about the possibilities of the preparation devices to act on various registration devices. Therefore, the laws will govern the possibilities of constructing registration devices (characterized by elements $b_0 \in \mathcal{R}_0$), but also the possibilities of the action carriers to act on these registration devices, described by the probability function $\lambda_{\mathcal{S}}(a \cap b_0,\ a \cap b)$ for the various $a \in \mathcal{Q}'$. Summarizing, we can briefly say: The axioms to be introduced in this section govern the possible *interactions* between action carriers and registration devices. Here, the word "possible" is always used in the sense of "physically possible" (introduced in [3] § 10.4). Hence, besides the mathematical formulation of the axioms we shall always give the physical interpretation which "results" due to the considerations from [3] § 10.4. The reader who has not studied [3] should accept the physical formulation as a "known", intuitively "customary" manner of speaking about mathematical relations in a physical theory.

§ 1.1 Increase in Sensitivity Relative to Two Effect Procedures

Registration device plus indication were characterized by the pair $(b_0, b) \in \mathcal{F}$. Very intuitively we call an effect procedure $f_1 \in \mathcal{F}$ more sensitive than an $f_2 \in \mathcal{F}$ if $\mu(a, f_1) \geqq \mu(a, f_2)$ for all those $a \in \mathcal{Q}'$ for which $\mu(a, f_1)$ as well as $\mu(a, f_2)$ is defined. One sees immediately that this is equivalent to $\mu(w, \psi(f_1)) \geqq \mu(w, \psi(f_2))$ for all $w \in \mathcal{K}$ and hence for all $w \in K$. But with respect to the order in $\mathcal{B}'$ this is equivalent with $\psi(f_1) \geqq \psi(f_2)$. Therefore we adopt

D 1.1.1 The effect procedure $f_1 \in \mathcal{F}$ is called *more sensitive* than $f_2 \in \mathcal{F}$ if $\psi(f_1) \geqq \psi(f_2)$. An effect $g_1 \in L$ is called more sensitive than $g_2 \in L$ if $g_1 \geqq g_2$.

In this sense, the effect procedures (b_0, b_0) are most sensitive since $\psi(b_0, b_0) = \mathbf{1}$. This is no physically profound assertion but only a trivial consequence of b_0 being the unit element of the Boolean ring $\mathcal{R}(b_0)$. Therefore, really physical structures are only described if one considers such $f \in \mathcal{F}$ that satisfy special supplementary conditions.

One such condition is suggested if one asks for those action carriers which cannot trigger a definite registration procedure $b \in \mathcal{R}$ $(b \neq \emptyset)$. This means asking for those $a \in \mathcal{Q}'$ which make $a \cap b = \emptyset$ although some b_0 (with $b_0 \supset b$) is combinable with a. If such an $a \in \mathcal{Q}'$ makes $a \cap b = \emptyset$ with some $b \neq \emptyset$ (by APS 5.2, for every $b \neq \emptyset$ an $a' \in \mathcal{Q}'$ exists with $a' \cap b \neq \emptyset$), this b therefore does not appear in the device b_0 "by itself", but only by its interaction with a preparing device. Of course, on devices b_0' there can also be such indications b' the response of which does not depend on the interaction with preparing devices. Then, however,

$\psi(b_0', b') = \lambda \mathbf{1}$ holds since this is the condition for $\lambda_{\mathscr{F}}(a \cap b_0', a \cap b') = \mu(\varphi(a), \psi(b_0', b)) = \lambda$ to hold for all $a \in \mathscr{Q}$ combinable with b_0'. In other words, the response probability of b' is *independent* of the preparation procedure a.

A $b \in \mathscr{R}$ (with $b \neq \emptyset$) for which $K_0(\psi(b_0, b))$ is not empty (K_0 as in IV (5.4)), is therefore an indication which can only be triggered by "real" interactions with action carriers. In this connection,

$$M_0(b_0, b) = \bigcup_{\substack{a \in \mathscr{Q}' \\ \varphi(a) \in K_0(\psi(b_0, b))}} a$$

is the set of all those action carriers which cannot trigger b. All other preparing procedures a' with $\varphi(a') \notin K_0(\psi(b_0, b))$ must also generate such systems $x \in a'$ (of course not only these) which make b occur, i.e. obey $x \in b$. This raises the experimentally obvious problem of finding devices b_0 with such signals $b \subset b_0$ that the sets $M_0(b_0, b)$ are not decreased while the response probability for (b_0, b) is made as large as possible, i.e. the problem of making effect procedures (b_0, b) more sensitive without decreasing $M_0(b_0, b)$. But it is *a priori* far from clear whether such an experimental problem is solvable. We shall step by step formulate axioms (natural laws) which say that it is "physically possible" (in the sense of [3] § 10.4) to solve such experimental problems of increasing the sensitivity of effect procedures. We begin with an axiom about increasing a sensitivity relative to two effect procedures, $f_1, f_2 \in \mathscr{F}$. We shall not in advance postulate that such an $f \in \mathscr{F}$ exist for which there is a $w \in \mathscr{K}$ obeying $\hat{\mu}(w, f) = 0$ exactly (with $\hat{\mu}$ as in III (5.1.1)). Rather we shall for an $f \in \mathscr{F}$ consider a $w_0 \in \mathscr{K}$ such that $\hat{\mu}(w_0, f) = \varepsilon$ holds with a number $\varepsilon \ll 1$. The mathematically exact requirement $\hat{\mu}(w, f) = 0$ could be an idealization (for the concept idealization, see [3] §§ 6 and 9), which is experimentally only approximately attainable (i.e. $\hat{\mu}(w_0, f) = \varepsilon$ with ε physically equivalent to zero). Therefore we tentatively consider the heuristic relation

A_1: For prescribed $f_1, f_2 \in \mathscr{F}$ and $w_0 \in K$, with small $\varepsilon_i = \mu(w_0, \psi(f_i))$, there is an $f \in \mathscr{F}$ with $\mu(w_0, \psi(f)) = \eta$ and small η and f more sensitive than f_1 and f_2.

The set K instead of only $\mathscr{K}$ has been used for convenience, since $K \subset \bar{\mathscr{K}}$ holds according to IV § 3 while the elements of $\bar{\mathscr{K}}$ are in *no* physical way distinguishable from those of $\mathscr{K}$ (also not by the topology of physical imprecision in $\mathscr{S}$!). For "ease of mind" the reader can check that one could have restricted A_1 to elements from $\mathscr{K}$, nevertheless obtaining the same relation for elements of K, since $\mathscr{K}$ is norm-dense in K.

Physically interpreted (by concepts from [3] §§ 10.4 and 11.4), relation A_1 with $\varepsilon_1, \varepsilon_2, \eta \ll 1$ means the following: Assume two effect procedures f_1, f_2 whose responses can practically not be evoked by systems prepared due to $a \in w_0$, i.e. these responses can physically "almost certainly" be excluded (see [3] § 11.4). Then it is physically possible to construct an effect procedure which likewise is almost certainly not triggered by systems from an $a \in w_0$ but is more sensitive than f_1 and f_2.

Therefore, A_1 concerns the practical possibility of constructing a registration device with contingent interactions with the action carriers. When we formulate A_1 as a physical law, of course one cannot deduce it from experience, but only intuitively guess from experiences (see [3] § 5). A physical law of the form A_1 can therefore be suggested by *special* experiences as well as by *intuitive notions* that transcend experience. That intuitive notions can also play a role is entirely legitimate, since the methods of physics do not prescribe how to arrive at physical laws ([3] § 5). One only must not mistake such a law as *necessarily* true, perhaps regarding certain intuitive notions as necessary although they are not.

Hence we shall describe some experimental experiences and intuitive notions which suggest a natural law of the form A_1. But in order not to interrupt the mathematical formulation of such a law "similar" to A_1, let us describe experimental and intuitive hints to A_1 afterwards in § 1.2.

Why doesn't A_1 satisfy us as mathematical formulation of a natural law? Because of the "physically small" numbers ε_1, ε_2 and η occuring there. Since we have no idea how small these numbers should be, as always in such cases we idealize (see [3] §§ 6 and 9).

For the numbers ε_1, ε_2, η this means not to prescribe bounds, but to let them "become" arbitrarily small. By such an idealization, we mathematically formulate A_1 as

A_2: For given numbers $\varepsilon_1,\ \varepsilon_2 > 0$ and given $f_1,\ f_2 \in \mathscr{F}$ and $w_0 \in K$ with $\mu(w_0, \psi(f_1)) < \varepsilon_1$, $\mu(w_0, \psi(f_2)) < \varepsilon_2$, there exist an $\eta(\varepsilon_1, \varepsilon_2)$ and an $f \in \mathscr{F}$ with $\mu(w_0, \psi(f)) < \eta$ and f more sensitive than f_1, f_2. Here $\eta(\varepsilon_1, \varepsilon_2)$ can be so chosen that $\varepsilon_1, \varepsilon_2 \to 0$ implies $\eta(\varepsilon_1, \varepsilon_2) \to 0$.

We also define the relation

AV 1.1 For each pair $g_1, g_2 \in L$, there is a $g \in L$ with $g \geqq g_1, g_2$ and $K_0(g_1) \cap K_0(g_2) \subset K_0(g)$.

We shall prove the theorem

T 1.1.1 $A_2 \Rightarrow$ AV 1.1.

Proof. Let g_1, $g_2 \in L$. Then sequences $g_1^\nu \in \mathscr{L}$ and $g_2^\nu \in \mathscr{L}$ exist with $g_1^\nu \to g_1$ and $g_2^\nu \to g_2$ in the (metrizable!) $\sigma(\mathscr{B}', \mathscr{B})$-topology. Hence any $w_0 \in K_0(g_1) \cap K_0(g_2)$ makes $\mu(w_0, g_1^\nu) \to 0$ and $\mu(w_0, g_2^\nu) \to 0$. With $\varepsilon_1^\nu = \mu(w_0, g_1^\nu)$ and $\varepsilon_2^\nu = \mu(w_0, g_2^\nu)$, from A_2 follows that for each ν there is a $g^\nu \in \mathscr{L}$ with $\mu(w_0, g^\nu) < \eta(\varepsilon_1^\nu, \varepsilon_2^\nu)$ and $g^\nu \geqq g_1^\nu$, g_2^ν such that $\nu \to \infty$ gives $\eta(\varepsilon_1^\nu, \varepsilon_2^\nu) \to 0$. Since L is $\sigma(\mathscr{B}', \mathscr{B})$-compact, there is a subsequence of the ν (again to be called ν) for which $g^\nu \to g \in L$ holds in the $\sigma(\mathscr{B}', \mathscr{B})$-topology. From this follows $\mu(w_0, g) = 0$ and $g \geqq g_1, g \geqq g_2$.

Since $K_0(g_1) \cap K_0(g_2)$ is a closed face of K (IV § 5), there exists (by T 5.4) a w_0 such that $C(w_0) = K_0(g_1) \cap K_0(g_2)$. We now assume that the above w_0 is just so chosen that $C(w_0) = K_0(g_1) \cap K_0(g_2)$. From $\mu(w_0, g) = 0$, i.e. $w_0 \in K_0(g)$, follows $C(w_0) \subset K_0(g)$. □

As the reader can easily see, AV 1.1 conversely implies a relation almost like A_2, only that one must replace "f_3 is more sensitive than f_1 and f_2" by "f_3 is *approximately* more sensitive than f_1 and f_2", what is more physical than A_2 itself. We have chosen A_2 in order not to obscure the formulation by many approximation estimates.

Because of $g \geqq g_1, g_2$ follows $K_0(g) \subset K_0(g_1)$, $K_0(g_2)$ and hence $K_0(g) \subset K_0(g_1) \cap K_0(g_2)$. Hence we can in AV 1.1 simply replace $K_0(g) \supset K_0(g_1) \cap K_0(g_2)$ by $K_0(g) = K_0(g_1) \cap K_0(g_2)$.

We regard AV 1.1 as the idealization of A_1, an idealization which just replaces "physically small" $\varepsilon_1, \varepsilon_2, \eta$ by the limit zero. In the form AV 1.1 we introduce the "first main law" of registering as an axiom. We saw clearly how this idealized AV 1.1 is so simple just because we mathematically completed $\mathscr{K}$ and $\mathscr{L}$ to K and L. Let us call AV 1.1 the *law of increase in sensitivity of the first kind* (another law AV 1.2 will follow in § 1.4).

§ 1.2 Some Experimental and Intuitive Indications for the Law of Increase in Sensitivity

In order to suggest the relation A_1 in § 1.1 by experience, it would for $f_1 = (b_{01}, b_1)$, $f_2(b_{02}, b_2)$ be nicest to derive from the physical construction of the registration devices corresponding to b_{01}, b_{02} a physical construction for a device b_0 with indication b, so that $f = (b_0, b)$ satisfies A_1. Here the "physical construction" of a device b_0 means what we have in III described by the preparation procedure $a_2 \in \mathscr{Q}_2$ of the systems 2 and the corresponding state space Z_2 (see the construction of $\mathscr{R}_0$, $\mathscr{R}$ and their connection with trajectory effects in III §§ 4 and 6). Since in III we agreed to conceive the state space Z_2 so broadly that all systems 2 could be described in it, our question of a physical construction amounts to this: Given two $a_{21}, a_{22} \in \mathscr{Q}_2$ and two trajectory effects $\psi_{2S}(b_{021}, b_{21})$, $\psi_{2S}(b_{022}, b_{22})$, we seek an $a_2 \in \mathscr{Q}_2$ and a trajectory effect $\psi_{2S}(b_{02}, b_2)$ so that for f_1, f_2, f with $f = (b_0, b)$, $f_1 = (b_{01}, b_1)$, $f_2 = (b_{02}, b_2)$ and

$$
\begin{aligned}
b_0 &= (M_1 \times a_2 \cap b_{02}) \cap M, & b &= (M_1 \times a_2 \cap b_2) \cap M; \\
b_{01} &= (M_1 \times a_{21} \cap b_{021}) \cap M, & b_1 &= (M_1 \times a_2 \cap b_{21}) \cap M; \\
b_{02} &= (M_1 \times a_{22} \cap b_{022}) \cap M, & b_2 &= (M_1 \times a_{22} \cap b_{22}) \cap M;
\end{aligned}
\tag{1.2.1}
$$

the relation A_1 holds with physically small $\varepsilon_1, \varepsilon_2, \eta$. Here one can allow, still in the sense of the remarks to the proof of T 1.1.1, that f is only approximately more sensitive than f_1 and f_2.

Since we assumed (see II § 3.1) that the measurement of trajectories is explained by pretheories for given $k_1(y) = \psi_{2S}(b_{021}, b_0)$ and $k_2(y) = \psi_{2S}(b_{022}, b_{22})$, it suffices to find a suitable $k(y)$, without worrying about the measuring methods. If in particular $a_{22} = a_{21}$, a suitable, more sensitive f is easy to find: One sets $a_2 = a_{21}$ and chooses $k(y) = \max\{k_1(y), k_2(y)\}$. Hence, the problem of greater sensitivity is of crucial importance only for $a_{22} \neq a_{21}$; then *another* construction rule a_2 is sought as well as a suitable trajectory effect k.

In order to express this more clearly, let us formulate this problem with the mapping β from III (6.4.14):

For given a_{21}, k_1 and a_{22}, k_2 and $w_0 \in K$ with $\mu(w_0, \beta(a_{21}, k_1)) \sim 0$, $\mu(w_0, \beta(a_{22}, k_2)) \sim 0$, we seek a pair a_2, k to fulfill $\mu(w_0, \beta(a_2, k)) \sim 0$ and $\beta(a_2, k) \gtrsim \beta(a_{21}, k_1)$, $\beta(a_2, k) \gtrsim \beta(a_{22}, k_2)$.

It appears entirely hopeless to solve this problem for $a_{22} \neq a_{21}$ *in generality*. It was just an advantage of the basic relations presented in III that no "complete survey" of the construction rules $a_2 \in \mathscr{Q}_2$ was needed. In the sense of the correspondence rules, we need *only* to know whether an actual production method of registration devices implements *some* $a_2 \in \mathscr{Q}_2$. Hence, in III we had no need, by pretheories to describe "comprehensively" "all" the possible devices. Just vice versa, a principal problem of the experimental physicist is to "find" especially appropriate construction rules $a_2 \in \mathscr{D}_2'$ for measuring devices. Here "appropriate" means that the trajectory effects on the devices lead to such $\psi(b_0, b) \in L$ (with b_0, b from (1.2.1)) which allow one to say especially much about the action carriers (see in [2] XVII § 2.3 the discussion of the information about the microsystem provided by the effects $g \in L$).

Hence we *cannot in generality* answer the question for a pair a_2, k for given a_{21}, k_i and a_{22}, k_2. We must in each concrete case leave the answer to intuition and to the wealth of ideas of the experimental physicist. Nevertheless let us at least by some *examples* (simplified for theorists!) demonstrate how such a construction a_2 and a corresponding k could appear.

For our example we choose a filter and impact counters (similar considerations were performed by Mielnik [36]). When in such examples speaking of microsystems or even quanta instead of saying "action carriers", we do so for the sake of physical intuition, an intuition quite permissible for "guessing" natural laws and often successful (see the figure at the end of § 5 in [3]). By a filter we here understand a "device part", (more or less complicated, e.g., a piece of painted glass) which lets microsystems pass through or "absorbs" them. An impact counter is a device part (see [2] XVI § 6.1) which responds as soon as a microsystem hits a sensitive surface. Figure 4 depicts a registration device composed of a filter and an impact counter. If microsystems pass through (from the left), they will be registered. Thus Fig. 4 symbolizes as device $a_2 \in \mathscr{D}_2'$. The trajectory effects consist in that the counter does or does not respond. When (b_{02}, b_2) denotes the effect of *non* response, one often calls the probability $\mu(w, \psi\varrho(a_2 \cap b_{20}, a_2 \cap b_2))$ the absorption coefficient of the filter for the ensemble w (for $\psi\varrho$ see III (6.4.9)), even if the microsystems not hitting the collector can also be reflected by the filter. The filter and the impact counter together form a registration device.

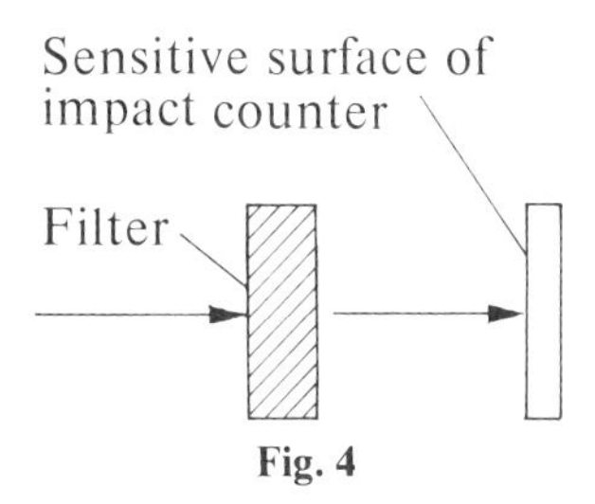

Fig. 4

Let a_{21} and a_{22} denote two registration devices similar to Fig. 4, only with different filters. Let w_0 be an ensemble with $\mu(w_0, \psi\varrho(a_{21} \cap b_{210}, a_{21} \cap b_{21}) \sim 0$ and $\mu(w_0, \psi\varrho(a_{22} \cap b_{220}, a_{22} \cap b_{22}) \sim 0$, i.e. an ensemble which passes the filter F_1 of a_{21} as well as the F_2 of a_{22}. How can one construct an a_2 with the properties desired in A_1? One often succeeds by the method from Fig. 5. One constructs a block of "very many", alternatingly inserted filters F_1, F_2 followed by the impact counter. This device a_2 of Fig. 5, with b_{02}, b_2 as the effect that a microsystem was absorbed by the filter block, then has the desired properties: w_0 passes practically without absorption through the whole block; each ensemble is absorbed more by the block than by one of F_1, F_2 alone. The example of polarization filters for light

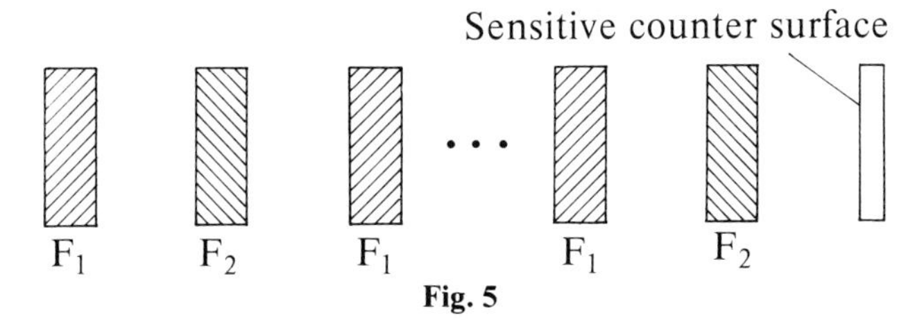

Fig. 5

shows that a block of just *two* filters F_1, F_2 occasionally absorb certain ensembles less than each filter F_1, F_2 alone!

The example of constructing a_2 with a block of many filters should not suggest that this construction succeeds for all possible filters. We only wanted to show a method which *frequently* yields the desired a_2.

Under the caption "semipermeable walls" the indicated example achieved a physical significance beyond the theory of microsystems (e.g. in thermodynamics). Nature itself constructs many such walls as biological membranes.

Such semipermeable walls resp. membranes (for the use of such walls in a foundation of quantum mechanics see [37]) are called "impassable" for an ensemble w_0 if $\mu(w_0, \psi\varrho(a_2 \cap b_{20}, a_2 \cap b_2)) \sim 1$, when (as above) b_2 indicates that the impact counter has *not* responded. Then $b_2^* = b_{20} \backslash b_2$ is the indicator that the counter has responded. Hence the semipermeable wall is impassable for w_0 if $\mu(w_0, \psi\varrho(a_2 \cap b_{20}, a_2 \cap b_2^*)) \sim 0$. This suggests the following sensitivity question: Let two given walls be impassable for w_0; we seek another semipermeable wall which likewise is impassable for w_0, but otherwise *more passable* than the first two. (This problem can of course also be posed for filters!).

A more passable wall is not so easily found as a more absorbing filter. But AV 1.1 says that it is experimentally possible to solve this problem.

Therefore it is not true that AV 1.1 can simply be read from the experiments. Deep physical laws are never read off from experience. Rather one can only expect that a *few* experiences can stimulate an intuitive step to a natural law. Hence it is quite legitimate if we now, after discussing experimental possibilities, also invoke purely intuitive notions which can lead to the natural law AV 1.1. The statement $\mu(w_0, \beta(a_2, k)) \sim 0$ means that the devices $x_1 \in a_1 \cap b_1$ with $a = (a_1 \cap b_1) \times M_2 \cap M \in w_0$ are unable, by interacting with the devices $x_2 \in a_2$, to evoke on these x_2 trajectories from the support of k. Something analogous holds for a_{21}, k_1 and a_{22}, k_2.

Thus one could imagine that the interaction possibilities described by w_0 are endowed with some "property" (characterized by w_0), namely of not being able to evoke certain trajectories on the systems $x_2 \in a_{21}$ resp. $x_2 \in a_{22}$. Concerning the action carriers determined by $a_1 \cap b_1$ one could say that w_0 determines a "property of the action carriers" $x \in a \in w_0$, namely of not being able to evoke certain trajectories on the $x_2 \in a_{21}$ resp. $x_2 \in a_{22}$.

If some w has the probabilities $\mu(w, \beta(a_2, k_1)) \neq 0$ and $\mu(w, \beta(a_{22}, k_2)) \neq 0$, then for realizations of w by preparation procedure $a \in w$ there should always *appear* action carriers $x \in a$ which do *not* have the property determined by w_0.

This "notion" of the action carriers having a property characterized by w_0 suggests to regard $\mu(w, \beta(a_{21}, k_1))$ and $\mu(w, \beta(a_{22}, k_2))$ as measures for the ability of the devices $x_2 \in a_{21}$ resp. $x_2 \in a_{22}$ from an $a \in w$ to separate such action carriers which do not have that property. But then there should also be devices which can separate at least as well as those from a_{21} and a_{22}. After all, the devices from a_{21} and a_{22} show that one can, at least with the probability $\max\{\mu(w, \beta(a_{21}, k_1)), \mu(w, \beta(a_{22}, k_2))\}$ separate those action carriers $x \in a \in w$ which do not have the property determined by w_0! But this means that there exists a pair a_2, k which makes

$$\mu(w_0, \beta(a_2, k)) \sim 0 \quad \text{but} \quad \mu(w, \beta(a_2, k)) \geqq \mu(w, \beta(a_{21}, k_1))$$

and

$$\mu(w, \beta(a_2, k)) \geqq \mu(w, \beta(a_{22}, k_2)) .$$

Not these intuitive notions, but the experimental experiences with filter blocks as in Fig. 5 suggest the following possibility for greater sensitivities: One chooses the two filters F_1 and F_2 in Fig. 5 to be equal. Experiment shows that the absorption coefficient increases, in fact for an appropriate ensemble up to the maximal value 1. Or, expressed otherwise: For $\gamma = \sup_{w \in K} \mu(w, \beta(a_2, k))$ and an ensemble w_s with $\mu(w_s, \beta(a_2, k)) \sim \gamma$, there ought to be a pair a_2', k' such that $\mu(w_s, \beta(a_2', k')) \sim 1$. But also for all other w we should have $\mu(w, \beta(a_2', k')) \geqq \mu(w, \beta(a_2, k_2))$. Nevertheless, for w_0 with $\mu(w_0, \beta(a_2, k_2) \sim 0$ also $\mu(w_0, \beta(a_2, k)) \sim 0$ should hold.

This assertion of a possible increase in sensitivity of *one* effect, suggested by experiments with filter blocks, is not contained in AV 1.1. We shall postulate this idealization as AV 1.2 in § 1.4.

The discussions in this subsection shows how difficult it can be to classify a natural law such as AV 1.1. On the one hand, it contains a challenge to construct a more sensitive device (under given constraints), on the other hand an "existing" structure, namely that "nature" does not erect unsurmountable limits to attempts to meet that challenge. But experiences in constructing devices might raise the suspicion that such limits could exist. Later such limits might even become recognizable, so that they can be formulated in laws. Then physics would progress toward a more comprehensive theory (see [3] § 8), while the present theory would remain useable in a certain fundamental domain, delimited by means of the more comprehensive theory.

§ 1.3 Decision Effects

Here let us deduce some consequences from axiom AV 1.1, but first show several equivalences.

T 1.3.1 The following relations are equivalent.

(i) AV 1.1.

(ii) The elements of $\mathscr{U}$ (IV (5.13)) are upward directed sets.

(iii) The sets $L_1(k) = \{g \mid g \in L \text{ and } \mu(w, g) = 1 \text{ for all } w \in k\}$ are downward directed.

(iv) The set $\mathscr{U}$ has a largest element.

(v) The sets $L_1(k)$ have a smallest element.

Proof. (i) $\Rightarrow$ (ii): Let $g_1, g_2 \in L_0(k)$. By AV 1.1 there is a g with $g \geqq g_1, g_2$ and $K_0(g) \supset K_0(g_1) \cap K_0(g_2)$.

From $g_1, g_2 \in L_0(k)$ follows $K_0(g_1) \cap K_0(g_2) \supset K_0 L_0(k)$ and hence $K_0(g) \supset K_0 L_0(k)$, i.e. $g \in L_0 K_0(g) \subset L_0 K_0 L_0(k) = L_0(k)$ (the last by IV T 5.4).

(ii) $\Rightarrow$ (iv): By IV T 6.2.

(iv) $\Rightarrow$ (i): For $k = K_0(g_1) \cap K_0(g_2)$ we have $g_1, g_2 \in L_0(k)$. The largest element g of $L_0(k)$ then makes $g \geqq g_1, g_2$ and by $g \in L_0(k)$ yields $K_0(g) \supset K_0 L_0(K_0(g_1) \cap K_0(g_2)) = K_0(g_1) \cap K_0(g_2)$.

(ii) $\Rightarrow$ (iii) and (iv) $\Rightarrow$ (v) follow from the fact that $g \rightarrow 1 - g$ is a bijection of $L_0(k)$ onto $L_1(k)$. □

When we impose AV 1.1 as an axiom, (ii) through (v) in T 1.3.1 thus hold as theorems. For the next discussions in this § 6.1, let us presume AV 1.1 as an axiom.

D 1.3.1 The largest element of $L_0(k)$ which due to T 1.3.1 exists, will be denoted by $e\, L_0(k)$ and called a *decision effect.* By G we denote the set of all decision effects.

T 1.3.2 The following relations hold:

$$L_0(k) = \{g \mid g \in L \text{ and } g \leqq e\, L_0(k)\}\,, \tag{1.3.1}$$

$$K_0 L_0(k) = K_0(e\, L_0(k))\,, \tag{1.3.2}$$

$$L_0(k) = L_0 K_0(e\, L_0(k))\,, \tag{1.3.3}$$

$$\{g \mid g \in L \text{ and } g \leqq e\} = L_0 K_0(e) \text{ for each } e \in G. \tag{1.3.4}$$

Proof. From $g \in L_0(k)$ immediately follows $g \leqq e\, L_0(k)$. From $g \leqq e\, L_0(k)$ follows

$$\mu(w, e\, L_0(k)) = 0 \Rightarrow \mu(w, g) = 0, \tag{1.3.5}$$

i.e. $\mu(w, g) = 0$ for all $w \in k$, since $e\, L_0(k) \in L_0(k)$. Thus $g \in L_0(k)$ holds, which proves (1.3.1).

From (1.3.1) follows (1.3.5) for all $g \in L_0(k)$ and hence $K_0(e\, L_0(k)) \subset K_0 L_0(k)$, which implies (1.3.2). By applying L_0 to (1.3.2), and with $L_0 K_0$ as the identity mapping (IV T 5.4), we get (1.3.3), which due to (1.3.1) is identical with (1.3.4). □

T 1.3.3 G is a complete lattice. The mapping $L_0(k) \rightarrow e\, L_0(k)$ is a lattice isomorphism $\mathscr{V} \xrightarrow{e} G$.

Proof. From (1.3.1) follows immediately that the mapping $L_0(k) \rightarrow e\, L_0(k)$ is an order isomorphism of $\mathscr{V}$ and G. Since $\mathscr{V}$ is a complete lattice (IV T 5.3), also G is a complete lattice. □

T 1.3.4 The mapping $e\, L_0(k) \rightarrow K_0(e\, L_0(k))$ is a dual isomorphism of the lattices G and $\mathscr{W}_L$.

Proof. $\mathscr{V} \xrightarrow{K_0} \mathscr{W}_L$ is a dual isomorphism due to IV T 5.4. Therefore (with e^{-1} as the mapping inverse to the e in T 1.3.3), $G \xrightarrow{e^{-1}} \mathscr{V} \xrightarrow{K_0} \mathscr{W}_L$ is a dual isomorphism of G onto

$\mathscr{W}_L$. With $e L_0(k) \in G$ we thus find that $e L_0(k) \to e^{-1} e L_0(k) = L_0(k) \xrightarrow{K_0} K_0 L_0(k) = K_0(e L_0(k))$ is a dual isomorphism (the last equality sign by (1.3.2)).

T 1.3.5 For $g \in L$ and $e \in G$, the relations

(i) $g \leqq e$,

(ii) $K_0(g) \supset K_0(e)$

are equivalent. For each subset $l \subset L$ there is exactly one $e \in G$ with $K_0(l) = K_0(e)$, i.e. a smallest $e \in G$ with $e \geqq g$ for all $g \in l$.

Proof. (i) $\Rightarrow$ (ii) follows immediately.

(ii) $\Rightarrow$ (i): From (ii) follows $L_0 K_0(g) \subset L_0 K_0(e)$ and hence $g \in L_0 K_0(e)$, from which (i) follows by (1.3.4).

By T 1.3.4 an element $K_0(l)$ of $\mathscr{W}_L$ is equal to a $K_0(e)$ for one and only one $e \in G$. From (ii) $\Rightarrow$ (i) follows $g \leqq e$ for all $g \in l$. If $e_2 \in G$ and $e_2 \geqq g$ for all $g \in l$, then $K_0(e_2) \subset K_0(l) = K_0(e)$ and thus $e_2 \geqq e$ due to T 1.3.4. □

T 1.3.6 $G \subset \partial_e L$.

Proof. From $e \in G$ and $e = \lambda g_1 + (1 - \lambda) g_2$ with $0 < \lambda < 1$ and $g_1, g_2 \in L$ follows $K_0(g_1) \supset K_0(e)$ and hence $g_1 \leqq e$ due to T 1.3.5. Likewise follows $g_2 \leqq e$. If we had $g_1 \neq e$, some $w \in K$ would exist with $\mu(w, g_1) \lneqq \mu(w, e)$. From this follows $\mu(w, e) = \lambda \mu(w, g_1) + (1 - \lambda) \mu(w, g_2) \lneqq \lambda \mu(w, e) + (1 - \lambda) \mu(w, e) = \mu(w, e)$ which is a contradiction. □

Since in $\mathscr{W}_L$ the lattice operation $\wedge$ is identical with set intersection, T 1.3.4 implies

T 1.3.7 With $e_1, e_2 \in G$ we get

$$K_0(e_1) \cap K_0(e_2) = K_0(e_1 \vee e_2),$$

i.e. from $\mu(w, e_1) = 0$ and $\mu(w, e_2) = 0$ follows $\mu(w, e_1 \vee e_2) = 0$.

T 1.3.8 With $e_1\, e_2 \in G$ and $g \in L$ we find $g \leqq e_1, e_2 \Rightarrow g \leqq e_1 \wedge e_2$.

Proof. Due to T 1.3.5, from $g \in L$ and $g \leqq e_1, e_2$ follows $K_0(g) \supset K_0(e_1)$ and $K_0(g) \supset K_0(e_2)$ and hence

$$K_0(g) \supset K_0(e_1) \vee K_0(e_2)$$

in the lattice $\mathscr{W}_L$.

Due to T 1.3.4 follows $K_0(e_1) \vee K_0(e_2) = K_0(e_1 \wedge e_2)$ and hence $K_0(g) \supset K_0(e_1 \wedge e_2)$, whence T 1.3.5 yields $g \leqq e_1 \wedge e_2$. □

T 1.3.9 For a subset $A \subset G$, the lower bound of A in L is $\bigwedge_{e \in A} e$. If A is directed downward, $\bigwedge_{e \in A} e$ is the limit of A in the $\sigma(\mathscr{B}', \mathscr{B})$-topology.

Proof. The first part of the theorem generalizes T 1.3.8. As there, for $g \in L$ with $g \leqq e$ for all $e \in A$ follows $K_0(g) \supset K_0(e)$ and hence $K_0(g) \supset \bigvee_{e \in A} K_0(e) = K_0(\bigwedge_{e \in A} e)$, whence $g \leqq \bigwedge_{e \in A} e$ follows; therefore $\bigwedge_{e \in A} e$ is also the lower bound of A in L.

If A is directed downward, by IV T 6.2 the limit g of A exists and g is the lower bound of A in $\mathscr{B}'_+$. Since L is compact, we find $g \in L$ and hence $g = \bigwedge_{e \in A} e$ by the first part of the theorem. □

§ 1.4 The Increase in Sensitivity of an Effect

For effects $g \in \mathscr{L}$ which have $\mu(w_0, g) \sim 0$ but *not* $\sup_{w \in K} \mu(w, g) \sim 1$, the discussion in § 1.2 suggests that one can construct a more sensitive effect $g' \in \mathscr{L}$ with $\mu(w_0, g') \sim 0$, which makes $\mu(w, g') \sim 1$ for those $w \in K$ for which $\mu(w, g)$ comes close to its supremum.

We could express this as follows: For $g \in \mathscr{L}$ with $\mu(w_0, g) \sim 0$ and $0 < \lambda^{-1} = \sup_{w \in K} \mu(w, g)$, there exists a $g' \in \mathscr{L}$ with $\mu(w, g') \sim 1$ for those $w \in K$ for which $\mu(w, \lambda g) \sim 1$.

L is convex and always contains $\mathbf{1} - g$ whenever it contains g. Each accumulation point can in the $\sigma(\mathscr{B}', \mathscr{B})$-topology be represented by a convergent sequence. This suggests to consider an "auxiliary set" $\bar{\bar{L}}$ defined as follows:

$\bar{\bar{L}} \subset [\mathbf{0}, \mathbf{1}]$ is the smallest convex $\sigma(\mathscr{B}', \mathscr{B})$-closed set with $L \subset \bar{\bar{L}}$, where $y \in \bar{\bar{L}}$ and $\|\lambda y\| \leqq 1$ for $\lambda > 0$ imply $\lambda y \in \bar{\bar{L}}$ and where $y \in \bar{\bar{L}}$ implies $\mathbf{1} - y \in \bar{\bar{L}}$.

Idealizing for the increase in sensitivity of *one* effect from L one could then formulate the axiom:

For $g_0 \in L$, $\mu(w_0, g_0) = 0$, $y \in \bar{\bar{L}}$, $\mu(w_0, y) = 0$, there is a $g \in L$ with $\mu(w_0, g) = 0$ which also makes $\mu(w, g) \geqq 1 - \delta(\varepsilon)$ for $\mu(w, y) \geqq 1 - \varepsilon$, where $\delta(\varepsilon) \to 0$ as $\varepsilon \to 0$.

To obtain the final formulation of this axiom, we take into account the *theorem* that $\bar{\bar{L}} = [\mathbf{0}, \mathbf{1}]$. This is *proven* once we show $2\bar{\bar{L}} - \mathbf{1} = [-\mathbf{1}, \mathbf{1}]$. We first show that $F = \bigcup_{\lambda > 0} \lambda(2\bar{\bar{L}} - \mathbf{1})$ is the subspace spanned by $2\bar{\bar{L}} - 1$. This follows easily if we take note that $y \in 2\bar{\bar{L}} - \mathbf{1}$ implies $-y \in 2\bar{\bar{L}} - \mathbf{1}$ and $2\bar{\bar{L}} - \mathbf{1}$ is convex.

The set $F \cap [-\mathbf{1}, \mathbf{1}]$ consists of all $y \in F$ with $\|y\| \leqq 1$. For $y = \lambda y_0$ with $y_0 \in 2\bar{\bar{L}} - \mathbf{1}$ follows $\lambda \|y_0\| \leqq 1$. Provided we show below that $\|y_0\|^{-1} y_0$ is also an element of $2\bar{\bar{L}} - \mathbf{1}$ if y_0 is, we get $F \cap [-\mathbf{1}, \mathbf{1}] = 2\bar{\bar{L}} - \mathbf{1}$. Thus, since $2\bar{\bar{L}} - \mathbf{1}$ is $\sigma(\mathscr{B}', \mathscr{B})$-closed, also F is $\sigma(\mathscr{B}', \mathscr{B})$-closed ([7] IV § 6.4). Since the $\sigma(\mathscr{B}', \mathscr{B})$-closed subspace of $\mathscr{B}'$ spanned by L is all of $\mathscr{B}'$, we obtain $F = \mathscr{B}'$ and hence $2\bar{\bar{L}} - \mathbf{1} = [-\mathbf{1}, \mathbf{1}]$.

If $y_0 \neq 0$ and $y_0 \in 2\bar{\bar{L}} - \mathbf{1}$, then $\mathbf{1}$ and y_0 span a two-dimensional subspace E of $\mathscr{B}'$, whose intersection $E \cap [\mathbf{0}, \mathbf{1}]$ must have the form given in Fig. 6. Since $\|g\|^{-1} g$ is an element of $\bar{\bar{L}}$ provided $g = (1/2)(y_0 - \mathbf{1})$ is, and $\mathbf{1} - g \in L$ provided $g \in L$, also $g_1 = \|\mathbf{1} - \|g\|^{-1} g\|^{-1} (\mathbf{1} - \|g\|^{-1} g)$ and $g_2 = \mathbf{1} - g_1$ are elements of L. But $g_1, g_2, \mathbf{1}, \mathbf{0}$ are the four extreme points of $E \cap [-\mathbf{1}, \mathbf{1}]$, i.e. $E \cap [\mathbf{0}, \mathbf{1}] = E \cap \bar{\bar{L}}$, whence $E \cap [-\mathbf{1}, \mathbf{1}] = E \cap (2\bar{\bar{L}} - \mathbf{1})$ follows. For y_0 thus $\|y_0\|^{-1} y_0 \in E \cap [-\mathbf{1}, \mathbf{1}]$ implies $\|y_0\|^{-1} y_0 \in 2\bar{\bar{L}} - \mathbf{1}$. □

Therefore, as final formulation of the axiom we choose

AV 1.2 From $g_0 \in L$, $w_0 \in K$, $\mu(w_0, g_0) = 0$, $y \in [\mathbf{0}, \mathbf{1}]$, $\mu(w_0, y) = 0$ follows the existence of a $g \in L$ with $\mu(w_0, g) = 0$, which makes $\mu(w, g) \geqq 1 - \delta(\varepsilon)$ for all $w \in K$ with $\mu(w, y) \geqq 1 - \varepsilon$, where $\delta(\varepsilon)$ can be chosen with $\delta(\varepsilon) \to 0$ for $\varepsilon \to 0$.

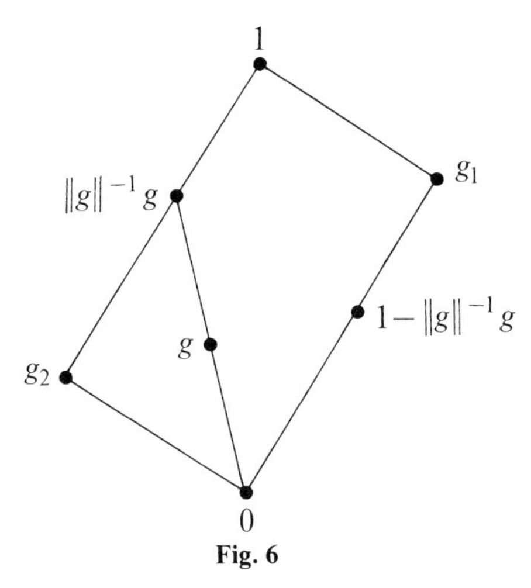

Fig. 6

There are arguments for sharpening this axiom to

AV 1.2 s $L = [\mathbf{0}, \mathbf{1}]$

(obviously stronger than AV 1.2). As an argument for AV 1.2 s one could invoke a method often used in physical theories: If *no experiments indicate* that L is not the *maximal possible* set $[\mathbf{0}, \mathbf{1}]$, one just chooses L as large as possible. Such a requirement brings us closer to the goal of a g. G.-closed theory (in the sense of [3] § 10.3). If it should later turn out by experience that not all $g \in [\mathbf{0}, \mathbf{1}]$ are approximately realizable by effect procedures (i.e. physically possible), then one must modify the theory. In the fundamental domain of microsystems (for macrosystems see X §§ 1, 2 and 3) there have not yet appeared experimental indications against the "physical possibilities" formulated by AV 1.2 s for constructing registration devices.

Therefore it is a matter of taste whether one adds AV 1.2 rather than AV 1.2 s as an axiom. Perhaps even AV 1.2 $\Leftrightarrow$ AV 1.2 s holds, but this could not be proven. In § 2.2 we shall prove other relations to make clearer which choice of axioms leads to equivalent theories. For that reason, we shall here deduce no further consequences from AV 1.2 or AV 1.2 s.

Let us only note

T 1.4.1 AV 1.2 is equivalent to the relation

A: For each $g_0 \in L$ and $y \in [\mathbf{0}, \mathbf{1}]$ with $K_0(y) \supset K_0(g_0)$, there is a $g \in L$ with $K_0(g) \supset K_0(g_0)$ and $\mu(w, g) \geqq 1 - \delta(\varepsilon)$ for all $w \in K$ with $\mu(w, y) \geqq 1 - \varepsilon$, where $\delta(\varepsilon)$ can be chosen with $\delta(\varepsilon) \to 0$ as $\varepsilon \to 0$.

Proof. Suppose AV 1.2 holds. If we choose the element w_0 in AV 1.2 so that $K_0(g_0) = C(w_0)$ (possible by V T 5.6), then $K_0(y) \supset K_0(g_0)$ implies $\mu(w_0, y) = 0$. Hence by AV 1.2 there is a g with $\mu(w_0, g) = 0$ etc; but from $\mu(w_0, g) = 0$ follows $w_0 \in K_0(g)$ and hence $K_0(g_0) = C(w_0) \subset K_0(g)$, whence AV 1.2 $\Rightarrow A$.

Now suppose A holds. From $\mu(w_0, g_0) = 0$ follows $w_0 \in K_0(g_0)$ and hence $w_0 \in K_0(y)$, i.e. $\mu(w_0 y) = 0$. By A there is a $g \in L$ with $K_0(g) \supset K_0(g_0)$ etc. But from $K_0(g) \supset K_0(g_0)$ follows $w_0 \in K_0(g)$, i.e. $\mu(w_0, g) = 0$, whence A $\Rightarrow$ AV 1.2. □

AV 1.2 resp. the equivalent relation A in T 1.4.1 is called the *law of increase in sensitivity of the second kind.* We shall call AV 1.2 s the "strong form" of this law. Let us emphasizes that AV 1.2 s does not imply AV 1.1. Even when one requires AV 1.2 s (which does not influence the convex set K!), AV 1.1 does *not* hold automatically, but presents a true restriction on the convex set K. Thus, one must view AV 1.1 as a *fundamental structure law* of the directed action by physical systems (as action carriers). But despite this we must warn against the opinion that AV 1.1 is typical for microsystems. Rather, it holds much more generally, e.g. for the directed actions by macroscopic electromagnetic waves ("radio waves").

Corresponding to T 1.4.1, in citing and applying AV 1.2 we shall not distinguish its first formulation and the relation A from T 1.4.1.

§ 2 Relations Between Preparation and Registration Procedures

Here let us discuss the relations between the de-mixing of ensembles and the registrable effects. This will lead us to an axiom which, together with AV 1.2, is equivalent to AV 1.2 s. Thus we have to ask whether and how the possibilities of de-mixing are connected with those of registering.

§ 2.1 Main Law for the De-mixing of Ensembles and Related Possibilities of Registering

In V §§ 6, 7 and 8, we discussed the mathematical possibilities of de-mixing an ensemble $w \in K$ and related it to the "physical possibilities" of de-mixing preparation procedures. Hence we may here limit ourselves to mixtures and de-mixings of ensembles. Since we consider only elements $w \in K$, and $K \subset \bar{\mathscr{K}}$ holds (IV T 3.17), we can always act as if the $w \in K$ are also produceable by preparation procedures.

In V § 8 we saw that the closed faces of K characterize just those sets which remain invariant under mixtures and de-mixings and are closed in the norm topology (hence also in the $\sigma(\mathscr{B}', \mathscr{B})$-topology).

Let F be a closed face of K. The set

$$M_F = \bigcup_{\substack{a \in \mathscr{Q}' \\ \varphi(a) \in F}} a \tag{2.1.1}$$

of systems is then characterized by the fact that it is not physically possible (physically excluded in the sense of [3] § 10.4) to construct ensembles outside F by decomposition into subsets and by joining subsets. In this sense the microsystems from M_F admit a *uniform* characterization, which one uses to call a "property". This uniform characterization belonging to an F is no longer appropriate for all systems from an $\tilde{a} \in \mathscr{Q}'$ with $\varphi(\tilde{a}) \notin F$.

This intuitively suggests that it should also be possible to construct an effect procedure f such that $\mu(\varphi(\tilde{a}), \psi(f))$ noticeably differs from zero, whereas $\mu(\varphi(a), \psi(f)) \sim 0$ holds for all the $a \in \mathscr{Q}'$ with $\varphi(a) \in F$. Such an effect procedure indeed does not respond to systems from M_F but can be triggered by systems from $\tilde{a}$, since not all systems from $\tilde{a}$ can have the characteristic belonging to F. These considerations intuitively suggest a law similar to:

For a closed face F of K, a $w_0 \in F$ and a $\tilde{w} \in K$ with $\tilde{w} \notin F$, there always exists a $g \in \mathscr{S}$ with $\mu(w_0, g) \sim 0$ and $\mu(\tilde{w}, g) > \delta$ with δ noticeably different from zero.

Again, it is obvious to idealize this to

AV 2 s: If F is a closed face of K, for $w_0 \in F$, $\tilde{w} \in K$ and $\tilde{w} \notin F$ there is a sequence $g_\nu \in \mathscr{S}$ with $\mu(w_0, g_\nu) \to 0$ and $\mu(\tilde{w}, g_\nu) > \delta$ for a fixed number $\delta > 0$.

AV 2 s could appear too sharp for the following reasons:

Choosing a certain closed face F, let us assume that to w_0 with $F = C(w_0)$ and $\tilde{w} \in K$, $\tilde{w} \notin F$ there belongs a sequence $g_\nu \in \mathscr{S}$ with $\mu(w_0, g_\nu) \to 0$ and $\mu(\tilde{w}, g_\nu) > \delta$ (δ fixed). Then the compactness of L implies that there is a $g \in L$ with $\mu(w_0, g) = 0$ and $\mu(\tilde{w}, g) \neq 0$. From this follows $K_0 L_0(w_0) = F$; hence F must be an exposed face. Therefore AV 2 s would imply that all closed faces of K are also exposed. One could thus try to demand AV 2 s only for exposed faces:

AV 2: If F is an exposed face of K, for $w_0 \in F$, $\tilde{w} \in K$ and $\tilde{w} \notin F$ there is a sequence $g_\nu \in \mathscr{S}$ with $\mu(w_0, g_\nu) \to 0$ and $\mu(\tilde{w}, g_\nu) > \delta$ ($\delta > 0$ a fixed number).

On the other hand, according to the considerations that led to AV 2 s, the postulate AV 2 could again appear too weak: If for a $w_0 \in K$ one considers $A(w_0) = \{w \mid w = w' \mu(w', 1)^{-1}$ with $w' \neq 0$, $w' \in [\mathbf{0}, w_0]\}$ (see V T 9.1), this $A(w_0)$ is the set of the mixture components of w_0. Therefore, it is "physically doable" to demix the ensemble w_0 into $w_0 = \mu(w', \mathbf{1})\, w + (1 - \mu(w', \mathbf{1}))\, w''$. If $A(w_0)$ is a closed face of K, then all $w \in A(w_0)$ are "produceable" mixture components of w_0. For a $\tilde{w}$ outside $A(w_0)$, one should then find a sequence $g_\nu \in \mathscr{S}$ with $\mu(w_0, g_\nu) \to 0$ and $\mu(\tilde{w}, g_\nu) > \delta > 0$. Due to V T 9.2, finite-dimensional faces of K make $A(w_0) = C(w_0)$. This would suggest AV 2 s at least for finite-dimensional faces. As we saw above, this implies that the finite-dimensional faces are exposed. Since we shall always require AV 2 (or since AV 2 by T 2.2.2 follows from AV 2.1 s) let us complement AV 2 by the next axiom, to be used much later:

AV 2 f: Each finite-dimensional face of K is exposed.

Since it is convenient also to use forms equivalent to AV 2, let us show:

T 2.1.1 The following relations are equivalent to AV 2:

(i) For each exposed face F of K and $w \in K$, $w \notin F$, there is a $g \in L$ with $K_0(g) \supset F$ and $\mu(w, g) \neq 0$.

(ii) The sets $\mathscr{W}_L$ and $\mathscr{W}_{B'}$ (from IV § 5) are equal.

(iii) For two exposed faces $F_1 = C(w_1)$, $F_2 = C(w_2)$ with $F_1 \supsetneqq F_2$, there is a $g \in L$ with $\mu(w_1, g) \neq 0$, $\mu(w_2, g) = 0$.

(iv) For two exposed faces $F_1 = C(w_1)$, $F_2 = C(w_2)$, from $F_1 \supsetneqq F_2$ follows $L_0(w_1) \neq L_0(w_2)$.

(v) For two exposed faces $F_1 = C(w_1)$ and $F_2 = C(w_2)$, from $L_0(w_1) = L_0(w_2)$ follows $F_1 = F_2$.

(vi) For each exposed face F of K there is a $g \in L$ with $F = K_0(g)$.

Proof. AV 2 $\Rightarrow$ (i): Since L is $\sigma(\mathscr{B}', \mathscr{B})$-compact, there is a subsequence of g_ν (which we again denote by g_ν) such that $g_\nu \to g \in L$. From this follows $\mu(w_0, g) = 0$ and $\mu(\tilde{w}, g) \neq 0$. Choosing w_0 so that $C(w_0) = F$, we get $K_0(g) \supset F$.

(i) $\Rightarrow$ AV 2: Choosing a $w_0 \in F$, we find $w_0 \in K_0(g)$. Because $g \in L$, there is a sequence $g_\nu \in \mathscr{S}$ which in the $\sigma(\mathscr{B}', \mathscr{B})$-topology converges to g. Therefore, $\mu(w_0, g_\nu) \to \mu(w_0, g) = 0$ and $\mu(w, g_\nu) \to \mu(w, g) = 2\delta$, so that from some ν on $\mu(w, g_\nu) > \delta$ must hold.

(i) $\Rightarrow$ (ii): Because of IV (5.6), we have only to show that each element of $\mathscr{W}_{B'}$ is an element of $\mathscr{W}_L$. But $\mathscr{W}_{B'}$ is just the set of all exposed faces. With $F \in \mathscr{W}_{B'}$, by T 5.5 we get $F \subset K_0 L_0(F)$, where $K_0 L_0(F)$ is an element of $\mathscr{W}_L$. If we had $F \neq K_0 L_0(F)$, a $w \in K_0 L_0(F)$ would exist with $w \notin F$. From (i) follows that there is a g with $K_0(g) \supset F$ and $\mu(w, g) \neq 0$. From $K_0(g) \supset F$ follows $g \in L_0(F)$ and hence $K_0(g) \subset K_0 L_0(F)$, contradicting the fact that $\mu(w, g) \neq 0$ for a $w \in K_0 L_0(F)$.

(ii) $\Rightarrow$ (iii): From (ii) follows $F_1 = K_0(g_1)$, $F_2 = K_0(g_2)$ with $g_1, g_2 \in L$. With $F_1 = C(w_1)$, $F_2 = C(w_2)$, we have $\mu(w_1, g_1) = 0$, $\mu(w_2, g_2) = 0$. If we had $\mu(w_1, g_2) = 0$, we would get $w_1 \in K_0(g_2)$ and hence $F_1 = C(w_1) \subset K_0(g_2) = F_2$. For $F_1 \supsetneqq F_2$ we thus must have $\mu(w_1, g_2) \neq 0$.

(iii) $\Rightarrow$ (iv) follows directly from the meaning of L_0.

(iv) $\Rightarrow$ (v): For $F_1 \neq F_2$, either $F_1 \subsetneqq F_1 \vee F_2$ or $F_2 \subsetneqq F_1 \vee F_2$ would hold. Let $F_2 \subsetneqq F_1 \vee F_2$. For some w_3 we get $F_1 \vee F_2 = C(w_3)$, while (iv) yields $L_0(w_2) \neq L_0(w_3)$. Because $L_0(w_3) = L_0 C(w_3) = L_0(C(w_1) \cup C(w_3)) = L_0 C(w_1) \wedge L_0 C(w_2)$, from $L_0(w_2) \neq L_0(w_3)$ follows $L_0(w_2) = L_0 C(w_2) \neq L_0 C(w_1) \wedge L_0 C(w_2)$ and hence $L_0(w_2) \neq L_0(w_1)$.

(v) $\Rightarrow$ (i): For $F_1 = C(w_1)$ and $w \notin F_1$ and $F_2 = K_0 L_0(w_1, w)$ we have $F_1 \subsetneqq F_2$. By (v) follows $L_0(F_1) \neq L_0(F_2)$. Because $F_1 \subset F_2$, we have $L_0(F_1) \supset L_0(F_2)$. Hence a $g \in L_0(F_1)$ exists with $g \notin L_0(F_2) = L_0 K_0 L_0(w_1, w) = L_0(w_1, w)$; therefore $\mu(w_1, g) = 0$ implies $\mu(w, g) \neq 0$.

(i) $\Rightarrow$ (vi): L convex implies that $L_0(F)$ is also convex. From (i) follows $K_0 L_0(F) = F$. For a set $\{g_\nu\}$ that is countable and $\sigma(\mathscr{B}', \mathscr{B})$-dense in $L_0(F)$, for $\lambda_\nu > 0$, $\sum_\nu \lambda_\nu = 1$ we have $g = \sum_\nu \lambda_\nu g_\nu \in L_0(F)$ and $K_0 L_0(F) = K_0(g)$.

(vi) $\Rightarrow$ (i) is trivial. □

§ 2.2 Some Consequences of Axiom AV 2

Here we shall draw consequences from AV 2 and discuss relations among the axioms AV 1.2, AV 1.2 s, AV 2, *not* assuming the axiom AV 1.1.

T 2.2.1 From AV 2 follows that the $\sigma(\mathscr{B}', \mathscr{B})$-closed cone generated by L equals $\mathscr{B}'_+$.

Proof. One obtains the $\sigma(\mathscr{B}', \mathscr{B})$-closed cone generated by L as the bipolar set of L. Hence, this cone equals $\mathscr{B}'_+$ provided the cone polar to L equals $\mathscr{B}_+$, i.e. if the set

$$L^0 = \{x \mid x \in \mathscr{B}, \mu(x, g) \geqq 0 \text{ for all } g \in L\}$$

equals $\mathscr{B}_+$. Because $L \subset [\mathbf{0}, \mathbf{1}]$, we have $L^\circ \supset \mathscr{B}_+$. Thus we must only show that for each $x \notin \mathscr{B}_+$ there is a $g \in L$ such that $\mu(x, g) < 0$.

Due to IV T 4.10, for x there is a decomposition $x = \alpha w_1 - \beta w_2$ (with $\beta > 0$ because $x \notin \mathscr{B}_+$) such that a $y_0 \in [-\mathbf{1}, \mathbf{1}]$ exist with $\mu(w_0, y_0) = 1$ and $\mu(w_2, y_0) \leqq -1 + \varepsilon$. Then $y = (1/2)(\mathbf{1} - y_0)$ is an element of $[\mathbf{0}, \mathbf{1}]$, so that $\mu(w_1, y) = 0$ and

$\mu(w_2, y) \geqq 1 - \frac{\varepsilon}{2}$ hold. Therefore, $w_1 \in K_0(y)$ and $w_2 \notin K_0(y)$. By (i) in T 2.1.1, a $g_0 \in L$ thus exists with $\mu(w_1, g_0) = 0$ and $\mu(w_2, g_0) > 0$, hence $\mu(x, g_0) = -\beta\mu(w_2, g_0) < 0$. □

T 2.2.2 AV 1.2 and AV 2 ⇔ AV 1.2s.

Proof. Very easily the preceding theorems yield AV 1.2s ⇒ AV 1.2 and AV 2. Conversely, now let AV 1.2 and AV 2 be satisfied. Due to the proof of T 2.2.1, there is a decomposition $x = \alpha w_1 - \beta w_2$ such that $\mu(w_1, y) = 0$ and $\mu(w_2, y) \geqq 1 - \frac{\varepsilon}{2}$ hold with a $y \in [\mathbf{0}, \mathbf{1}]$, while $\mu(w_1, g_0) = 0$ with a $g_0 \in L$. Also a $g \in L$ with $\mu(w_1, g) = 0$ and $\mu(w_2, g) \geqq 1 - \delta(\frac{\varepsilon}{2})$ then exists by AV 1.2. With $y' = \mathbf{1} - 2g$ we have $y' \in \mathbf{1} - 2L \subset [-\mathbf{1}, \mathbf{1}]$ and

$$\begin{aligned}\mu(x, y') &\geqq \alpha - \beta(-1 + 2\delta(\tfrac{\varepsilon}{2})) = \alpha + \beta - 2\beta\delta(\tfrac{\varepsilon}{2}) \\ &\geqq \alpha + \beta - 2\beta\delta(\tfrac{\varepsilon}{2}) - 2\alpha\delta(\tfrac{\varepsilon}{2}) = (\alpha + \beta)(1 - 2\delta(\tfrac{\varepsilon}{2})).\end{aligned}$$

Therefore, $\|x\| \leqq \alpha + \beta$ yields

$$\mu(x, y') \geqq \|x\|(1 - 2\delta(\tfrac{\varepsilon}{2})),$$

so that $\mathbf{1} - 2L \subset [-\mathbf{1}, \mathbf{1}]$ implies

$$\|x\| \geqq \sup_{y \in \mathbf{1} - 2L} \mu(x, y) \geqq \|x\|(1 - 2\delta(\tfrac{\varepsilon}{2})).$$

For $\varepsilon \to 0$, also $\delta(\frac{\varepsilon}{2}) \to 0$ holds and hence

$$\sup_{y \in \mathbf{1} - 2L} \mu(x, y) = \|x\|.$$

Therefore the set polar to $\mathbf{1} - 2L$ is the unit ball in $\mathscr{B}$; hence we obtain $\mathbf{1} - 2L = [-\mathbf{1}, \mathbf{1}]$, since $\mathbf{1} - 2L$ is absolutely convex and $\sigma(\mathscr{B}', \mathscr{B})$-closed. □

Therefore T 2.2.2 shows that it really is a matter of taste whether we take AV 1.2 *and* AV 2 as an axiom or rather require AV 1.2 s. When we assume AV 1.2 s as an axiom, one therefore can also presume AV 1.2 and AV 2, and conversely. Thus in the sequel we shall conveniently state that we presume AV 1.2 s although we could also say that AV 1.2 and AV 2 are presumed.

T 2.2.3 Assume AV 1.2s. Then $\mathscr{D} \cap [\mathbf{0}, \mathbf{1}]$ is $\sigma(\mathscr{B}', \mathscr{B})$-dense in $[\mathbf{0}, \mathbf{1}]$; likewise, $\mathscr{D} \cap [-\mathbf{1}, \mathbf{1}]$ is dense in $[-\mathbf{1}, \mathbf{1}]$. Hence $\mathscr{B}$ is a Banach subspace of $\mathscr{D}'$. Also $\bar{\mathscr{K}} = K$ and $\bar{\mathscr{L}} = \mathscr{D} \cap [\mathbf{0}, \mathbf{1}]$ hold.

Proof. Because $\mathscr{L} \cap \mathscr{D}$ and $\mathscr{L}$ is $\sigma(\mathscr{B}', \mathscr{B})$-dense in $L = [\mathbf{0}, \mathbf{1}]$, the first part of the theorem follows immediately. From IV T 3.19 follows $\bar{\mathscr{K}} = K$, while $\bar{\mathscr{L}} = \mathscr{D} \cap [\mathbf{0}, \mathbf{1}]$ follows from IV T 3.16. □

§ 3 The Lattice G

Here let us assemble several fundamental properties yielded by AV 1.1 *and* AV 1.2s. We thus continue the considerations of § 1.3, in addition assuming $L = [\mathbf{0}, \mathbf{1}]$.

T 3.1 We have $\overline{\text{co}}^{\sigma} G = [\mathbf{0}, \mathbf{1}] = L$, which implies $\partial_e L \subset \bar{G}^{\sigma}$. For $e \in G$ we get $L_0 K_0(e) = [\mathbf{0}, e]$.

Proof. It suffices to show

$$\sup_{e \in G} \mu(x, e) = \sup_{g \in [\mathbf{0}, \mathbf{1}]} \mu(x, g).$$

Due to IV T 4.10, there is a decomposition $x = \alpha w_1 - \beta w_2$ with a $y_0 \in [-\mathbf{1}, \mathbf{1}]$, with $\mu(w_1, y_0) \geqq 1 - \varepsilon$, $\mu(w_2, y_0) = -1$. For $g_0 = \frac{1}{2}(\mathbf{1} + y_0)$ we have $g_0 \in [\mathbf{0}, \mathbf{1}]$ and $\mu(w_1, g_0) \geqq 1 - \frac{\varepsilon}{2}$, $\mu(w_2, g_0) = 0$. From this follows $\mu(x, g_0) \geqq \alpha(1 - \frac{\varepsilon}{2})$. With $K_0(g_0) = K_0(e_0)$, $e_0 \in G$, by T 1.3.5 we get $\mu(w_1, e_0) \geqq \mu(w_1, g_0) \geqq 1 - \frac{\varepsilon}{2}$. Because $K_0(e_0) = K_0(g_0)$, we have $\mu(w_2, e_0) = 0$ and hence $\mu(x, e_0) \geqq \alpha(1 - \frac{\varepsilon}{2})$. Because $\mu(w, g) \leqq 1$ for all $w \in K$, $g \in [\mathbf{0}, \mathbf{1}]$, we have $\sup_{g \in [\mathbf{0}, \mathbf{1}]} \mu(x, g) \leqq \alpha$, hence

$$\mu(x, e_0) \geqq (1 - \tfrac{\varepsilon}{2}) \sup_{g \in [\mathbf{0}, \mathbf{1}]} \mu(x, g).$$

Since ε was arbitrary, we obtain

$$\sup_{e \in G} \mu(x, e) \geqq \sup_{g \in [\mathbf{0}, \mathbf{1}]} \mu(x, g).$$

Because $G \subset [\mathbf{0}, \mathbf{1}]$, this proves

$$\sup_{e \in G} \mu(x, e) = \sup_{g \in [\mathbf{0}, \mathbf{1}]} \mu(x, g).$$

$\partial_e L \subset \bar{G}^{\sigma}$ follows by [7] II 10.5 from $\overline{\text{co}}^{\sigma} G = L$.
With $L = [\mathbf{0}, \mathbf{1}]$, from (1.3.4) we get $L_0 K_0(e) = [\mathbf{0}, e]$. □

By T 1.3.6 we have $G \subset \partial_e [\mathbf{0}, \mathbf{1}]$, so that G is $\sigma(\mathscr{B}', \mathscr{B})$-dense in $\partial_e [\mathbf{0}, \mathbf{1}]$. Without further assumptions, the equality $G = \partial_e [\mathbf{0}, \mathbf{1}]$ has not be proven.

T 3.2 For $e \in G$ and $e \neq 0$ we have $\| e \| = 1$, i.e. $\sup_{w \in K} \mu(w, e) = 1$.

Proof. With $g = \| e \|^{-1} e \in [\mathbf{0}, \mathbf{1}]$ we have $K_0(g) = K_0(e)$, so that T 1.3.5 gives $g \leqq e$, i.e. $\| e \| \geqq 1$. Because $e \in [\mathbf{0}, \mathbf{1}]$, we have $\| e \| \leqq 1$. □

According to T 3.2 the probabilities $\mu(w, e)$ for an $e \in G$ with $e \neq 0$ approach the value 1 arbitrarily closely. But it has not been proven that $\mu(w, e)$ really attains the value 1 on K. Physically, the case that $\mu(w, e)$ reaches the value 1 on K cannot be distinguished from $\sup_{w \in K} \mu(w, e) = 1$ (also see [3] § 11.4).

Therefore, we change nothing in the physical structure of $K \times L \xrightarrow{\mu} [0, 1]$ if we require as an axiom that the function $\mu(w, e)$ attains its supremum on K for each $e \in G$ with $e \neq 0$. Such an axiom is a *purely* mathematical idealization. Whereas mathematical idealizations entered the axiom AV 1.1 in the way described in § 1.1, an axiom that the function $\mu(w, e)$ reaches its supremum is *only* a mathematical idealization. Therefore, one cannot invoke any physical arguments against such an axiom. The only counter argument could be that this axiom contradicts later mathematical idealizations. For the case of quantum mechanics, such contradictions (to the assumption that $\mu(w, e)$ reaches its supremum) are not known. In order to express the fact of a pure idealization, we denote this axiom by AVid.

AVid For $e \in G$, the function $\mu(w, e)$ attains its supremum on K.

Without presuming this axiom, let us first prove some equivalences, following from AV 1.1 and AV 1.2s.

T 3.3 The following relations are equivalent:

(i) AVid.
(ii) $\mathbf{1} - e \in G$ for all $e \in G$, i.e. $\mathbf{1} - G \subset G$.
(iii) $\mathbf{1} - G = G$.
(iv) $e \in G, e \neq 0 \Rightarrow K_0(\mathbf{1} - e) \neq \emptyset$.
(v) $e_1, e_2 \in G$ and $e_1 \leqq e_2 \Rightarrow e_2 - e_1 \in G$.

Proof. From T 3.2 follows (i) $\Leftrightarrow$ (iv). (iii) $\Rightarrow$ (ii) is trivial. (ii) $\Rightarrow$ (iii) follows from $\mathbf{1}-(\mathbf{1}-g)=g$ so that $e \to \mathbf{1}-e$ is bijective. With $e_2 = \mathbf{1}$ follows (v) $\Rightarrow$ (ii). (ii) $\Rightarrow$ (iv) follows because $K_0(\mathbf{1}-e) \neq \emptyset$ for $\mathbf{1}-e \neq \mathbf{1}$.

(iv) $\Rightarrow$ (v): From $e_2 \geqq e_1$ follows $\mathbf{1} \geqq e_2 \geqq e_2 - e_1 \geqq \mathbf{0}$ and hence $e_2 - e_1 \in [\mathbf{0}, \mathbf{1}]$. Due to T 1.3.5, an $e' \in G$ with $e_2 - e_1 \leqq e'$ is determined by $K_0(e_2 - e_1) = K_0(e')$. Because $e_2 \geqq e_2 - e_1$, we get $K_0(e') = K_0(e_2 - e_1) \supset K_0(e_2)$ so that T 1.3.5 yields $e' \leqq e_2$. From $\mathbf{0} \leqq e_2 - e_1 \leqq e'$ follows $\mathbf{0} \leqq e' - (e_2 - e_1) \leqq e'$, and because $e' \leqq e_2$ we also have $e' - (e_2 - e_1) \leqq e_2 - (e_2 - e_1) = e_1$. For $g = e' + e_1 - e_2$ we therefore get $g \leqq e_1, e'$ so that T 1.3.8 gives $g \leqq e_1 \wedge e'$. $\tilde{e} \overset{\text{def}}{=} e_1 \wedge e'$ is an element of G. Because $\tilde{e} \leqq e_1$ and $\tilde{e} \leqq e'$, from $\mu(w, \tilde{e}) = 1$ follows $\mu(w, e_1) = 1$ and $\mu(w, e') = 1$. From $\mu(w, e_1) = 1$ follows $w \in K_0(e_2 - e_1) = K_0(e')$ and thus $\mu(w, e') = 0$, which contradicts $\mu(w, e') = 1$. By (iv) this implies $\tilde{e} = \mathbf{0}$ and hence $g = \mathbf{0}$, i.e. $e' = e_2 - e_1 \in G$. □

T 3.4 From AVid follows that all elements of G are exposed points of $L = [\mathbf{0}, \mathbf{1}]$.

Proof. We must for each $e \in G$ show that there is an $x \in \mathscr{B}$ such that $\mu(x, g)$ with $g \in L = [\mathbf{0}, \mathbf{1}]$ attains the value $\sup_{g \in L} \mu(x, g)$ only at $g = e$. Let $e \neq \mathbf{0}, \mathbf{1}$ and set $x = w_2 - w_1$ with w_1, w_2 given by $C(w_1) = K_0(e)$ and $C(w_2) = K_0(\mathbf{1} - e)$ (by T 3.3 (iv) we have $K_0(\mathbf{1} - e) \neq \emptyset$). From $\mu(x, g) = \mu(w_2, g) - \mu(w_1, g) \leqq 1$ and $\mu(x, e) = \mu(w_2, e) - \mu(w_1, e) = 1$ follows $\sup_{g \in L} \mu(x, g) = 1$ and this supremum is reached for $g = e$. Let $g \in L$ with $\mu(x, g) = \mu(w_2, g) - \mu(w_1, g) = 1$; then we get $\mu(w_2, g) = 1$ and $\mu(w_1, g) = 0$. From $\mu(w_1, g) = 0$ follows $g \in L_0 K_0(e)$ and hence $g \leqq e$. From $\mu(w_1, g) = 1$ follows $\mu(w_1, \mathbf{1} - g) = 0$ and hence $\mathbf{1} - g \in L_0 K_0(\mathbf{1} - e)$. This gives $\mathbf{1} - g \leqq \mathbf{1} - e$ and hence $e \leqq g$, so that $e = g$. □

If $\mathscr{B}$ possesses the minimal decomposition property, then G equals the set of exposed points of $L = [\mathbf{0}, \mathbf{1}]$ (see A III).

T 3.5 If $\mathbf{1} - G = G$, then the mapping $\perp$ of G onto itself, defined by $e \to e^{\perp} = \mathbf{1} - e$, is an orthocomplementation.

Proof. It follows immediately that $\perp$ is a dual automorphism of G with $e^{\perp\perp} = e$. For $e \vee e^{\perp}$ we have $K_0(e \vee e^{\perp}) = K_0(e) \cap K_0(e^{\perp}) = K_0(e) \cap K_0(1 - e) = \emptyset$ and hence $e \vee e^{\perp} = 1$. With the mapping $\perp$, from this follows $e^{\perp} \wedge e = \mathbf{1}^{\perp} = \mathbf{1} - \mathbf{1} = \mathbf{0}$. □

Thus AVid makes G an orthocomplemented lattice. Henceforth we presume AV 1.1, AV 1.2s and AVid.

D 3.1 $e_1, e_2 \in G$ are called *mutually orthogonal* if $e_1 \leqq e_2^{\perp}$; for this we write $e_1 \perp e_2$.

The relation $e_1 \perp e_2$ is symmetric since $e_1 \leqq e_2^{\perp}$ by the mapping $\perp$ implies $e_1^{\perp} \geqq e_2$. When we write $K_1(e) = K_0(\mathbf{1} - e) = K_0(e^{\perp})$ the mappings $G \xrightarrow{K_1} \mathscr{W}_{\mathscr{B}'} = \mathscr{W}_L$ forms a lattice isomorphism. This follows immediately from $K_1 = K_0 \perp$, i.e. K_1 equals the mapping composed of $\perp$ and K_0.

T 3.6 Two exposed faces F_1, F_2 of K are orthogonal in the sense of IV D 5.1, if and only if $F_1 = K_1(e_1)$, $F_2 = K_1(e_2)$ with $e_1, e_2 \in G$ and $e_1 \perp e_2$.

Proof. With $y = \mathbf{1} - e_1$ and $e_1 \perp e_2$ follows $F_1 \subset K_0(y)$ and (because $e_1 \leqq e_2^{\perp}$) $F_2 = K_0(e_2^{\perp}) \subset K_0(e_1) = K_0(\mathbf{1} - y)$.

With $F_1 = K_1(e_1) \subset K_0(y')$ and $F_2 = K_1(e_2) \subset K_0(\mathbf{1} - y')$ follows $K_0(\mathbf{1} - e_1) \subset K_0(y')$ and $K_0(\mathbf{1} - e_2) \subset K_0(\mathbf{1} - y')$. This gives $y' \leqq \mathbf{1} - e_1$ and $\mathbf{1} - y' \leqq \mathbf{1} - e_2$ and hence $e_2 \leqq y' \leqq \mathbf{1} - e_1$, so that $e_2 \perp e_1$. □

The orthocomplemention, transferred from G to $\mathscr{W}_{\mathscr{B}'}$ by the isomorphic mapping K_1, therefore makes $K_1(e)^{\perp} = K_1(e^{\perp})$, i.e. $K_0(e^{\perp})^{\perp} = K_0(e)$ and hence $K_0(e^{\perp}) = K_0(e)^{\perp}$.

T 3.7 From $e_1, e_2 \in G$ and $e_1 + e_2 \leqq \mathbf{1}$ follows $e_1 \perp e_2$. From $e_1, e_2 \in G$ and $e_1 \perp e_2$ follows $e_1 \vee e_1 = e_1 + e_2$.

Proof. From $e_1 + e_2 \leqq \mathbf{1}$ follows $e_1 \leqq \mathbf{1} - e_2 = e_2^{\perp}$, i.e. $e_1 \perp e_2$. From $e_1 \perp e_2$ follows $e_1 \leqq \mathbf{1} - e_2$ and hence $e_1 + e_2 \leqq \mathbf{1}$. From this follows $K_0(e_1 + e_2) = K_0(e_1) \cap K_0(e_2) = K_0(e_1 \vee e_2)$ and thus $e_1 + e_2 \leqq e_1 \vee e_2$. By T 3.3, from $e_1 \leqq e_2^{\perp}$ follows $e_2^{\perp} - e_1 \in G$ and $\mathbf{1} - (e_2^{\perp} - e_1) = e_1 + e_2 \in G$. Because $e_1 \leqq e_1 + e_2$, $e_2 \leqq e_1 + e_2$, this also gives $e_1 \vee e_2 \leqq e_1 + e_2$. □

This T 3.7 can be extended to many summands. But to this end let us first extend T 1.3.9:

T 3.8 For a subset $A \subset G$, the upper bound of A in $L = [\mathbf{0}, \mathbf{1}]$ is $\bigvee_{e \in A} e$.

If A is directed upward, this $\bigvee_{e \in A} e$ is the limit of A in the $\sigma(\mathscr{B}', \mathscr{B})$-topology.

Proof. Apply T 1.3.9 to the set $A^{\perp} = \{e^{\perp} \mid e \in A\}$. □

T 3.9 From $A \subset G$ and $\sum_{e \in A} e \leqq \mathbf{1}$ follows that the $e \in A$ are pairwise orthogonal and at most countably many e are in A, $e \neq 0$ (here, $\sum_{e \in A} e$ is meant in the $\sigma(\mathscr{B}', \mathscr{B})$-topology). If the elements of A are pairwise orthogonal, then A has at most countably many $e \neq 0$. With the positive integer ν as index of the elements $e \neq 0$ in A, we then have $\sum_\nu e_\nu = \bigvee_\nu e_\nu$.

Proof. Since there is an effective $w \in K$, for $e \neq \mathbf{0}$ we also have $\mu(w, e) \neq 0$. Then $\sum_{e \in A} \mu(w, e) \leqq 1$ shows that there are at most countably many $e \neq 0$. From $\sum_\nu e_\nu \leqq \mathbf{1}$ then follows $e_{\nu_i} \leqq \mathbf{1} - \sum_{\nu \neq \nu_i} e_\nu \leqq \mathbf{1} - e_{\nu_j}$, i.e. $e_{\nu_i} \perp e_{\nu_j}$ for $\nu_i \neq \nu_j$.

Conversely, if two distinct e in A are always orthogonal, T 3.7 by induction implies that each finite subset $A_f \subset A$ yields

$$\sum_{e \in A_f} e = \bigvee_{e \in A_f} e \overset{\text{def}}{=\!=} e_{A_f}.$$

The e_{A_f} form an upward directed subset of G with $\bigvee_{A_f} e_{A_f} = \bigvee_{e \in A} e$.

An effective $w \in K$ therefore gives

$$\mu(w, e_{A_f}) = \sum_{e \in A_f} \mu(w, e) \leqq \mu(w, \bigvee_{e \in A} e).$$

Therefore, only countably many $e \in A$ can be different from zero. Hence T 3.8 yields $\sum_{\nu=1}^{N} e_\nu \to \bigvee_{e \in A} e$ in the $\sigma(\mathscr{B}', \mathscr{B})$-topology. □

T 3.10 In every subset $A \supset G$ there is a countable set $\{e_\nu\} \subset A$ so that $\bigvee_{e \in A} = \bigvee_\nu e_\nu$; the corresponding result holds for the intersection $\wedge$.

Proof. Let Φ be the set of all finite subsets of A. For $\varphi \in \Phi$ we write $e_\varphi = \bigvee_{e \in \varphi} e$, such that $\bigvee_{e \in A} e = \bigvee_{\varphi \in \Phi} e_\varphi$. The set of e_φ is directed upward. Due to T 3.8 thus $\bigvee_{e \in A} e$ is also the $\sigma(\mathscr{B}', \mathscr{B})$-limit of the directed set of the e_φ. Since the $\sigma(\mathscr{B}', \mathscr{B})$-topology on $[\mathbf{0}, \mathbf{1}]$ is metrizable, there is a countable subsequence $e_{\varphi_\nu} \to \bigvee_{e \in A} e$. With $\varphi_n = \bigcup_{\nu=1}^{n} \varphi_\nu$, by T 3.8 we have

$$e_{\varphi_n} \to \bigvee_n e_{\varphi_n} \leqq \bigvee_{\varphi \in \Phi} e_\varphi = \bigvee_{e \in A} e.$$

For each e_{φ_ν}, there is an e_{φ_n} with e_{φ_ν} with $e_{\varphi_\nu} \leqq e_{\varphi_n}$. From $e_{\varphi_\nu} \to \bigvee_{e \in A} e$ then follows $e_{\varphi_n} \to e' \geqq e_{\varphi_\nu} \to \bigvee_{e \in A} e$, i.e. $e' = \bigvee_n e_{\varphi_n} \geqq \bigvee_{e \in A} e$, such that $\bigvee_n e_{\varphi_n} = \bigvee_{e \in A} e$. Since the φ_ν are finite subsets of A, the set $\bigvee_\nu \varphi_\nu$ is countable and yields $\bigvee_{e \in \bigvee_\nu \varphi_\nu} e = \bigvee_{e \in A} e$. By the mapping $\perp$, an analogous theorem follows for the intersection. □

T 3.11 The lattice operations $\wedge$, $\vee$, $\perp$ for G are continuous in the $\sigma(\mathscr{B}', \mathscr{B})$-topology. More rigorously,

$$e_\nu \to e \text{ implies } e_\nu^\perp \to e^\perp$$

(since the $\sigma(\mathscr{B}', \mathscr{B})$-topology on $[\mathbf{0}, \mathbf{1}]$ is metrizable, it suffices to consider sequences).

From $e_\nu \to e$, $e'_\mu \to e'$ follows $e_\nu \vee e'_\mu \to e \vee e'$ *in case* there is a number $\beta > 0$ such that

$$\inf_{w \in K} \{\mu(w, (e_\nu \vee e'_\mu)^\perp + \tfrac{1}{2}(e_\nu + e'_\mu))\} \geqq \beta$$

for all μ, ν, up to finitely many; and $e_\nu \wedge e'_\mu \to e \wedge e'$ *in case* there is a number $\gamma > 0$ such that

$$\inf_{w \in K} \{\mu(w, e_\nu \wedge e'_\mu + \tfrac{1}{2}(e_\nu^\perp + e'^\perp_\mu))\} \geqq \gamma$$

for all μ, ν up to finitely many.

Proof. From $e_\nu \to e$ follows $e_\nu^\perp = \mathbf{1} - e_\nu \to \mathbf{1} - e = e^\perp$. From $\frac{1}{2}(e_\nu + e'_\mu) \leqq \mathbf{1}$ and $K_0(\frac{1}{2}e_\nu + \frac{1}{2}e'_\mu) = K_0(e_\nu \vee e'_\mu)$ follows $\frac{1}{2}(e_\nu + e'_\mu) \leqq e_\nu \vee e'_\mu$. Let

$$\sup_{w \in K} \mu(w, e_\nu \vee e'_\mu - \tfrac{1}{2}(e_\nu + e'_\mu)) = 1 - \beta_{\nu\mu} \quad \text{with} \quad 1 - \beta_{\nu\mu} \neq 0\,;$$

then $g_{\nu\mu} \overset{\text{def}}{=} (1 - \beta_{\nu\mu})^{-1}[e_\nu \vee e'_\mu - \frac{1}{2}(e_\nu + e'_\mu)] \in [\mathbf{0}, \mathbf{1}]$ holds, hence $g_{\nu\mu} \leqq e_\nu \vee e'_\mu$ because $K_0(g_{\nu\mu}) \supset K_0(e_\nu \vee e'_\mu)$. For $\beta_{\nu\mu} \neq 0$ this implies

$$e_\nu \vee e'_\mu \leqq \frac{1}{2\beta_{\nu\mu}}(e_\nu + e'_\mu)\,.$$

This also holds for $\beta_{\nu\mu} = 1$ since $1 - \beta_{\nu\mu} = 0$ implies $e_\nu \vee e'_\mu = \frac{1}{2}(e_\nu + e'_\mu)$. If there is a $\beta > 0$ with $\beta_{\nu\mu} > \beta$ for all ν, μ, it makes

$$e_\nu \vee e'_\mu \leqq \frac{1}{2\beta}(e_\nu + e'_\mu)\,.$$

Since $L = [\mathbf{0}, \mathbf{1}]$ is compact, a subsequence ν_i, μ_i can be so chosen that $e_{\nu_i} \vee e'_{\mu_i} \to g \in L$ converges. Together with $e_\nu \to e$, $e'_\mu \to e'$ this yields

$$g \leqq \frac{1}{2\beta}(e + e')\,,$$

so that $K_0(g) \supset K_0(\frac{1}{2}(e + e')) = K_0(e \vee e')$ and hence $g \leqq e \vee e'$.

From $e_{\nu_i} \vee e'_{\mu_i} \geqq e_{\nu_i}$ and $e_{\nu_i} \vee e'_{\mu_i} \geqq e'_{\mu_i}$, in the limit follows $g \geqq e, e'$ and hence $\mathbf{1} - g \leqq e^\perp$, $\mathbf{1} - g \leqq e'^\perp$. By T 1.3.8 follows $\mathbf{1} - g \leqq e^\perp \wedge e'^\perp$ and hence

$$g \geqq \mathbf{1} - (e^\perp \wedge e'^\perp) = (e^\perp \wedge e'^\perp)^\perp = e \vee e'\,,$$

therefore $g = e \vee e'$. Since g was any accumulation point of $e_\nu \vee e'_\mu$, thus $e_\nu \vee e'_\mu \to e \vee e'$ must hold.

$\beta_{\nu\mu} > \beta$ means

$$\sup_{w \in K} \mu(w, e_\nu \vee e'_\mu - \tfrac{1}{2}(e_\nu + e_\mu)) \leqq 1 - \beta\,,$$

which due to $(e_\nu \vee e'_\mu)^\perp = \mathbf{1} - e_\nu \vee e'_\mu$ is equivalent to

$$\inf_{w \in K} \mu(w, (e_\nu \vee e'_\mu)^\perp + \tfrac{1}{2}(e_\nu + e'_\mu)) \geqq \beta\,.$$

The last part of the theorem follows easily with $e_\nu \wedge e'_\mu = (e_\nu^\perp \vee e'^\perp_\mu)^\perp$ and $e_\nu^\perp \to e^\perp$, $e'^\perp_\mu \to e'^\perp$. □

§ 4 Commensurable Decision Effects

Here let us give some complements to the considerations in V which result when AV 1.1, AV 1.2s and AVid are valid. Due to AV 1.2s we can everywhere in V identify L with $[\mathbf{0}, \mathbf{1}]$. In particular now all the relations derived in V "under the assumption $L=[\mathbf{0}, \mathbf{1}]$" become valid. Thus AV 1.1 (complemented by the idealization AVid) becomes the crucial axiom for the following considerations.

In this section let us consider some concepts introduced in V for the subset G of $L=[\mathbf{0}, \mathbf{1}]$. Corresponding to V D 1.2.2, for a subset A of G we adopt

D 4.1 A set $A \subset G$ (a set of decision effects) is called *commensurable* if there is a Boolean ring Σ with an additive measure $\Sigma \xrightarrow{F} G$ such that $A \subset F\Sigma$.

Observe that in this definition the mapping F proceeds from Σ into G. According to V D 1.2.2, a set $A \subset G$ is coexistent if there is a Boolean ring Σ' with an additive measure $\Sigma' \xrightarrow{F'} L$ so that $A \subset F'\Sigma$. Therefore, a set of commensurable decision effects is always coexistent. Does the converse also hold?

T 4.1 For $g \in L$ and $e \in G$, the following conditions are equivalent:

(i) $\{g, e\}$ is coexistent;

(ii) $g = g_1 + g_2$ with $g_1, g_2 \in L$ and $g_1 \leqq e$, $g_2 \leqq e^{\perp}$ (in this decomposition, g_1 and g_2 are uniquely determined);

(iii) $e = g_1' + g_3'$ with $g_1', g_3' \in L$ and $g_1' \leqq g_1$, $g_3' \leqq \mathbf{1} - g_1$ (in this decomposition, g_1' and g_3' are uniquely determined and $g_1' = g_1$ with g_1 from (ii)).

Proof. See [2] IV Th 1.3.1.

This theorem can be specialized further:

T 4.2 The coexistence of two decision effects $e_1, e_2 \in G$ is equivalent to the decompositions $e_1 = g_1 + g_2$, $e_2 = g_1 + g_3$ with $g_1, g_2, g_3 \in L$ and $g_1 + g_2 + g_3 \in L$. Here g_1, g_2, g_3 are *uniquely* determined by e_1, e_2, and in fact $g_1 = e_1 \wedge e_2$, $g_2 = e_1 \wedge e_2^{\perp}$, $g_3 = e_1^{\perp} \wedge e_2$ and hence $g_1, g_2, g_3 \in G$. If Σ is a Boolean ring with the additive measure $\Sigma \xrightarrow{F} L$ for which $F(\sigma_1) = e_1$, $F(\sigma_2) = e_2$, we therefore obtain $F(\sigma_1 \wedge \sigma_2) = e_1 \wedge e_2$, $F(\sigma_1 \wedge \sigma_2^*) = e_1 \wedge e_2^{\perp}$, $F(\sigma_2 \wedge \sigma_1^*) = e_2 \wedge e_1^{\perp}$.

Proof. By T 4.1 (ii) we get $g_1 \leqq e_2$, $g_2 \leqq e_2^{\perp}$ and $g_1 \leqq e_1$, $g_3 \leqq e_1^{\perp}$. By T 1.3.8 follows $g_1 \leqq e_1 \wedge e_2$, $g_2 \leqq e_1 \wedge e_2^{\perp}$, $g_3 \leqq e_1^{\perp} \wedge e_2$. Therefore T 3.7 yields $e_1 = g_1 + g_2 \leqq (e_1 \wedge e_2) + (e_1 \wedge e_2^{\perp}) = (e_1 \wedge e_2) \vee (e_1 \wedge e_2^{\perp}) \leqq e_1$. Because $g_1 \leqq e_1 \wedge e_2$ and $g_2 \leqq e_1 \wedge e_2^{\perp}$, this implies $g_1 = e_1 \wedge e_2$, $g_2 = e_1 \wedge e_2^{\perp}$. Analogously follows $g_3 = e_1^{\perp} \wedge e_2$. The remainder of the theorem then follows from the additivity of the measure $\Sigma \xrightarrow{F} L$. □

T 4.3 If $\Sigma \xrightarrow{F} G$ is an additive, effective measure on the Boolean ring Σ, then F is an isomorphic mapping of the Boolean ring Σ onto the Boolean sublattice $F\Sigma$ of G.

Proof. See [2] IV Th 1.3.3.

T 4.4 The following conditions are equivalent for decision effects:

(i) $\{e_1, e_2\}$ is coexistent.

(ii) $\{e_1, e_2\}$ is commensurable.

(iii) The orthocomplemented sublattice Γ of G generated by e_1, e_2 is a Boolean ring.

(iv) $e_1 = (e_1 \wedge e_2) \vee (e_1 \wedge e_2^{\perp})$.

Proof. See [2] IV Th 1.3.4.

Let us list some special cases of coexistent resp. commensurable effects. From $g_1 \leqq g_2$ follows that g_1, g_2 are commensurable since $g_1 = g_1 + \mathbf{0}$ and $g_2 = g_1 + (g_2 - g_1)$. From $e_1, e_2 \in G$ and $e_1 \leqq e_2$ thus follows that e_1, e_2 are commensurable. If $e_1 \perp e_2$, then e_1, e_2 are commensurable. If $e_1 + e_2 \leqq \mathbf{1}$, then $e_1 \perp e_2$ (T 3.7) and hence e_1, e_2 are commensurable. If e_1 and e_2 are commensurable and $e_1 \wedge e_2 = 0$, we get $e_1 \perp e_2$ and hence $e_1 \vee e_2 = e_1 + e_2$.

T 4.5 If $A_1, A_2 \subset G$ and each $e_1 \in A_1$ is commensurable with each $e_2 \in A_2$, also $\bigwedge_{e \in A_2} e$ and $\bigvee_{e \in A_2} e$ are commensurable with each $e_1 \in A_1$. If e is commensurable with each $e_1 \in A_1$, also $e^\perp$ is so.

Proof. See [2] IV Th 1.3.5.

T 4.6 The following conditions are equivalent for $A \subset G$:

(i) A is coexistent.
(ii) A is commensurable.
(iii) Each two $e_1, e_2 \in A$ are coexistent.
(iv) Each two $e_1, e_2 \in A$ are commensurable.
(v) The orthocomplemented sublattice Γ_A of G generated by A is a Boolean ring.
(vi) The orthocomplemented and complete sublattice $\bar{\Gamma}_A$ of G generated by A is a Boolean ring.

Proof. See [2] IV Th 1.3.6.

T 4.7 If e_ν is a sequence of commensurable decision effects which in the $\sigma(\mathscr{B}', \mathscr{B})$-topology converge to an $e \in G$, then e is commensurable with all $\bar{e} \in G$ which are commensurable with the e_ν. A subsequence e_{ν_k} can then be chosen so that

$$e = \bigwedge_{n=1}^{\infty} \tilde{e}_n \quad \text{with} \quad \tilde{e}_n = \bigvee_{k=n}^{\infty} e_{\nu_k}.$$

(In particular, every complete Boolean subring of G is $\sigma(\mathscr{B}', \mathscr{B})$-closed in G.)

Proof. See [2] IV Th 1.3.7.

§ 5 The Orthomodularity of G

Let V be an orthocomplemented lattice. For brevity we define the relations:

A pair of elements $a, b \in V$ is called a *modular* pair if the following relation $M(a, b)$ holds:

$$M(a, b)\text{: For all } c \text{ with } c < b \text{ follows } (c \vee a) \wedge b = c \vee (a \wedge b).$$

The last relation is nothing else than the distributive relation $(c \vee a) \wedge b = (c \wedge b) \vee (a \wedge b)$ in case $c < b$.

A pair $a, b \in V$ is said to be *compatible* if the following relation $\mathscr{C}(a, b)$ holds:

$$\mathscr{C}(a, b): \; a = (a \wedge b) \vee (a \wedge b^{\perp}) .$$

This immediately gives $a \perp b \Rightarrow \mathscr{C}(a, b)$; $\mathscr{C}(a, b) \Rightarrow \mathscr{C}(a, b^{\perp})$; $a < b \Rightarrow \mathscr{C}(a, b)$, and we get the important theorem:

T 5.1 In an orthocomplemented lattice the following relations are equivalent:

(i) $M(a, b)$ holds for all pairs a, b with $a \perp b$.

(ii) $M(a, a^{\perp})$ holds for all a.

(iii) $\mathscr{C}(b, a)$ holds for all pairs a, b with $a < b$, i.e. $b = a \vee (b \wedge a^{\perp})$.

(iv) All pairs a, b with $a < b$ give $a = b \wedge (a \vee b^{\perp})$.

(v) All pairs a, b give $\mathscr{C}(a, b) \Rightarrow \mathscr{C}(b, a)$.

(vi) Each triple a, b, c with $a \perp b$ and $a \perp c$ gives $a \vee b = a \vee c \Rightarrow b = c$.

(vii) Each triple a, b, c with $a \perp b$ and $a \perp c$ gives $a \vee b = a \vee c$ and $b < c$ $\Rightarrow b = c$.

(viii) For all a, from $b \perp a$ and $a \vee b = 1$ follows $b = a^{\perp}$.

Proof. (i) $\Rightarrow$ (ii) follows directly since $a \perp a^{\perp}$ holds.

(ii) $\Rightarrow$ (iii): From $a < b$ follows $b^{\perp} < a^{\perp}$, which by $M(a, a^{\perp})$ yields $(b^{\perp} \vee a) \wedge a^{\perp} = b^{\perp} \vee (a \wedge a^{\perp}) = b^{\perp}$. Applying $\perp$, from this we obtain $b = a \vee (b \wedge a^{\perp})$.

(iii) $\Rightarrow$ (v): From $\mathscr{C}(a, b)$, i.e. $a = (a \wedge b) \vee (a \wedge b^{\perp})$, follows $a^{\perp} = (a^{\perp} \vee b^{\perp}) \wedge (a^{\perp} \vee b)$ and thus $b \wedge a^{\perp} = b \wedge (a^{\perp} \vee b^{\perp})$ which gives $(b \wedge a) \vee (b \wedge a^{\perp}) = (b \wedge a) \vee [b \wedge (a^{\perp} \vee b^{\perp})] = (b \wedge a) \vee [b \wedge (a \wedge b^{\perp})]$. Because $b \wedge a < b$, by (iii) we conclude $(b \wedge a) \vee [b \wedge (b \wedge a^{\perp})] = b$.

(v) $\Rightarrow$ (vi): If (v) holds, we get $\mathscr{C}(a, b) \Rightarrow \mathscr{C}(a, b^{\perp}) \Rightarrow \mathscr{C}(b^{\perp}, a) \Rightarrow \mathscr{C}(b^{\perp}, a^{\perp}) \Rightarrow \mathscr{C}(a^{\perp}, b^{\perp})$. From $a \vee b = a \vee c$ with $b \perp a$ and $c \perp a$ follows $a^{\perp} \wedge b^{\perp} = a^{\perp} \wedge c^{\perp}$. Because $a \perp b \Rightarrow \mathscr{C}(a, b) \Rightarrow \mathscr{C}(b^{\perp}, a^{\perp})$, we find $b^{\perp} = (b^{\perp} \wedge a) \vee (b^{\perp} \wedge a^{\perp}) = a \vee (b^{\perp} \wedge a^{\perp}) = a \vee (c^{\perp} \wedge a^{\perp})$. Similarly follows $c^{\perp} = a \vee (c^{\perp} \wedge a^{\perp})$ and hence $b^{\perp} = c^{\perp}$, i.e. $b = c$.

(vi) $\Rightarrow$ (vii) is trivial.

(vii) $\Rightarrow$ (viii): From $a \vee b = 1 = a \vee a^{\perp}$ and $b \perp a$, i.e. $b \leqq a^{\perp}$, in particular follows $b = a^{\perp}$.

(viii) $\Rightarrow$ (iii): Due to (iii) we assume $a < b$. We define $d = a \vee (b \wedge a^{\perp})$. From this follows $d < b$ and hence $d \wedge b^{\perp} = (d \wedge b) \wedge b^{\perp} = 0$. Since also $b^{\perp} \vee d = b^{\perp} \vee a \vee (b \wedge a^{\perp}) = (b \wedge a^{\perp})^{\perp} \vee (b \wedge a^{\perp}) = 1$, from (viii) follows $d = (b^{\perp})^{\perp} = b$, i.e. $b = a \vee (b \wedge a^{\perp})$.

(iii) $\Rightarrow$ (i): By (i) we assume $c < b$ and $a \perp b$, which makes $M(a, b)$ equivalent to $(c \vee a) \wedge b = c$. Because $c < b$ and $c < c \vee a$, we have $(c \vee a) \wedge b > c$. Therefore (iii) with $g = [(c \vee a) \wedge b] \wedge c^{\perp}$ yields $(c \vee a) \wedge b = c \vee g$. Because $(c \vee a) \wedge b < b$, we have $g < b \wedge c^{\perp}$, i.e. $b^{\perp} \vee c < g^{\perp}$. From $(c \vee a) \wedge b < c \vee a$ follows $g < c \vee a$, which by $a \perp b$ gives $g < c \vee b^{\perp} < g^{\perp}$. This implies $g = 0$, i.e. $c = (c \vee a) \wedge b$.

(iii) $\Rightarrow$ (iv): For $a < b$ we get $b^{\perp} < a^{\perp}$; thus (iii) gives $a^{\perp} = b^{\perp} \vee (a^{\perp} \wedge b)$, hence $a = b \wedge (a \vee b^{\perp})$. Analogously one shows (iv) $\Rightarrow$ (iii). □

If an orthocomplented lattice V obeys one (and hence all) the relations (i) through (viii) of T 5.1, then V is called *orthomodular*.

Still presuming AV 1.1, AV 1.2s and AVid, we find

T 5.2 *G* is orthomodular.

Proof. Let us show (vi) from T 5.1. From e_1, $e_2 \perp e$ and $e \vee e_1 = e \vee e_2$, with T 3.7 follows $e + e_1 = e + e_2$ and hence $e_1 = e_2$. □

All the relations (i) through (viii) from T 5.1 thus hold as theorems in G. Since this is isomorphic to the lattice $\mathscr{W}_{\mathscr{F}'} = \mathscr{W}_L$ of the exposed faces of K, the same relations hold for that lattice.

The relation (vi) used in T 5.2, i.e. $e_1 \perp e$, $e_2 \perp e$, $e \vee e_1 = e \vee e_2 \Rightarrow e_1 = e_2$, has by § 4 the intuitive meaning that e and $\tilde{e} = e_1 \vee e = e_2 \vee e$ are commensurable (because $e_1 \vee e \geqq e_1$). Therefore, we may by V § 5 think of a registration method which measures e and $\tilde{e}$ "approximately". In order to simplify the language let us pretend that some measuring method b_0 registers the e and $\tilde{e}$ "exactly", i.e. that there are a $b \in \mathscr{R}(b_0)$ and a $\tilde{b} \in \mathscr{R}(b_0)$ with $\psi(b_0, b) = e$ and $\psi(b_0, \tilde{b}) = \tilde{e}$. Because $\tilde{e} \geqq e$, we have $\tilde{b} \supset b$. Then $\psi(b_0, \tilde{b} \backslash b) = \psi(b_0, \tilde{b}) - \psi(b_0, b) = \tilde{e} - e$ is necessarily *uniquely* determined (i.e. $e_1 = e_2$).

The relations (vii) and (viii) are only special cases of (vi). It is mathematically interesting that the validity of these special cases suffices to guarantee (vi).

The relation $\mathscr{C}(e_1, e_2)$ coincides with (iv) in T 4.4. Since the relations (i) and (ii) in T 4.4 are symmetric in e_1, e_2, the proof of T 4.4 contains an implicit proof of (v) in T 5.1. Therefore one can replace (iv) in T 4.4 by "e_1, e_2 are compatible".

§ 6 The Main Law for Not Coexistent Registrations

Although we could derive interesting structures in the theory set forth up till now (i.e. practically on the basis of AV 1.1), this theory turns out as rather weakly restricted. Hence one cannot assume to have already a g. G.-closed theory ([3] § 10.3), at least not for microsystems. Therefore, the two axioms AV 3 and AV 4 yet to be set up shall strengthen the structure type; they will have essential significance just for microsystems as action carriers.

The axioms AV 1.1, AV 1.2 s and AVid are also satisfied for so called "classical systems" which one can characterize by "all decision effects being commensurable" (see VII § 5.3). For this case, the main law AV 3 now to be formulated for not coexistent registrations is "without meaning", as we shall recognize more precisely below.

Therefore, summarizing, let us emphasize again that by the next axioms AV 3 and AV 4 we shall restrict the domain of application, i.e. the fundamental domain (see [3] §§ 3 and 5) of the theory to "microsystems as action carriers".

§ 6.1 Experimental Hints for Formulating the Main Law for Not Coexistent Registrations

We consider (straightaway idealizing) the registration of three decision effects, e_1, e_2, e_3 with $e_2 \perp e_3$. Our language is abbreviated much if we now "after some practice" speak of registration methods b_0 which permit us to register decision effects. Strictly speaking, we always ought to add that b_0 registers decision effects "approximately" and that idealizing we can let these approximations become arbitrarily good in the limit. Having assumed $e_2 \perp e_3$, we find e_2 and e_3 commensurable (see § 4), i.e. we can think of a registration method b_0 with such registration procedures $b_2, b_3 \in \mathscr{R}(b_0)$ that $\psi(b_0, b_i) = e_i$ $(i = 2, 3)$. From $e_2 \perp e_3$ then follows $b_2 \cap b_3 = \emptyset$. Moreover we must have $\psi(b_0, b_0 \backslash (b_2 \cup b_3)) = \mathbf{1} - (e_2 + e_3)$.

Therefore, since e_1 need *not* necessarily be commensurable with e_2, e_3, we think of a *second* registration method $\tilde{b}_0$ with $\tilde{b}_1, \tilde{b}_2 \in \mathscr{R}(\tilde{b}_0)$ and $\psi(\tilde{b}_0, \tilde{b}_1) = e_1$, $\psi(\tilde{b}_0, \tilde{b}_2) = e_1 \vee e_2$ (since $e_1 \leqq e_1 \vee e_2$ holds, e_1 and $e_1 \vee e_2$ are commensurable; see § 4). It yields $\tilde{b}_2 \supset \tilde{b}_1$ and $\psi(\tilde{b}_0, \tilde{b}_2 \backslash \tilde{b}_1) = (e_1 \vee e_2) - e_1$ and $\psi(\tilde{b}_0, \tilde{b}_0 \backslash \tilde{b}_2) = \mathbf{1} - (e_1 \vee e_2)$.

Of course there are important relations between the responses of the two devices b_0 and $\tilde{b}_0$ for the various ensembles $w \in K$. We are interested in the response of the device b_0 to ensembles $w \in K_1(e_1 \vee e_2)$ and that of $\tilde{b}_0$ to the $w \in K_1(e_3)$. The $w \in K_1(e_1 \vee e_2)$ are just those ensembles for which the indication $\tilde{b}_2$ on $\tilde{b}_0$ occurs with certainty; the $w \in K_1(e_3)$ are those for which the b_3 on b_0 occurs with certainty.

About the response of the registrations of the device b_0 to the $w \in K_1(e_1 \vee e_2)$, resp. of the device $\tilde{b}_0$ to the $w \in K_1(e_3)$, we can immediately say something in the case $e_1 \perp e_2 \vee e_3 = e_2 + e_3$: Namely, we than have $e_1 \vee e_2 = e_1 + e_2 \perp e_3$ and hence $\mu(w, \psi(b_0, b_3)) = 0$ for all $w \in K_1(e_1 \vee e_2)$ and $\mu(w, \psi(\tilde{b}_0, \tilde{b}_2)) = 0$ for all $w \in K_1(e_3)$.

We are now interested in the response for the case that e_1 is no longer "quite orthogonal" to $e_2 + e_3$ and thus also no longer commensurable with e_2, e_3. Indeed $e_1 \perp e_2 + e_3$ shall no longer hold, but despite that e_1 shall not be "close" to $e_2 + e_3$. What should be understood by that?

$e_1 \perp e_2 + e_3$ is equivalent to $\mu(w, e_2 + e_3) = 0$ for all $w \in K_1(e_1)$ and to $\mu(w, e_1) = 0$ for all $w \in K_1(e_1 + e_3)$. For e_1 to be not close to $e_2 + e_3$, let us understand that at least one of the following relations holds:

$$\sup_{w \in K_1(e_1)} \mu(w, e_2 + e_3) \neq 1 \text{ resp. } \sup_{w \in K_1(e_2 + e_3)} \mu(w, e_1) \neq 1 .$$

These two conditions can be tested by means of the two devices $b_0, \tilde{b}_0$.

$K_1(e_2 + e_3)$ is the set of all ensembles w for which $b_2 \cup b_3$ (on the device b_0) occurs with certainty.

Hence

$$\sup_{w \in K_1(e_2 + e_3)} \mu(w, e_1) = 1 - \alpha \neq 1$$

means that at least the fraction α for those ensembles triggers the indications $\tilde{b}_0 \backslash \tilde{b}_1$ on $\tilde{b}_0$. For $C(w_3) = K_1(e_3)$, from $w_3 \in K_1(e_2 + e_3)$ thus follows $\mu(w_3, \psi(\tilde{b}_0, \tilde{b}_0 \backslash \tilde{b}_1)) \geqq \alpha \neq 0$. Can in $\mu(w_3, \psi(\tilde{b}_0, \tilde{b}_0 \backslash \tilde{b}_1) = \mu(w_3, \psi(\tilde{b}_0, \tilde{b}_2 \backslash \tilde{b}_1)) + \mu(w_3, \psi(\tilde{b}_0, \tilde{b}_0 \backslash \tilde{b}_2))$ the last summand vanish?

Intuitively, one expects the answer "no", because w_3 due to $e_3 \perp e_2$ characterizes an ensemble, in whose realization by preparation procedures only such systems are prepared that are "totally different" from e_2 (with certainty do not trigger the indications b_2 on b_0). Hence, at least some of these systems should trigger the indication $\tilde{b}_0 \backslash \tilde{b}_2$ even when $\alpha \neq 0$, i.e. not yet *all* systems trigger the indication $\tilde{b}_1$. Experimental experience appears not to contradict this. We thus arrive at a relation which we might axiomatically postulate as follows:

$$\sup_{w\in K_1(e_2+e_3)} \mu(w, e_1) \neq 1 \Rightarrow \{\text{for } w_3 \text{ with } C(w_3) = K_1(e_3) \text{ we have } \mu(w_3, \mathbf{1} - e_1 \vee e_2) \neq 0\}. \quad (6.1.1)$$

Quite analogously we can consider what follows: $K_1(e_1)$ is the set of all ensembles for which $\tilde{b}_1$ (on the device $\tilde{b}_0$) occurs with certainty. Hence

$$\sup_{w\in K_1(e_1)} \mu(w, e_1 + e_2) = 1 - \beta \neq 1$$

means that for each of those ensembles the indication $b_0 \backslash b_2 \cup b_3$ occurs at least with the frequency β. If we now consider ensembles $w \in K_1(e_1 \vee e_2)$ (for which $\tilde{b}_2$ on $\tilde{b}_0$ occurs with certainty), one intuitively conjectures that such a w due to $e_3 \perp e_2$ has a frequency $\mu(w, \psi(b_0, b_3)) \neq 1$. Experiments again corroborate this. We thus arrive at a further relation which we might require axiomatically:

$$\sup_{w\in K_1(e_1)} \mu(w, e_2 + e_3) \neq 1 \Rightarrow K_1(e_1 \vee e_2) \cap K_1(e_3) = \emptyset. \quad (6.1.2)$$

With $C(w_3) = K_1(e_3)$, from (6.1.2) we obtain the apparently weaker relation:

$$\sup_{w\in K_1(e_1)} \mu(w, e_2 + e_3) \neq 1 \Rightarrow \{\text{for } w_3 \text{ with } C(w_3) = K_1(e_3) \text{ we have } w_3 \notin K_1(e_1 \vee e_2)\}. \quad (6.1.3)$$

The relation $\mu(w_3, \mathbf{1} - e_1 \vee e_2) \neq 0$ in (6.1.1) is identical with $\mu(w_3, e_1 \vee e_2) \neq 1$, i.e. with $w_3 \notin K_1(e_1 \vee e_2)$. Thus (6.1.1) can also be written

$$\sup_{w\in K_1(e_2+e_3)} \mu(w, e_1) \neq 1 \Rightarrow \{\text{for } w_3 \text{ with } C(w_3) = K_1(e_3) \text{ we have } w_3 \notin K_1(e_1 \vee e_2)\}. \quad (6.1.4)$$

Because $C(w_3) = K_1(e_3)$, we find $w_3 \in K_1(e_1 \vee e_2)$ equivalent to $C(w_3) = K_1(e_3) \not\subset K_1(e_1 \vee e_2)$, i.e. to $e_3 \not\leqq e_1 \vee e_2$. Therefore (6.1.3) and (6.1.4) can also be written

$$\sup_{w\in K_1(e_1)} \mu(w, e_2 + e_3) \neq 1 \Rightarrow e_3 \not\leqq e_1 \vee e_2; \quad (6.1.5)$$

$$\sup_{w\in K_1(e_2+e_3)} \mu(w, e_1) \neq 1 \Rightarrow e_3 \not\leqq e_1 \vee e_2. \quad (6.1.6)$$

From $e_4 \leqq e_3$ follows $\mu(w, e_2 + e_4) \leqq \mu(w, e_2 + e_3)$ and $K_1(e_2 + e_4) \subset K_1(e_2 + e_3)$, so that (6.1.5) applied to e_4 rather than to e_3 yields

$$\sup_{w\in K_1(e_1)} \mu(w, e_2 + e_3) \neq 1 \Rightarrow e_4 \not\leqq e_1 \vee e_2 \text{ for all } e_4 \leqq e_3 \quad (6.1.7)$$

while (6.1.6) yields

$$\sup_{w\in K_1(e_2+e_3)} \mu(w, e_1) \neq 1 \Rightarrow e_4 \not\leqq e_1 \vee e_2 \text{ for all } e_4 \leqq e_3. \quad (6.1.8)$$

$e_4 \not\leqq e_1 \vee e_2$ for all $e_4 \leqq e_3$ is equivalent to $e_3 \wedge (e_1 \vee e_2) = 0$. Therefore (6.1.3) is equivalent to (6.1.5) and also to

$$\sup_{w\in K_1(e_1)} \mu(w, e_2 + e_3) \neq 1 \Rightarrow e_3 \wedge (e_1 \vee e_2) = 0, \quad (6.1.9)$$

whereas (6.1.4) is equivalent to (6.1.6) and also to

$$\sup_{w\in K_1(e_2+e_3)} \mu(w, e_1) \neq 1 \Rightarrow e_3 \wedge (e_1 \vee e_2) = 0. \quad (6.1.10)$$

In all these relations we assumed $e_2 \perp e_3$ so that $e_4 \leqq e_3$ implies $e_2 \perp e_4$. Since (6.1.9) is identical with (6.1.2), also (6.1.3) is so.

In order to summarize (6.1.5) and (6.1.9) resp. (6.1.9) and (6.1.10) in one relation between two elements of G we define the "distance":

$$\Delta(e_1, e_2) = \max\left\{\inf_{w \in K_1(e_1)} \mu(w, \mathbf{1} - e_2);\ \inf_{w \in K_1(e_2)} \mu(w, \mathbf{1} - e_1)\right\}. \tag{6.1.11}$$

Then the relation "(6.1.5) and (6.1.6)" is equivalent to

$$\Delta(e_1, e_2 + e_3) \neq 0,\ e_2 \perp e_3 \Rightarrow e_3 \not\leqq e_1 \vee e_2, \tag{6.1.12}$$

whereas "(6.1.9) and (6.1.10)" is equivalent to

$$\Delta(e_1, e_2 + e_3) \neq 0,\ e_2 \perp e_3 \Rightarrow e_3 \wedge (e_1 \vee e_2) = 0. \tag{6.1.13}$$

Moreover, these relations (6.1.12) and (6.1.13) are equivalent.

In order to "visualize" the important relation (6.1.12) or (6.1.13) even better, let us by an example illustrate the general considerations of registration devices and ensembles. As registration method let us use a somewhat modified Stern-Gerlach experiment. Since this example requires a knowledge of quantum mechanics we invoke it *only to illustrate* the general considerations (it falls out of the present construction of quantum mechanics).

We think of two Stern-Gerlach devices into which an atomic beam (of hydrogen atoms in the ground state) falls in the 1-direction (in a rectangular system of axes 1, 2, 3). One device shall decompose the beam according to the 3-component of spin, the other according to an r-direction (which lies in the 2–3-plane). Moreover, let the two devices be equipped to measure the energy of the atoms in the outgoing beam.

For the above e_1, e_2, e_3 let us now in particular assume: The first device measures

e_2: spin in the (+ 3)-direction, energy greater than ε;
e_3: spin in the (+ 3)-direction, energy smaller than ε:
$\mathbf{1} - (e_2 + e_3)$: spin in the (− 3)-direction.

The second device measures

e_1: spin in the (+ r)-direction;
$e_1 \vee e_2 = e_1 + (e_1 \vee e_2 - e_1)$ with $e_1 \vee e_2 - e_1$: spin in the (− r)-direction, energy greater than ε;
$\mathbf{1} - e_1 \vee e_2$: spin in the (− r)-direction, energy smaller than ε.

The reader should check that e_1, e_2 imply the given meaning of measuring $e_1 \vee e_2$!

If the r-direction is the (− 3)-direction, then both devices do the same and $\Delta(e_1, e_2 + e_3) = 1$. Then (6.1.1) and (6.1.3) hold trivially, since $\mathbf{1} - e_1 \vee e_2 = e_3$ immediately causes "$C(w_3) = K_1(e_3) \Rightarrow \mu(w_3, \mathbf{1} - e_1 \vee e_2) = 1$".

If one slowly rotates the r-direction out of the (− 3)-direction, $\mu(w_3, \mathbf{1} - e_1 \vee e_2)$ should not jump to zero but continuously vary away from 1. Very pictorially, $\Delta(e_1, e_2 + e_3)$ is a measure for the deviation of the r-direction from the 3-direction, becoming zero as the r-direction approaches the 3-direction. Then $\mu(w_3, \mathbf{1} - e_1 \vee e_2)$ can also tend to zero.

Therefore, (6.1.12) is a sort of continuity law for the probability function μ. In generality (not only for the example) this can be seen as follows.

Let $C(w_3) = K_1(e_3)$. For $\Delta(e_1, e_2 + e_3) = 1$ we have $\mu(w_3, \mathbf{1} - e_1 \vee e_2) = 1$. As $\Delta(e_1, e_2 + e_3)$ decreases, $\mu(w_3, \mathbf{1} - e_1 \vee e_2)$ also decreases but tends to zero *only if* $\Delta(e_1, e_2 + e_3)$ does.

We might introduce (6.1.12) or (6.1.13) as a further main law. Before doing so, let us yet prove theorems intimately connected with (6.1.12) and (6.1.13).

§ 6.2 Some Important Equivalences

First let us in theorems collect some assertions about $\Delta(\ldots)$:

Besides $\Delta(e_1, e_2)$ we can introduce the following distances derivable from the norms in $\mathscr{B}$ and $\mathscr{B}'$: First $\| e_1 - e_2 \|$ with the norm in $\mathscr{B}'$, second the distance

$$\delta(e_1, e_2) = \tfrac{1}{2} \inf_{\substack{w_1 \in K_1(e_1) \\ w_2 \in K_1(e_2)}} \| w_1 - w_2 \|, \tag{6.2.1}$$

with $\| w_1 - w_2 \|$ the norm in $\mathscr{B}$.

From $\| e_1 - e_2 \| = \sup_{w \in K} |\mu(w, e_1 - e_2)|$ follows

$$\begin{aligned} \| e_1 - e_2 \| &\geqq \sup_{w \in K_1(e_1) \cup K_1(e_2)} |\mu(w, e_1 - e_2)| \\ &= \max\left\{ \sup_{w \in K_1(e_1)} |\mu(w, e_1 - e_2)|, \sup_{w \in K_1(e_2)} |\mu(w, e_1 - e_2)| \right\} \\ &= \max\left\{ \sup_{w \in K_1(e_1)} |\mu(w, 1 - e_2)|, \sup_{w \in K_1(e_2)} |\mu(w, 1 - e_1)| \right\} = \Delta(e_1, e_2). \end{aligned}$$

Analogously,

$$\begin{aligned} \| w_1 - w_2 \| &= \sup_{y \in [-\mathbf{1}, \mathbf{1}]} |\mu(w_1 - w_2, y)| \\ &= \sup_{g \in L} |\mu(w_1 - w_2, 2g - \mathbf{1})| \\ &= 2 \sup_{g \in L} |\mu(w_1 - w_2, g)| \end{aligned}$$

implies the relations

$$\| w_1 - w_2 \| \geqq 2 |\mu(w_1 - w_2, \mathbf{1} - e_1)|,$$

$$\| w_1 - w_2 \| \geqq 2 |\mu(w_1 - w_2, \mathbf{1} - e_2)|,$$

and hence

$$\delta(e_1, e_2) \geqq \inf_{\substack{w_1 \in K_1(e_1) \\ w_2 \in K_1(e_2)}} |\mu(w_1 - w_2, \mathbf{1} - e_1)| = \inf_{w_2 \in K_2(e_2)} \mu(w_2, \mathbf{1} - e_1),$$

$$\delta(e_1, e_2) \geqq \inf_{\substack{w_1 \in K_1(e_1) \\ w_2 \in K_1(e_2)}} |\mu(w_1 - w_2, \mathbf{1} - e_2)| = \inf_{w_1 \in K_1(e_1)} \mu(w_1, \mathbf{1} - e_2).$$

From these immediately follows $\delta(e_1, e_2) \geqq \Delta(e_1, e_2)$.

Because $0 \leqq \mu(w, e) \leqq 1$, we obtain $\| e_1 - e_2 \| \leqq 1$. From $0 \leqq \mu(w, g) \leqq 1$ similarly follows $\delta(e_1, e_2) \leqq 1$, such that

$$0 \leqq \Delta(e_1, e_2) \leqq \delta(e_1, e_2) \leqq 1, \tag{6.2.2}$$

$$0 \leqq \Delta(e_1, e_2) \leqq \| e_1 - e_2 \| \leqq 1. \tag{6.2.3}$$

T 6.2.1 The following relations are equivalent:

(i) $e_1 \perp e_2$,
(ii) $\Delta(e_1, e_2) = 1$,
(iii) $\delta(e_1, e_2) = 1$.

Proof. (i) $\Rightarrow$ (ii): Because $e_1 \perp e_2$ makes $e_1 \leqq \mathbf{1} - e_2$ and $e_2 \leqq \mathbf{1} - e_1$, from T 3.6 follows $\Delta(e_1, e_2) = 1$. From (6.2.2) follows (ii) $\Rightarrow$ (iii).

(iii) $\Rightarrow$ (i): (6.2.1) and $\delta(e_1, e_2) = 1$ imply $\|w_1 - w_2\| \geqq 2$ for all $w_i \in K_1(e_i)$ with $i = 1, 2$, hence

$$\sup_{g \in L} |\mu(w_1 - w_2, g)| = 1 \text{ for all } w_i \in K_1(e_i).$$

We choose $w_i = w_{i0}$ so that $C(w_{i0}) = K_1(e_i)$. Since L is $\sigma(\mathscr{B}', \mathscr{B})$-compact, there is a $g \in L$ with $(w_{10} - w_{20}, g) = 1$ (for $\mu(w_{10} - w_{20}, g) = -1$ follows $\mu(w_{10} - w_{20}, \mathbf{1} - g) = 1$, so that it suffices to consider $\mu(w_{10} - w_{20}, g) = 1$), whence $\mu(w_{10}, g) = 1$ and $\mu(w_{20}, g) = 0$ follow. Thus we get $\mu(w_1, g) = 1$ for all $w_1 \in C(w_{10}) = K_1(e_1)$, and $\mu(w_2, g) = 0$ for all $w_2 \in C(w_{20}) = K_1(e_2)$. This gives $g \in L_1 K_1(e_1)$ and $g \in L_0 K_1(e_2)$ and hence $\mathbf{1} - g \in L_0 K_1(e_1) = L_0 K_0(\mathbf{1} - e_1)$, $g \in L_0 K_0(\mathbf{1} - e_2)$. By (1.3.4) follows $\mathbf{1} - g \leqq \mathbf{1} - e_1$ and $g \leqq \mathbf{1} - e_2$, i.e. $e_1 \leqq g \leqq \mathbf{1} - e_2$ and thus $e_1 \perp e_2$. □

T 6.2.2 If e_1 is commensurable with e_2, we have $\Delta(e_1, e_2) = 0$ or 1 and likewise $\delta(e_1, e_2) = 0$ or 1 as well as $\|e_1 - e_2\| = 0$ or 1.

For e_1, e_2 commensurable, T 4.4 (iv) makes $e_1 = (e_1 \wedge e_2) \vee (e_1 \wedge e_2^{\perp})$. For $e_1 \wedge e_2 \neq 0$ we get $K_1(e_1) \cap K_1(e_2) \neq \emptyset$ and hence $\Delta(e_1, e_2) = 0$. From $K_1(e_1) \cap K_1(e_2) \neq 0$ follows $\delta(e_1, e_2) = 0$. For $e_1 \wedge e_2 = 0$ we obtain $e_1 = e_1 \wedge e_2^{\perp}$, hence $e_1 \perp e_2$. By T 6.2.1 this gives $\Delta(e_1, e_2) = 1$ and $\delta(e_1, e_2) = 1$. □

When all $e \in G$ are mutually commensurable ($G = Z$ with Z as in VII D 1.2), T 6.2.2 makes (6.1.13) trivially satisfied, because $\Delta(e_1, e_2 + e_2) \neq 0$ then gives $e_1 \perp e_2 + e_3$ and hence $e_1 \perp e_2$ and $e_3 \perp e_1 + e_2 = e_1 \vee e_2$, whence $e_3 \wedge (e_1 \wedge e_2) = 0$. Our intended main law AV 3, to be equivalent to (6.1.13), hence is meaningless when *all* decision effects are commensurable (see VIII § 5.3).

T 6.2.3 For elements $e_i \in G$, the following relations are equivalent:

(i) $e_2 \perp e_3, \Delta(e_1, e_2 + e_3) \neq 0 \Rightarrow e_3 \nleqq e_1 \vee e_2$,
(ii) $e_2 \perp e_3, \Delta(e_1, e_2 + e_3) \neq 0 \Rightarrow e_3 \wedge (e_1 \vee e_2) = 0$,
(iii) $e_2 \leqq e_4, \Delta(e_1, e_4) \neq 0, e_1 \vee e_2 = e_1 \vee e_4 \Rightarrow e_2 = e_4$,
(iv) $e_2 \leqq e_4, \Delta(e_1, e_4) \neq 0, e_2 \neq e_4 \Rightarrow e_4 \nleqq e_1 \vee e_2$.
(v) $e_3 \nleqq e_2, \Delta(e_1, e_2 + e_3) \neq 0 \Rightarrow e_3 \nleqq e_1 \vee e_2$.
(vi) $e_3 \leqq e_1 \vee e_2, \Delta(e_1, e_2 \vee e_3) \neq 0 \Rightarrow e_3 \leqq e_2$.
(vii) $e_3 \leqq e_1 \vee e_2, e_3 \leqq 1 - e_2, \Delta(e_1, e_2 \vee e_3) \neq 0 \Rightarrow e_3 = 0$.
(viii) $e_2 \leqq e_1, \Delta(e_1 \wedge (e_2 \vee e_3), e_3 - e_1 \wedge e_3) \neq 0 \Rightarrow (e_2 \vee e_3) \wedge e_1 = e_2 \vee (e_3 \wedge e_1)$.
(ix) $e_2 \leqq e_1 \leqq e_2 \vee e_3, \Delta(e_1, e_3) \neq 0 \Rightarrow e_1 = e_2$.

Proof. (i) $\Rightarrow$ (ii) has been shown in § 6.1, since (i) and (ii) are identical with (6.1.12) and (6.1.13).

(ii) $\Rightarrow$ (iii): If $e_4 - e_2 = e_3$, then $e_2 \perp e_3$ and $e_2 + e_3 = e_4$. By (ii) thus follows $e_3 \wedge (e_1 \vee e_2) = 0$. From $e_3 \leqq e_4$ and $e_1 \vee e_2 = e_1 \vee e_4$ follows $e_3 \wedge (e_1 \vee e_2) = e_3 \wedge (e_1 \vee e_4) = e_3$ and hence $e_3 = 0$.

(iii) $\Rightarrow$ (iv): If we had $e_4 \leqq e_1 \vee e_2$, we would get $e_1 \vee e_2 = e_1 \vee e_2 \vee e_4 = e_1 \vee e_4$ (because $e_2 \leqq e_4$) so that (iii) would give $e_2 = e_4$.

(iv) $\Rightarrow$ (v): With $e_4 = e_2 + e_3$ we have $e_2 \leqq e_4$. Because $e_3 \not\leqq e_2$, we have $e_3 \neq \mathbf{0}$ and thus $e_2 \neq e_4$. From (iv) follows $e_4 \not\leqq e_1 \vee e_2$. If we had $e_3 \leqq e_1 \vee e_2$, we would get $e_4 = e_2 \vee e_3 \leqq e_2 \vee e_1 \vee e_2 = e_1 \vee e_2$ which contradicts (iv).

(v) $\Rightarrow$ (i) follows immediately since for $e_3 \perp e_2$ we have *a fortiori* $e_3 \not\leqq e_2$.

(iii) $\Rightarrow$ (vi): With $e_4 = e_2 \vee e_3$ we have $e_2 \leqq e_4$. Because $e_3 \leqq e_1 \vee e_2$, we have $e_1 \vee e_4 = e_1 \vee e_2 \vee e_3 = e_1 \vee e_2$. With $\Delta(e_1, e_2 \vee e_3) \neq 0$, from (iii) follows $e_2 = e_2 \vee e_3$ and thus $e_3 \leqq e_2$.

(vi) $\Rightarrow$ (vii) follows directly because $e_3 \leqq e_2$, $e_3 \leqq \mathbf{1} - e_2 = e_2^\perp \Rightarrow e_3 \leqq e_2 \wedge e_2^\perp = \mathbf{0}$.

(vii) $\Rightarrow$ (viii): From $e_3 \wedge e_1 \leqq e_3$ and $e_3 \wedge e_1 \leqq e_1$ follows $e_2 \vee (e_3 \wedge e_1) \leqq e_2 \vee e_3$ and $e_2 \vee (e_3 \wedge e_1) \leqq e_2 \vee e_1 = e_1$ (because $e_2 \leqq e_1$), hence $e_2 \vee (e_3 \wedge e_1) \leqq (e_2 \vee e_3) \wedge e_1$. Putting $(e_2 \vee e_3) \wedge e_1 - e_2 \vee (e_3 \wedge e_1) = \tilde{e}_3$ we get $\tilde{e}_3 \leqq \mathbf{1} - \tilde{e}_2$ with $\tilde{e}_2 = e_2 \vee (e_3 \wedge e_1)$ and $\tilde{e}_3 \leqq e_1 \wedge (e_2 \vee e_3)$. With $\tilde{e}_1 = e_3 - e_3 \wedge e_1$ we have $\tilde{e}_1 \wedge e_3 = \tilde{e}_1$ and $\tilde{e}_1 \perp e_3 \wedge e_1$, i.e. $\tilde{e}_1 \wedge (e_3 \wedge e_1) = \mathbf{0}$, whence $\tilde{e}_1 \wedge e_1 = \tilde{e}_1 \wedge e_2 \wedge e_1 = 0$.

From $\tilde{e}_3 \leqq e_1 \wedge (e_2 \vee e_3) \leqq e_2 \vee e_3 = e_2 \vee [(e_3 \wedge e_1) \vee \tilde{e}_1] = \tilde{e}_2 \vee \tilde{e}_1$ and (vii), for $\Delta(\tilde{e}_1, \tilde{e}_2 \vee \tilde{e}_3) \neq 0$ follows $\tilde{e}_3 = \mathbf{0}$. With $\tilde{e}_2 \vee \tilde{e}_3 = (e_2 \vee e_3) \wedge e_1$ and $\tilde{e}_1 = e_3 - e_3 \wedge e_1$, finally (viii) follows. Let us yet remark that always $\tilde{e}_1 \wedge (\tilde{e}_2 \vee \tilde{e}_3) = \tilde{e}_1 \wedge e_1 \wedge (e_2 \vee e_3) = \mathbf{0}$ because $\tilde{e}_1 \wedge e_1 = 0$. Thus $\Delta(\tilde{e}_1, \tilde{e}_2 \wedge \tilde{e}_3) \neq 0$ sharpens the always satisfied relation $\tilde{e}_1 \wedge (\tilde{e}_2 \vee e_3) = \mathbf{0}$!

(viii) $\Rightarrow$ (ix): With $e_2 \leqq e_1 \leqq e_2 \vee e_3$ follows $e_1 \wedge (e_2 \vee e_3) = e_1$. From $\Delta(e_1, e_3) \neq 0$ follows $e_1 \wedge e_3 = \mathbf{0}$ and hence $e_2 \vee (e_1 \wedge e_3) = e_2$. Thus (viii) in particular gives $e_1 = e_2$.

(ix) $\Rightarrow$ (iii): $e_2 \leqq e_4$ and $e_1 \vee e_2 = e_1 \vee e_4$ imply $e_4 \leqq e_1 \vee e_4 = e_2 \vee e_1$; hence (ix) with $\Delta(e_1, e_4) \neq 0$ finally proves $e_2 = e_4$. □

Mathematically especially interesting are (viii) and (ix), called the generalized modular relation and the generalized covering condition. Physically, (i) and (ii) can be interpreted most clearly, as we tried to do in § 6.1.

§ 6.3 Formulation of the Main Law and Some Consequences

Due to T 6.2.3, we can as main law formulate the axiom

AV 3 One of the relations (i) through (ix) in T 6.2.3.

All the remaining relations in T 6.2.3 then hold as theorems. Approaching from experiments, one could choose (i) in T 6.2.3 as AV 3.

As remarked in § 6.1, for "classical systems as action carriers" (the case $G = Z$ with Z as in VII D 1.2), AV 3 is satisfied as a theorem and hence without any meaning. A mathematical counterexample (in which AV 1.1, AV 1.2s and AVid are satisfied but AV 3 is not) can apparently be constructed since there are examples of orthocomplemented, orthomodular lattices which do not satisfy (viii) in T 6.2.3. A rigorous proof for the independence of the axioms AVid and AV 3 is still lacking.

AV 3 does not exclude classical systems as action carriers, but it is essential for non-classical systems.

T 6.3.1 For $\Delta(e_1, e_3 - e_1 \wedge e_3) \neq 0$, the modular relation $M(e_3, e_1)$ holds, so that

$$(e \vee e_3) \wedge e_1 = e \vee (e_3 \wedge e_1) \quad \text{for all} \quad e \leqq e_1 .$$

Proof. According to (viii) of T 6.2.3, we need only show that $\Delta(e_1, e_3 - e_1 \wedge e_3) \neq 0$ implies $\Delta(e_1 \wedge (e \vee e_3), e_3 - e_1 \wedge e_3) \neq 0$. This is true because $e_1 \wedge (e \vee e_3) \leqq e_1$. □

If a face $K_1(e)$ of K is finite dimensional as an affine space, then on $K_1(e)$ all separating topologies of the vector space $\mathscr{B}$ coincide with the "euclidean" topology; hence in particular each finite-dimensional $K_1(e)$ is compact in the $\sigma(\mathscr{B}, \mathscr{B}')$-topology.

T 6.3.2 If $K_1(e_1)$ or $K_1(e_2)$ is compact in the $\sigma(\mathscr{B}, \mathscr{B}')$-topology, then

$$\Delta(e_1, e_2) \neq 0 \Leftrightarrow e_1 \wedge e_2 = \mathbf{0}.$$

Proof. That $\Delta(e_1, e_2) \neq 0$ implies $e_1 \wedge e_2 = \mathbf{0}$ is trivial. Let $K_1(e_1)$ be $\sigma(\mathscr{B}, \mathscr{B}')$-compact and let $\inf_{w \in K_1(e_1)} \mu(w, \mathbf{1} - e_2) = 0$. Since $K_1(e_1)$ is compact, there is a $w_0 \in K_1(e_1)$ with $\mu(w_0, \mathbf{1} - e_2) = 0$, i.e. with $w_0 \in K_1(e_2)$, so that $K_1(e_1) \cap K_1(e_2) = K_1(e_1 \wedge e_2) \neq \emptyset$. Since this would contradict $e_1 \wedge e_2 = 0$, we conclude $\Delta(e_1, e_2) \neq 0$. □

T 6.3.3 If $K_1(e_1)$ or $K_1(e_3)$ is $\sigma(\mathscr{B}, \mathscr{B}')$-compact, $M(e_3, e_1)$ holds and hence also $M(e_1, e_3)$.

Proof. Because $K_1(e_3 - e_1 \wedge e_3) \subset K_1(e_3)$, also $K_1(e_3 - e_1 \wedge e_3)$ is $\sigma(\mathscr{B}, \mathscr{B}')$-compact if $K_1(e_3)$ is, since $K_1(e_3 - e_1 \wedge e_3)$ is always $\sigma(\mathscr{B}, \mathscr{B}')$-closed. Since $e_1 \wedge (e_3 - e_1 \wedge e_3) = \mathbf{0}$, the theorem thus follows from T 6.3.1 and T 6.3.2. □

T 6.3.4 If $K_1(e_1)$ or $K_1(e_3)$ is $\sigma(\mathscr{B}, \mathscr{B}')$-compact, we find

$$e_2 \leqq e_1 \leqq e_2 \vee e_3, e_1 \wedge e_3 = \mathbf{0} \Rightarrow e_1 = e_2 .$$

Proof. This follows from T 6.2.3 (ix) and T 6.3.2. □

Therefore T 6.3.4 holds in particular when $K_1(e_3)$ is null-dimensional, i.e. an extreme point of K. Then e_3 is an atom of the lattice G. If moreover $K_1(e_3)$ is $\sigma(\mathscr{B}, \mathscr{B}')$-compact (not necessarily zero-dimensional), then T 6.3.4 for $e_1 \neq e_2$ implies $e_1 \wedge e_3 \neq \mathbf{0}$. Since e_3 is an atom, therefore $e_1 \wedge e_3 = e_3$ holds, hence $e_3 \leqq e_1$. With $e_2 \leqq e_1$ we obtain $e_2 \vee e_3 \leqq e_1 \vee e_3 = e_1$. Together with $e_1 \leqq e_2 \vee e_3$ this finally gives $e_1 = e_2 \vee e_3$; hence between e_2 and $e_2 \vee e_3$ there is no $e_1 \in G$, distinct from e_2 and $e_2 \vee e_3$.

T 6.3.5 If $K_1(e)$ is $\sigma(\mathscr{B}, \mathscr{B}')$-compact, the order interval $[\mathbf{0}, e] \subset G$ is a complete orthocomplemented and modular lattice.

Proof. That $[\mathbf{0}, e]$ is a complete lattice follows directly from the fact that G is complete. With $e - e_1$ as the orthocomplement of $e_1 \in [\mathbf{0}, e]$ in $[\mathbf{0}, e]$, this lattice is orthocomplemented.

From $e_1 \leqq e$ follows $K_1(e_1) \subset K_1(e)$. Since $K_1(e_1)$ is $\sigma(\mathscr{B}, \mathscr{B}')$-closed, also $K_1(e_1)$ is compact. Hence T 6.3.3 proves $M(e_1, e_2)$ for any two elements in $[\mathbf{0}, e]$, which therefore is modular. □

§ 7 The Main Law of Quantization

The typical pecularity of microsystems to occupy "discretely distinguishable states", as seen in the case of atoms, has found no expression in the previous axioms. But what are really the experimental experiences one uses to label with captions such as quantized states? On the historical path of heuristically guessing quantum mechanics, one perceived Planck's quantum of action as the prominent element of quantization. However, the value of a dimensioned constant cannot be essential in a theory. Hidden behind Planck's action quantum there must be some "finiteness" which he first encountered in the discretely quantized states of the harmonic oscillator. But the pure scale value of the energy of an harmonic oscillator must not be decisive, if we conjecture a central law of microsystems. If we now try to formulate such a general law, we cannot expect that one could directly grasp it in generality. Rather, it must just express a structure not familiar to us from classical systems.

§ 7.1 Intuitive Indications for Formulating the Main Law of Quantization

What is the significance of "discrete states" of atomic systems (e.g. of the harmonic oscillator first described by Planck), if we try to express it by the concepts of preparing and registering? It shows that there are preparations (i.e. ensembles) which no longer can be de-mixed into "arbitrarily" many different subensembles. The "ground state" of an atom (one of the typically quantum mechanical structures of microsystems) is an *easily* produced ensemble which no longer can be de-mixed properly, i.e. into different ensembles.

On the contrary, one uses to describe classical macrosystems (e.g. a system of mass points) in a continuous state space. Ensembles can never be produced as point measures, but rather only finer and finer. Never to reach a limit of maximal precision is a typical feature of experimenting with classical systems. On the contrary, with quantum systems there is the fact of easily producable ensembles which nonetheless cannot be made more uniform.

This manifests itself as follows in the concepts of the set K and its faces. For classical systems, every face can be arbitrarily refined, i.e. subdivided into smaller faces. On the other hand, for quantum systems there are faces which cannot be further refined, or allow only refinements in *finitely* (!) many steps. We can also express this as follows: The unexpected in quantum systems (as action carriers) just are those faces F of K whose ensembles can be distinguished by *finitely* many registrations, i.e by finitely many effects. On the other hand, from classical systems (described in continuous state spaces) one was accustomed to the situation that finitey many effects never suffice to distinguish the ensembles of a face. The finer one wishes to distinguish the ensembles of a face experimentally, the more registrations one needs. The continuous scales typical for classical systems are the striking indication of measurements that can be more and more refined. For quantum systems, there are faces where one gets along with a finite, discrete set of measurement points in order to distinguish the element of a face.

Therefore we adopt

D 7.1.1 A set $l \subset L$ is called *separating* with respect to a face F of K, if $w_1, w_2 \in F, \mu(w_1, g) = \mu(w_2, g)$ for all $g \in l$ imply $w_1 = w_2$.

The smallest cardinality of a separating set l for a face F is precisely the (affine) dimension of F. Of course, one could define this dimension by this smallest cardinality.

Therefore, continuously described classical systems are distinguished just by the fact that the dimension of all faces of K is infinite. For quantum systems, experiments yielded the initially strange fact that there are faces of *finite* dimension. But as a mathematical axiom, i.e. as mathematical formulation of a general law, a relation of the form: "There are faces of finite dimension" is unsuitable. It would not be a "universal" structure, since the existence of a single finite-dimensional face of K would satisfy it. This suggests, as a "main law of quantum systems" to try out

AV 4 For *each* exposed face F of K, there is another exposed face F_1 of *finite dimension* such that $F_1 \subset F$.

One could sharpen this "finiteness axiom" to

AV 4s Each exposed face F of K is the upper bound of a sequence $\{F_\nu\}$ of increasing exposed faces of finite dimension.

This AV 4s postulates that $F = \bigvee_\nu F_\nu$ with $\bigvee_\nu$ as the lattice-theoretical union in $\mathscr{W}_{\mathscr{B}'} = \mathscr{W}_L$ has $F_\nu \subset F_{\nu+1}$ and each F_ν of finite dimension.

In a certain sense, AV 4s expresses that any F has finite-dimensional approximations. Physically, not much speaks against a further sharpening of AV 4s to the form that K (and hence *every* face F) is finite dimensional, except for the fact that one does *not* know how to determine a finite dimension of K. As usual, one then replaces this lack of knowledge (see [3] §§ 7.3 and 9) by the requirement that K be infinite dimensional. In this sense the infinite dimensionality of K is an idealization. But it becomes "necessary" if one wishes to introduce yet further idealizations such as the Galileo (or Poincaré) group as the transport group of the registration devices (see III § 7; IX § 1). Namely, if we would assume K finite dimensional and introduce the idealization of the translation group (as a subgroup of the Galileo resp. Poincaré group) up to "infinity", then only the trivial "solution" of the axiom system would remain: There the preparation devices do not act at all on the registration devices. In the language of physics, there would be "no action-carriers at all" (one does not call the "vacuum" an action carrier; for "vacuum" see IX § 1).

We have stated the two formulations AV 4 and AV 4s since on one hand we need AV 4s in order to attain the "Hilbert space structure", but on the other hand many theorems can be proved with AV 4 alone.

We declare the "microsystems" to form the fundamental domain of those physical systems (as action carriers) for which the axioms AV 1.1, AV 1.2s, AVid, AV 3, AV 4 (perhaps sharpened by AV 2s and AV 4s) yield a useable theory.

Similarly one can give the word "classical systems" a precise sense: They form that fundamental domain of physical systems (as action carriers) for which (besides the axioms AV 1.1, AV 1.2s, AVid) the following axiom AVkl leads to a useable theory.

AVkl: Any two decision effects are commensurable and each exposed face of K is infinite dimensional.

§ 7.2 Simple Consequences of the Main Law of Quantization

T 7.2.1 From AV 4 follows: The lattices G and $\mathscr{W}_L$ are atomic. An atom in $\mathscr{W}_L$ is a finite-dimensional face of K.

Proof. G is isomorphic to $\mathscr{W}_{\mathscr{B}'} = \mathscr{W}_L$. Due to AV 4, each face of $\mathscr{W}_L$ contains a finite-dimensional face of $\mathscr{W}_L$. If it is not an atom, it contains a smaller face and hence a face of smaller dimension. After *finitely* many steps one must obtain an atom. □

T 7.2.2 From AV 4 and AV 2f follows that the atoms of $\mathscr{W}_L$ are just the extreme points of K.

Proof. By AV 2f, an extreme point of K is an element of $\mathscr{W}_L$ and obviously an atom of $\mathscr{W}_L$.

Due to T 7.2.1, an atom of $\mathscr{W}_L$ must be a finite-dimensional face. Since each finite-dimensional face possesses an extreme point (by AV 2f an element of $\mathscr{W}_L$), an atom of $\mathscr{W}_L$ must always be an extreme point. □

T 7.2.3 From AV 4 and AV 3 follows: If p is an atom of the lattice G, then $e_2 \leqq e_1 \leqq e_2 \vee p$ implies $e_1 = e_2$ or $e_1 = e_2 \vee p$.

Proof. Follows easily from T 7.2.1 and the remarks after T 6.3.4. □

One often calls the relation in T 7.2.3 the *covering* condition.

T 7.2.4 From AV 4 and AV 3 follows: If $e \in G$, $e \neq \mathbf{0}$, $e \neq \mathbf{1}$ and p is an atom, there are two atoms $q_1 \leqq e$, $q_2 \leqq e^{\perp}$ with $p \leqq q_1 \vee q_2$.

Proof. The theorem is trivial for $p < e$ and $p < e^{\perp}$. If $p \not\leqq e$ and $p \not\leqq e^{\perp}$, we have $e \neq e \vee p$ and $e^{\perp} \neq e^{\perp} \vee p$. With $q_2 = e \vee p - e$ we get $q_2 \leqq \mathbf{1} - e = e^{\perp}$.

If q_2 were not an atom, there would exist $e_1, e_2 \neq 0$ and $q_2 = e_1 + e_2$ and hence $e \not\geqq e + e_1 \not\geqq e + e_1 + e_2 = e + q_2 = e \vee p$, contradicting T 7.2.3. Therefore q_2 is an atom. Similarly follows that $q_1 = e^{\perp} \vee p - e^{\perp}$ is an atom with $q_1 \leqq e$. Because $p \leqq e \vee p$ and $p \leqq e^{\perp} \vee p$, we obtain $p \leqq (e \vee p) \wedge (e^{\perp} \vee p) = (e \vee q_2) \wedge (e^{\perp} \vee q_1)$. By T 4.2 we easily find (see the remarks after T 4.4) that $e, e^{\perp}, q_1, q_2$ are pairwise commensurable. Hence the orthocomplemented sublattice generated by these elements is a Boolean ring (see T 4.4 iii), such that $(e \vee q_2) \wedge (e^{\perp} \vee q_1) = (e \wedge e^{\perp}) \vee (e \wedge q_1) \vee (q_2 \wedge e^{\perp}) \vee (q_1 \wedge q_2) = q_1 \vee q_2$. □

VII Decision Observables and the Center

The following sections will extend and specialize the discussions in V for the case that certain main laws of preparing and registering are presumed. In this chapter, we shall in general presume only AV 1.1, AV 1.2s and AVid. Therefore, the results to follow are not only valid for microsystems as action carriers.

§ 1 The Commutator of a Set of Decision Effects

In VI § 4 we have introduced the concept of commensurability and proved theorems on commensurable decision effects. We will supplement these discussions by structure assertions about special subsets of decision effects.

D 1.1 We call the set $A' = \{e' \mid e' \in G$ and e' is commensurable with each $e \in A\}$ the *commutator* of the subset $A \subset G$.

T 1.1 The set A' is a complete, orthocomplemented sublattice of G. The set A' is $\sigma(\mathscr{B}', \mathscr{B})$-closed in G.

Proof. The first part of the theorem follows directly from VI T 4.5. The second part of the theorem follows in a way similarly to the proof of VI T 4.7: Since $\sigma(\mathscr{B}', \mathscr{B})$ is metrizable on L, we need only show that $e'_\nu \in A'$ and $e'_\nu \to e' \in G$ in the $\sigma(\mathscr{B}', \mathscr{B})$-topology implies $e' \in A'$. From e'_ν commensurable to $e \in A$ follows $e'_\nu = (e'_\nu \wedge e) + (e'_\nu \wedge e^\perp)$. Since L is compact, one can choose a subsequence ν_i such that

$$e'_{\nu_i} \wedge e \to g_1, \quad e'_{\nu_i} \wedge e^\perp \to g_2,$$

hence $e' = g_1 + g_2$. From $e'_{\nu_i} \wedge e \leqq e$ follows $g_1 \leqq e$ and likewise $g_2 \leqq e^\perp$. Thus VI T 4.1 and VI T 4.4 show that e' is commensurable with e. □

One easily finds $A'' \supset A$, $A''' = A'$, etc.

D 1.2 The set $Z = G'$ is called the *center* of G.

From T 1.1 and VI T 4.6 (vi) immediately follows

T 1.2 Z is a complete Boolean sublattice of G.

D 1.3 A set A of commensurable decision effects is called *maximal* if $A' = A$.

T 1.3 Every set A of commensurable decision effects is a subset of a maximal set B of commensurable decision effects. A maximal set of commensurable decision effects is a complete, maximal Boolean sublattice of G.

Proof. Due to VI T 4.6 (iv), the set of the $\tilde{B} \subset G$ with $\tilde{B} \supset A$ and $\tilde{B}$ commensurable satisfies the presumptions of Zorn's Lemma. Hence there is a (set-theoretic) maximal B in the set of the $\tilde{B}$, which one easily proves $B' = B$.

The remainder of the theorem follows easily from T 1.1 and VI T 4.6 (vi). □

§ 2 Decision Observables

It is obvious to specialize the general definition V D 1.1.1 of an observable to

D 2.1 A *decision observable* is a Boolean ring Σ with an effective additive measure $\Sigma \xrightarrow{F} G$ such that Σ is complete relative to the uniform structure $\mathscr{U}_g$ (that Σ is $\mathscr{U}_g$-separable follows as a theorem; see T 2.2).

We first prove:

T 2.1 If an additive measure $\Sigma \xrightarrow{F} G$ is given for a Boolean ring, this also holds for the *extension* $\hat{\Sigma}_g \xrightarrow{F} G$ of F onto $\hat{\Sigma}_g$, which also makes $F\hat{\Sigma}_g \subset \overline{F\Sigma^\sigma} \cap G$.

Proof. See [2] IV Th 1.4.7.

Due to T 2.1 it is therefore uninteresting to consider such measures $\Sigma \xrightarrow{F} G$ for which $\Sigma = \hat{\Sigma}_g$ does not hold, i.e. for which $\Sigma \xrightarrow{F} G$ is not a decision observable.

T 2.2 For a decision observable, Σ_g is separable. F is an isomorphism onto the image $F\Sigma \subset G$. Therefore one can identify a decision observable Σ with *a* $\mathscr{U}_g$-closed Boolean sublattice of G. Every lattice-theoretically closed Boolean sublattice of G is also $\mathscr{U}_g$-complete, and conversely; hence we can also identify the decision observables with the (lattice-theoretically) complete Boolean sublattices of G. The topology defined on Σ by $\mathscr{U}_g$ is identical with the $\sigma(\mathscr{B}', \mathscr{B})$-topology on Σ (the uniform structures might differ). From this also follows that Σ, as a subset of G, is $\sigma(\mathscr{B}', \mathscr{B})$-closed in G since Σ_g is complete.

Proof. See [2] IV Th 1.4.8.

From T 1.2 follows that the center Z is a decision observable. From T 1.3 follows that any maximal set of commensurable decision effects is a decision observable.

D 2.2 A decision observable is called *maximal* if, considered as a subset of G, it is a maximal set of commensurable decision effects.

From T 1.3 follows that every set A of commensurable decision effects is a subset of a maximal decision observable. Since the elements of Z are commensurable with all the elements of G, each maximal decision observable contains the center Z as a subset.

§ 3 Structures in That Class of Observables Whose Range also Contains Elements of G

All the considerations in V § 3.2 and § 3.3 remain intact. Because of $G \subset \partial_e L$, from V T 3.3.1 (iii) in particular follows

$$G \cap \overline{\text{co}}^{\sigma}(F\Sigma) = G \cap \partial_e \overline{\text{co}}^{\sigma}(F\Sigma) = G \cap (F\Sigma),$$

i.e. that part of the range of an observable that lies in G is $\sigma(\mathscr{B}', \mathscr{B})$-closed in G. From VI T 4.2 and VI T 4.3 follows that there is a complete Boolean subring Σ_G of Σ such that $F\Sigma_G = G \cap (F\Sigma)$, where F is an isomorphic mapping of Σ_G onto $G \cap (F\Sigma)$. In particular, if $\Sigma \xrightarrow{F} G$ is decision observable, we get

$$G \cap \overline{\text{co}}^{\sigma}(F\Sigma) = G \cap \partial_e \overline{\text{co}}^{\sigma}(F\Sigma) = F\Sigma .$$

By V T 3.3.1 (ii) follows $\partial_e \overline{\text{co}}^{\sigma}(F\Sigma) \subset F\Sigma$ and thus $F\Sigma = \partial_e \overline{\text{co}}^{\sigma}(F\Sigma)$. Hence, the kernel of a decision observable is identical with Σ, so that every decision observable is a kernel observable in the sense of D 3.3.5.

Since $G \subset \partial_e L$ holds, from V D 3.4.3 immediately follows that every decision observable is also irreducible (see V § 3.4).

But it is not generally true that an irreducible kernel observable is a decision observable. In particular, in quantum mechanics one can give examples for other irreducible kernel observables than decision observables ([2] IV § 2.4).

For decision observables the features of measurement scales become especially simple; this follows from

T 3.1 In the case of a decision observable, $\mathscr{B}'(\Sigma) \xrightarrow{S'} \mathscr{B}'$ is a norm isomorphism of $\mathscr{B}'(\Sigma)$ onto the subspace $S'\mathscr{B}'(\Sigma)$ of $\mathscr{B}'$. Here S is the mapping from V § 3.2, and $S'\mathscr{B}'(\Sigma)$ is $\sigma(\mathscr{B}', \mathscr{B})$-closed in $\mathscr{B}'$.

Proof. For $y \in \mathscr{B}'(\Sigma)$, since S' is norm continuous, V (3.5.1) yields

$$S' y = \int_{\alpha_1 - \delta}^{\alpha_2} \lambda \, de(\lambda), \tag{3.1}$$

where $e(\lambda) \in G$; $\delta > 0$ can be chosen arbitrarily. Here we have $\|y\| = \max(|\alpha_1|, |\alpha_2|)$ with $\alpha_1 = \sup\{\lambda \mid \sigma(y \leqq \lambda) = 0\}$ and $\alpha_2 = \inf\{\lambda \mid \sigma(y \leqq \lambda) = \varepsilon\}$. From (3.1) thus follows

$$\sup_{w \in K} \mu(w, S'y) = \alpha_2$$

and

$$\inf_{w \in K} \mu(w, S'y) = \alpha_1,$$

and hence $\|y\| = \|S'y\|$. Therefore S' is a norm isomorphism.

If $S'\mathscr{B}'(\Sigma) \cap \mathscr{B}'_{|1|}$ is $\sigma(\mathscr{B}', \mathscr{B})$-closed, $S'\mathscr{B}'(\Sigma)$ is so. Due to the norm isomorphism we have

$$S'\mathscr{B}'(\Sigma) \cap \mathscr{B}'_{|1|} = S'\mathscr{B}'_{|1|}(\Sigma) .$$

Since S' is σ-continuous, it maps the compact set $\mathscr{B}'_{|1|}$ on a compact set. Hence $S'\mathscr{B}'(\Sigma) \cap \mathscr{B}'_{|1|}$ is $\sigma(\mathscr{B}', \mathscr{B})$-compact and thus $\sigma(\mathscr{B}', \mathscr{B})$-closed. $\square$

Because of this isomorphism S', one can identify not only Σ with a complete Boolean sublattice of G, but all of $\mathscr{B}'(\Sigma)$ with the $\sigma(\mathscr{B}', \mathscr{B})$-closed subspace $\mathscr{B}'_\Sigma$ spanned in $\mathscr{B}'$ by Σ (as a subset of G).

If one identifies $S' \mathscr{B}'(\Sigma)$ with $\mathscr{B}'_\Sigma$ in this way, one can regard S' as the injection of $\mathscr{B}'_\Sigma$ in $\mathscr{B}'$. Since $\mathscr{B}'_\Sigma$ is a $\sigma(\mathscr{B}', \mathscr{B})$-closed subspace, one can interpret S as a surjective mapping of $\mathscr{B}$ onto $\mathscr{B}/\mathscr{B}'^{\perp}_\Sigma$ (see [10] § 22.3), i.e. one can identify $\mathscr{B}(\Sigma)$ with $\mathscr{B}/\mathscr{B}'^{\perp}_\Sigma$. Thus we have proved

T 3.2 For a decision observable, $\mathscr{B} \xrightarrow{S} \mathscr{B}(\Sigma)$ is surjective.

Because $\mathscr{B}'(\Sigma)$ is isomorphic with $\mathscr{B}'_\Sigma \subset \mathscr{B}'$ (where Σ is identified with a Boolean subring of G), a measuring scale y belonging to Σ (see V § 3.5) can be identified with an element $S' y$ from (3.1). Thus if Σ is a complete Boolean subring of G, and $\mathscr{B}'_\Sigma$ the $\sigma(\mathscr{B}', \mathscr{B})$-closed subspace in $\mathscr{B}'$ spanned by Σ, then a scale of the decision observable determined by Σ is uniquely determined by an element y' of B'_Σ, the spectral representation (3.1) with $y' = S' y$ being determined by prescribing Σ (hence $\mathscr{B}'_\Sigma$) and y'.

D 3.1 A decision observable $\Sigma \subset G$ having a scale $y \in \mathscr{B}'_\Sigma$ with $\Sigma(y) = \Sigma$ is often called briefly a "scale observable".

It cannot in generality be shown whether each $y \in \mathscr{B}'$ possesses a unique spectral representation of the form (3.1). In this case, already an element $y \in \mathscr{B}'$ itself would determine the whole scale observable. The representation theorem derived in VIII § 4 follows from the axioms AV 1.1, AV 1.2s, AV 2f, AVid, AV 3, AV 4s. Together with the decomposition from § 5.4 it will imply (by known theorems on operators in Hilbert space) that each $y \in \mathscr{B}'$ possesses a unique spectral representation of the form (3.1).

§ 4 Commensurable Decision Observables

Applying V D 4.1 to two decision observables $\Sigma_i \xrightarrow{F_i} G$ $(i = 1, 2)$, we find that $F_1(\Sigma_1) \cup F_2(\Sigma_2)$ must be a set of co-existent decision effects. By VI T 4.6 this is just the case when $F_1 \Sigma_1$ and $F_2 \Sigma_2$ are subsets of a complete Boolean ring $\Sigma \subset G$. Identifying Σ_1 with $F_1 \Sigma_1$ and Σ_2 with $F_2 \Sigma_2$, we get $\Sigma_1 \subset \Sigma$ and $\Sigma_2 \subset \Sigma$ so that the diagram in V D 4.1 is trivially satisfied with the $h_{1,2}$ as canonical injections of $\Sigma_{1,2}$ in Σ. Therefore, two decision observable are coexistent if and only if all decision effects of one of them are commensurable with those of the other.

D 4.1 Two coexistent decision observables are also called *commensurable.*

From V D 4.3 follows that two decision observables are complementary if and only if $e_i \in F_i \Sigma_i$ with $e_i = e_1, e_2$ commensurable implies either $e_1 \in Z$ or $e_2 \in Z$ (Z the center of G).

§ 5 Decomposition of $\mathscr{B}$ and $\mathscr{B}'$ Relative to the Center Z

If the center Z consists of only the two elements **0** and **1**, then G and also the pair $(\mathscr{B}, \mathscr{B}')$ are called "irreducible".

If the center Z does not only consist of the two elements $\mathbf{0}$ and $\mathbf{1}$, we can try to decompose the structures K, L into simpler components so that for each "part" the center has the trivial form $\{\mathbf{0}, \mathbf{1}\}$, i.e. each part is irreducible.

§ 5.1 Reduction of the Elements of $\mathscr{B}'$ by the Elements of G

We start with a definition suggested by VI T 4.1 (ii).

D 5.1.1 An $e \in \partial_e L$ (specifically $e \in G$) "reduces" a $y \in \mathscr{B}'$ if there exist a decomposition $y = y_1 + y_2$ and a $c > 0$ such that

$$-c\,e \leqq y_1 \leqq c\,e\,, \quad -c(\mathbf{1}-e) \leqq y_2 \leqq c(\mathbf{1}-e)\,.$$

T 5.1.1 If e reduces the element y then the decomposition in D 5.1.1 is unique.

Proof. For $y = y_1 + y_2 = y_1' + y_2'$ and

$$-c'\,e \leqq y_1' \leqq c'\,e, \quad -c'(\mathbf{1}-e) \leqq y_2' \leqq c'(\mathbf{1}-e)$$

we obtain $\mathbf{0} = (y_1 - y_1') + (y_2 - y_2')$, where

$$-(c+c')\,e \leqq y_1 - y_1' \leqq (c+c')\,e\,, \quad -(c+c')(\mathbf{1}-e) \leqq y_2 - y_2' \leqq (c+c')(\mathbf{1}-e)\,.$$

Hence it suffices to show that $\bar{y}_1 + \bar{y}_2 = \mathbf{0}$ and

$$-\bar{c}\,e \leqq \bar{y}_1 \leqq \bar{c}\,e\,, \quad -\bar{c}(\mathbf{1}-e) \leqq \bar{y}_2 \leqq \bar{c}(\mathbf{1}-e)$$

imply $\bar{y}_1 = \bar{y}_2 = \mathbf{0}$. Because $\bar{y}_2 = -\bar{y}_1$ we first get

$$-e \leqq \frac{1}{\bar{c}}\,\bar{y}_1 \leqq e\,, \quad -(\mathbf{1}-e) \leqq \frac{1}{\bar{c}}\,\bar{y}_1 \leqq (\mathbf{1}-e)\,.$$

From this follow

$$\mathbf{0} \leqq e + \frac{1}{\bar{c}}\,\bar{y}_1 \leqq e + (\mathbf{1}-e) = \mathbf{1}$$

and

$$\mathbf{0} \leqq e - \frac{1}{\bar{c}}\,\bar{y}_1 \leqq e - (e - \mathbf{1}) = \mathbf{1}\,,$$

hence $e \pm (1/\bar{c})\,\bar{y}_1 \in L$. Then $e = \frac{1}{2}(e + (1/\bar{c})\,\bar{y}_1) + \frac{1}{2}(e - (1/\bar{c})\,\bar{y}_1)$, which implies $\bar{y}_1 = \mathbf{0}$ since $e \in \partial_e L$. □

By this theorem, condition (ii) in VI T 4.1 also demands that e reduces g. Hence "e and g are coexistent" is equivalent to "e reduces g".

T 5.1.2 If $e \in G$ reduces the element y, the y_1 and y_2 from $y = y_1 + y_2$ obey $\alpha_1\,e \leqq y_1 \leqq \beta_1\,e$, $\alpha_2\,e^{\perp} \leqq y_2 \leqq \beta_2\,e^{\perp}$ with $\alpha_i = \inf\limits_{w \in K} \mu(w, y_i)$, $\beta_i = \sup\limits_{w \subset K} \mu(w, y_i)$.

The latter yield

$$\sup_{w \in K} \mu(w, y) = \max(\beta_1, \beta_2)\,,$$

$$\inf_{w \in K} \mu(w, y) = \min(\alpha_1, \alpha_2)$$

and

$$\| y \| = \max (|\alpha_1|, |\alpha_2|, |\beta_1|, |\beta_2|) .$$

Proof. From

$$-c\,e \leqq y_1 \leqq c\,e\,, \quad -c\,e^{\perp} \leqq y_2 \leqq c\,e^{\perp}$$

follows

$$y + c\,\mathbf{1} = (y_1 + c\,e) + (y_2 + c\,e^{\perp}) \geqq \mathbf{0}\,.$$

With $y' = y + c\,\mathbf{1}$, $y_1' = y_1 + c\,e$ and $y_2' = y_2 + c\,e^{\perp}$, we thus have

$$y' = y_1' + y_2' \geqq \mathbf{0} \quad \text{and} \quad \mathbf{0} \leqq y_1' \leqq 2\,c\,e\,, \quad \mathbf{0} \leqq y_2' \leqq 2\,c\,e^{\perp}\,.$$

From this follows

$$\tfrac{1}{2}\,e + \tfrac{1}{2}\,\frac{y_1'}{\| y_1' \|} \in [\mathbf{0}, \mathbf{1}]$$

and

$$K_0\left(\tfrac{1}{2}\,e + \tfrac{1}{2}\,\frac{y_1'}{\| y_1' \|}\right) \supset K_0(e)\,,$$

hence

$$\tfrac{1}{2}\,e + \tfrac{1}{2}\,\frac{y_1'}{\| y_1' \|} \leqq e$$

and thus

$$y_1' \leqq \| y_1' \|\, e\,.$$

Analogously follows $y_2' \leqq \| y_2' \|\, e^{\perp}$, hence $y' = y_1' + y_2' \leqq \| y_1' \|\, e + \| y_2' \|\, e^{\perp}$. Because $\mu(w, e) + \mu(w, e^{\perp}) = 1$ for all w, we have

$$\| y' \| = \sup_{w \in K} \mu(w, y') \leqq \max\{\| y_1' \|, \| y_2' \|\}\,.$$

In particular choosing $w \in K_1(e)$ and $w \in K_1(e^{\perp})$, we finally get

$$\| y' \| = \max\{\| y_1' \|, \| y_2' \|\}\,.$$

With $y' = y + c\,\mathbf{1}$, we have

$$\| y' \| = \sup_{w \in K} \mu(w, y') = \gamma + c\,,$$

where

$$\gamma = \sup_{w \in K} \mu(w, y)\,.$$

With $\| y_i' \| = \sup_{w \in K} \mu(w, y_i') = \beta_2 + c \ (i = 1, 2)$, from $\| y' \| = \max\{\| y_1' \|, \| y_2' \|\}$ follows $\gamma = \max\{\beta_1, \beta_2\}$. From $y_1' \leqq \| y_1' \|\, e$, $y_2' \leqq \| y_2' \|\, e^{\perp}$ follows $y_1 \leqq \beta_1\, e$, $y_2 \leqq b_2\, e^{\perp}$.

Repeating these considerations for $-y = (-y_1) + (-y_2)$, we obtain $y_1 \geqq \alpha_1\, e$, $y_2 \geqq \alpha_2\, e^{\perp}$. □

D 5.1.2 An $e \in G$ "decomposes" a y into a positive part y_+ and a negative part y_- if it reduces y to $y = y_+ + y_-$ with $y_+ \geqq 0$, $y_- \leqq 0$.

The preceding theorems show that y_+ and y_- are uniquely determined by e.

§ 5.2 Reduction by Center Elements

Due to T 1.2, the center Z is a complete Boolean sublattice of G, by § 2 representing a decision observable. By § 4 this Z is commensurable with all decision observables. Let us show that Z coexists with all observables. To this end, we first prove

T 5.2.1 An $r \in Z$ reduces each $y \in \mathscr{B}'$.

Proof. Since for each y there is a $c \geqq 0$ with $y + c\,\mathbf{1} \geqq 0$ and hence

$$\| y + c\,\mathbf{1} \|^{-1} (y + c\,\mathbf{1}) \in [\mathbf{0}, \mathbf{1}],$$

it suffices to prove the theorem for $g \in [\mathbf{0}, \mathbf{1}]$.

Since r is commensurable with each $e \in G$, due to T 1.4 (iv) we have

$$e = (e \wedge r) + (e \wedge r^{\perp}), \tag{5.2.1}$$

i.e. r reduces each $e \in G$.

Let l_r be the set of all $g \in L$ reduced by r, such that $G \subset l_r$. We easily see that l_r is convex since $K_0(r)$ and $K_0(\mathbf{1} - r)$ are convex. Let us show that l_r is $\sigma(\mathscr{B}', \mathscr{B})$-closed.

It suffices to consider sequences since the $\sigma(\mathscr{B}', \mathscr{B})$-topology is metrizable on L. Let $g_\nu \in l_r$ and $g_\nu \to g$, so that $g \in L$. From $g_\nu \in l_r$ follows $g_\nu = g_\nu^1 + g_\nu^2$ where $g_\nu^1 \leqq r$ and $g_\nu^2 \leqq r^{\perp} = \mathbf{1} - r$. Since L is compact, there is a subsequence (indexed again by ν) so that $g_\nu^i \to g^i \in L$ $(i = 1, 2)$ and thus $g = g^1 + g^2$ holds. From $g_\nu^1 \leqq r$ also follows $g^1 \leqq r$ and analogously we get $g^2 \leqq \mathbf{1} - r$, so that $g \in l_r$.

Since l_r is convex and $\sigma(\mathscr{B}', \mathscr{B})$-closed, from $G \subset l_r \subset L$ and VI T 3.1 follows $l_r = L$. □

In the decomposition $y = y_1 + y_2$ of a $y \in \mathscr{B}'$ according to the r that reduces y, due to T 5.1.1 the components y_1, y_2 are uniquely determined. To each $r \in Z$ thus corresponds a mapping T_r of $\mathscr{B}'$ into itself, with $T_r y = y_1$. This implies $T_r^2 = T_r$, i.e. T_r is a projector. For $e \in G$, from (5.2.1) follows

$$T_r e = e \wedge r. \tag{5.2.2}$$

From T 5.1.2 follows easily that T_r is norm continuous. But T_r is even $\sigma(\mathscr{B}', \mathscr{B})$-continuous. First we prove

T 5.2.2 The subspace $\mathscr{B}_r' = T_r \mathscr{B}'$ is $\sigma(\mathscr{B}', \mathscr{B})$-closed.

Proof. That $T_r \mathscr{B}'$ is a subspace follows from the linearity of T_r. Here, $\mathscr{B}_r'$ is $\sigma(\mathscr{B}' \mathscr{B})$-closed if $\mathscr{B}_r' \cap \mathscr{B}_{|1|}'$ is so. By T 5.1.2 we have

$$\mathscr{B}_r' \cap \mathscr{B}_{|1|}' = [-r, r],$$

which hence is $\sigma(\mathscr{B}', \mathscr{B})$-closed.

T 5.2.3 T_r is $\sigma(\mathscr{B}', \mathscr{B})$-continuous.

Proof. T_r is $\sigma(\mathscr{B}', \mathscr{B})$-continuous if each $x \in \mathscr{B}$ makes $l(y) = \mu(x, T_r y)$ a $\sigma(\mathscr{B}', \mathscr{B})$-continuous linear form on $\mathscr{B}'$. This happens when the space $\{y \mid l(y) = 0\}$ is $\sigma(\mathscr{B}', \mathscr{B})$-closed. In turn, this occurs if $\{y \mid l(y) = 0\} \cap \mathscr{B}'_{|1|} = \{y \mid l(y) = 0$ and $\|y\| \leqq 1\}$ is $\sigma(\mathscr{B}', \mathscr{B})$-closed. Due to T 5.1.2, with $X = \{y \mid \mu(x, y) = 0\}$ we have

$$\begin{aligned}\{y \mid l(y) = 0 \text{ and } \|y\| \leqq 1\} \\ = \{y \mid \|T_r y\| \leqq 1, \|T_{\mathbf{1}-r} y\| \leqq 1, \mu(x, T_r y) = 0\} \\ = \mathscr{B}'_r \cap \mathscr{B}'_{|1|} \cap X + \mathscr{B}'_{\mathbf{1}-r} \cap \mathscr{B}'_{|1|}.\end{aligned}$$

Since X is $\sigma(\mathscr{B}', \mathscr{B})$-closed while $\mathscr{B}'_r$ and $\mathscr{B}'_{\mathbf{1}-r}$ are so (by T 5.2.2) we find that $\mathscr{B}'_r \cap X \cap \mathscr{B}'_{|1|}$ and $\mathscr{B}'_{\mathbf{1}-r} \cap \mathscr{B}'_{|1|}$ are $\sigma(\mathscr{B}', \mathscr{B})$-compact. Hence $\{y \mid l(y) = 0, \|y\| \leqq 1\}$ is also $\sigma(\mathscr{B}', \mathscr{B})$-compact. □

T 5.2.4 There is a positive mapping $\mathscr{B} \xrightarrow{S_r} \mathscr{B}$ with $S_r K \subset \check{K}$, such that $T_r = S'_r$.

Proof. From T 5.2.3 follows that the T_r' adjoint to T_r exists as a norm-continuous mapping of $\mathscr{B}$ into itself. Therefore, with $S_r = S_r'$ we have $T_r = S'_r$.

For $g \in L = [\mathbf{0}, \mathbf{1}]$, from $w \in K$ follows

$$\mu(S_r w, g) = \mu(w, S'_r g) = \mu(w, T_r g) \geqq 0,$$

i.e. $S_r w \in \mathscr{B}_+$. For $g = \mathbf{1}$ follows

$$\mu(S_r w, \mathbf{1}) = \mu(w, T_r \mathbf{1}) = \mu(w, r) \leqq 1,$$

and hence $S_r w \in \check{K}$. □

A mapping S of $\mathscr{B}$ into itself with $SK \subset \check{K}$ (equivalent to $SK \subset \check{K}$) is often called an operation ([2] V § 4.1).

From $T_r^2 = T_r$ follows $S_r^2 = S_r$; hence S_r is also a projector.

T 5.2.5 For $r, r_1, r_2 \in Z$, we have $T_{r_1 \wedge r_2} = T_{r_1} T_{r_2} = T_{r_2} T_{r_1}$ and $T_{r^\perp} = T_{\mathbf{1}-r} = \mathbf{1} - T_r$. If $\{r_\nu\} \subset Z$ and $r_\nu \wedge r_\mu = 0$ for $\nu \neq \mu$, we get $\bigvee_\nu r_\nu = \sum_\nu r_\nu$ and $T_{\bigvee_\nu r_\nu} = \sum_\nu T_{r_\nu}$, where the convergence of the sum $\sum_\nu T_{r_\nu}$ is pointwise convergence in $\mathscr{B}'$ with the $\sigma(\mathscr{B}', \mathscr{B})$-topology.

Proof. From the reduction $y = y_1 + y_2$, according to r follows $T_r y = y_1$ and $T_{\mathbf{1}-r} y = y_2$, hence $T_r + T_{\mathbf{1}-r} = \mathbf{1}$.

From $T_r \mathbf{1} = r$ follows $T_{r_1} T_{r_2} \mathbf{1} = T_{r_1} r_2 = r_1 \wedge r_2 = T_{r_1 \wedge r_2} \mathbf{1}$. Thus it suffices to verify $T_{r_1} T_{r_2} g = T_{r_1 \wedge r_2} g$ for $g \in [\mathbf{0}, \mathbf{1}]$. First we obtain $T_{r_1} T_{r_2} g \leqq r_1$ and $T_{r_1} T_{r_2} g \leqq T_{r_2} g \leqq r_2$, which by VI T 1.2.8 gives $T_{r_1} T_{r_2} g \leqq r_1 \wedge r_2$. From

$$g = T_{r_1} T_{r_2} g + (\mathbf{1} - T_{r_1}) T_{r_2} g + T_{r_1} (\mathbf{1} - T_{r_2}) g + (\mathbf{1} - T_{r_1})(\mathbf{1} - T_{r_2}) g$$

then follows

$$(\mathbf{1} - T_{r_1}) T_{r_2} g \leqq r_1^\perp \wedge r_2, \quad T_{r_1} (\mathbf{1} - T_{r_2}) g \leqq r_1 \wedge r_2^\perp, \quad (\mathbf{1} - T_{r_1})(\mathbf{1} - T_{r_2}) g \leqq r_1^\perp \wedge r_2^\perp$$

and hence

$$(\mathbf{1} - T_{r_1} T_{r_2}) g \leqq r_1^\perp \wedge r_2 + r_1 \wedge r_2^\perp + r_1^\perp \wedge r_2^\perp = (r_1 \wedge r_2)^\perp,$$

such that $T_{r_1} T_{r_2} g = T_{r_1 \wedge r_2} g$.

From $r_1, r_2 \in Z$ and $r_1 \wedge r_2 = 0$ follows $r_1 \perp r_2$ because Z is a Boolean ring. By VI T 3.9 this permits at most a countable number of r_ν with $r_\nu \wedge r_\mu = 0$ for $\nu \neq \mu$. Since the r_ν are then pairwise orthogonal, from VI T 3.9 follows

$$\bigvee_\nu r_\nu = \sum_\nu r_\nu .$$

For $r_1 \perp r_2$ and $r = r_1 + r_2$ we have

$$T_r = T_r T_{r_1} + T_r (\mathbf{1} - T_{r_1}) = T_{r \wedge r_1} + T_{r \wedge r_1^\perp} = T_{r_1} + T_{r_2} .$$

By induction for $r = \sum_{\nu=1}^{N} r_\nu$ this yields

$$T_r = \sum_{\nu=1}^{N} T_{r_\nu} .$$

With $r_N = \sum_{\nu=1}^{\infty} r_\nu - \sum_{\nu=1}^{N} r_\nu$ and $r = \sum_{\nu=1}^{\infty} r_\nu$ we have

$$T_r = \sum_{\nu=1}^{N} T_{r_\nu} + T_{r_N} .$$

Thus we need only show that $T_{r_N} g \to \mathbf{0}$ with $g \in [\mathbf{0}, \mathbf{1}]$ holds in the $\sigma(\mathscr{B}', \mathscr{B})$-topology. Due to $\mathbf{0} \leqq T_{r_N} g \leqq r_N$ it suffices to show $r_N \to \mathbf{0}$; but just this is meant by the convergence of $\sum_{\nu=1}^{\infty} r_\nu$. □

From T 5.2.5 follows that the mapping $r \to T_r$ is a σ-additive measure on the Boolean ring Z. If $\Sigma \xrightarrow{F} L$ is any observable, we can consider the free Boolean algebra (Σ, Z) generated by Σ and Z, whose elements are finite sums $\sum_\nu \dot{+}\, \eta_\nu$ with $\eta_\nu \cdot \eta_\mu \, (= \eta_\nu \wedge \eta_\mu) = 0$ for $\nu \neq \mu$ (hence $\sum_\nu \dot{+}\, \eta_\nu = \bigvee_\nu \eta_\nu$) and $\eta_\nu = \sigma_\nu \cdot r_\nu$ with $\sigma_\nu \in \Sigma$ and $r_\nu \in Z$. An additive measure on $(\Sigma, Z) \xrightarrow{\tilde{F}} L$ is defined by $\sigma \cdot r \xrightarrow{\tilde{F}} T_r F(\sigma)$. For two homomorphisms $\Sigma \xrightarrow{h_1} (\Sigma, Z)$ and $Z \xrightarrow{h_2} (\Sigma, Z)$, defined by $h_1(\sigma) = (\sigma, \mathbf{1})$ and $h_2(r) = (\varepsilon, r)$, the diagram in V D 4.1 is then satisfied with $\tilde{F}$ instead of F, this F instead of F_1, and with F_2 as the identity mapping of Z into itself. Therefore we have shown that each observable coexists with the decision observable defined by Z.

Theorems corresponding to T 5.2.3, T 5.2.5 hold for $S_r = T_r'$.

T 5.2.6 $\mathscr{B}_r = S_r \mathscr{B}$ is a norm-closed subspace of $\mathscr{B}$; thus $\mathscr{B}_r$ is a Banach space. Its dual can be identified with $\mathscr{B}_r' = S_r' \mathscr{B}'$.

Proof. $\mathscr{B}_r$ consists of all the $x \in \mathscr{B}$ with $S_r x = x$. Since S_r is norm-continuous, from $x_\nu \to x$ and $S_r x_\nu = x_\nu$ follows $S_r x = x$, hence $x \in \mathscr{B}_r$.

For $x \in \mathscr{B}_r$ and $y \in S_r' \mathscr{B}$, obviously $\mu(x, y)$ is a norm-continuous linear form on $\mathscr{B}_r$. From $y_1, y_2 \in S_r' \mathscr{B}'$ and $\mu(x, y_1) = \mu(x, y_2)$ for all $x \in \mathscr{B}_r$, for $x = S_r \tilde{x}$ and $\tilde{x}$ arbitrary in $\mathscr{B}$ follows $0 = \mu(S_r \tilde{x}, y_1 - y_2) = \mu(\tilde{x}, S_r' y_1 - S_r' y_2) = \mu(\tilde{x}, y_1 - y_2)$ and hence $y_1 = y_2$. If $l(x)$ is a norm-continuous linear form on $\mathscr{B}_r$, then $\tilde{l}(\tilde{x}) = l(S_r \tilde{x})$ defines a norm-continuous linear form on $\mathscr{B}$ because $|\tilde{l}(\tilde{x})| = |l(S_r \tilde{x})|$

$\leqq C \| S_r \tilde{x} \| \leqq C \| \tilde{x} \|$. Hence there is a $\tilde{y}$ with $\tilde{l}(\tilde{x}) = \mu(\tilde{x}, \tilde{y})$. Because $S_r^2 = S_r$, we have $\tilde{l}(S_r \tilde{x}) = \tilde{l}(\tilde{x})$, i.e. $\mu(\tilde{x}, \tilde{y}) = \mu(S_r \tilde{x}, y) = \mu(\tilde{x}, S_r' \tilde{y})$. This gives $\tilde{y} = S_r' \tilde{y}$, hence $\tilde{y} \in \mathscr{B}_r'$. □

T 5.2.7 Corresponding to T 5.2.5 we have $S_{r_1 \wedge r_2} = S_{r_1} S_{r_2} = S_{r_2} S_{r_1}$ and $S_{r^\perp} = \mathbf{1} - S_r$. Furthermore, $S_{\bigvee_\nu r_\nu} = \sum_\nu S_{r_\nu}$ converges pointwise in $\mathscr{B}$ relative to the norm topology.

Proof. This follows from T 5.2.5 up to the asserted convergence of $\sum_\nu S_{r_\nu}$. With r_N from the proof of T 5.2.5 and with $w \in K$ we get

$$\| S_{r_N} w \| = \mu(S_{r_N} w, \mathbf{1}) = \mu(w, r_N) \to 0\,,$$

which also yields $\| S_{r_N} x \| \to 0$ for $x = \alpha w_1 - \beta w_2$. □

For a sequence $r_\nu \in Z$ with $\sum_\nu r_\nu = \mathbf{1}$, T 5.2.5 and T 5.2.7 in particular give $\sum_\nu T_{r_\nu} = \mathbf{1}$ and $\sum_\nu S_{r_\nu} = \mathbf{1}$.

For $y \in \mathscr{B}'$ and $y_\nu = T_{r_\nu} y$ together with $x \in \mathscr{B}$ and $x_\nu = S_{r_\nu} x$, we then have

$$x = \sum_\nu x_\nu\,, \qquad y = \sum_\nu y_\nu \tag{5.2.3}$$

and $\mu(x, y) = \sum_{\nu, \mu} \mu(x_\nu, y_\mu)$. From $x_\nu = S_{r_\nu} x$, $y_\mu = T_{r_\mu} y$ and $T_{r_\nu} T_{r_\mu} = T_{r_\nu \wedge r_\mu} = 0$ for $\nu \neq \mu$ (since $r_\nu \wedge r_\mu = 0$ for $\nu \neq \mu$!), we thus obtain

$$\mu(x, y) = \sum_\nu \mu(x_\nu, y_\nu)\,. \tag{5.2.4}$$

In particular, for $x = w \in K$ and $y = g \in L$ we have $w_\nu = S_{r_\nu} w \in \check{K}$ and $g_\nu = T_{r_\nu} g \in L$. For $e \in G$, from (5.2.2) in particular follows

$$e = \sum_\nu e_\nu\,, \quad \text{where} \quad e_\nu = T_{r_\nu} e = e \wedge r_\nu\,. \tag{5.2.5}$$

Most important about the dual pair $\mathscr{B}_r$, $\mathscr{B}_r'$ is that it has the same structure properties as $\mathscr{B}$, $\mathscr{B}'$; we shall show this in the next theorems.

T 5.2.8 $\mathscr{B}_r$ is a base normed space with the basis $K_r = \{w \mid w \in K \text{ and } S_r w = w\} = K_1(r)$. The dual $\mathscr{B}_r'$ is an order unit space with r as the order unit. We have $\mathscr{B}_{r+} = \mathscr{B}_r \cap \mathscr{B}_+$, $\mathscr{B}_{r+}' = \mathscr{B}_r' \cap \mathscr{B}_+'$ and $L_r = [0, r] = \mathscr{B}_r' \cap [\mathbf{0}, \mathbf{1}] = \mathscr{B}_r' \cap L = L_0 K_0(r)$.

Proof. The set $\mathscr{B}_r \cap \mathscr{B}_+$ is just the set of all $x \in \mathscr{B}_r$ with $x \geqq 0$, i.e. the positive cone $\mathscr{B}_{r+}$ in $\mathscr{B}_r$ generated by $\mathscr{B}_+$. This implies that $\{w \mid w \in K \text{ and } S_r w = w\}$ is just a basis of the cone $\mathscr{B}_{r+}$. From $w \in K$ and $S_r w = w$ follows $\mu(w, r) = \mu(w, S_r' \mathbf{1}) = \mu(w, \mathbf{1}) = 1$, i.e. $w \in K_1(r)$. From $w \in K_1(r)$ follows $\mu(S_r w, \mathbf{1}) = 1$, i.e. $S_r w \in K$. From $w = S_r w + S_{r^\perp} w$ then follows $\mu(S_{r^\perp} w, \mathbf{1}) = 0$, so that $S_{r^\perp} w \geqq \mathbf{0}$ gives $S_{r^\perp} w = \mathbf{0}$, hence $w = S_r w$. We have thus proved $\{w \mid w \in K \text{ and } S_r w = w\} = K_1(r)$.

In order to recognize $\mathscr{B}_r$ as base normed with the basis K_r, we need only show that each $x \in \mathscr{B}_r$ can be written $x = \alpha w_1 - \beta w_2$, where $w_1, w_2 \in K_r$ and $\| x \| \geqq \alpha + \beta - \varepsilon$ (with $\varepsilon > 0$ arbitrarily small). From $x \in \mathscr{B}$ follows $x = \alpha' w_1' - \beta' w_2'$ with $w_1', w_2' \in K$ and $\| x \| \geqq \alpha' + \beta' - \varepsilon$. Because $S_r x = x$ we get $x = \alpha' S_r w_1' - \beta' S_r w_2'$.

Because $S_r K \subset \check{K}$, we have $S_r w_1' = \lambda_1 w_1$, $S_r w_2' = \lambda_2 w_2$, where $w_1, w_2 \in K_r$ and $\lambda_1 \leqq 1$, $\lambda_2 \leqq 1$, which for $\alpha \leqq \alpha', \beta \leqq \beta'$ implies $x = \alpha w_1 - \beta w_2$ and hence $\| x \| \geqq \alpha + \beta - \varepsilon$.

Therefore, $\overline{\text{co}}\,(K_r \cup (-K_r))$ is the unit ball of $\mathscr{B}_r$, i.e. $\overline{\text{co}}\,(K_r \cup (-K_r)) = \mathscr{B}_r \cap \overline{\text{co}}\,(K \cup (-K))$. Since $\mathscr{B}_r$ is base-normed, $\mathscr{B}_r'$ is an order unit space. From $K_r = K_1(r)$ follows that r is the order unit in $\mathscr{B}_r'$. But one also finds

$$\mathscr{B}_r' \cap [-\mathbf{1}, \mathbf{1}] = [-r, r],$$

for we have $-\mathbf{1} \leqq y \leqq \mathbf{1}$ and $S_r' y = y$ (since S_r' preserves the ordering!) so that $S_r' \mathbf{1} = r$ implies $-r \leqq y \leqq r$.

From $\mu(w, y) \geqq 0$ for all $w \in K_r$ and a $y \in \mathscr{B}_r'$, with $\tilde{w} \in K$ and $S_r \tilde{w} \in \check{K}_r$ follows $\mu(\tilde{w}, y) \geqq 0$ for all $\tilde{w} \in K$ and hence $y \in \mathscr{B}_+'$. The relation $\mu(w, y) \geqq 0$ for all $w \in K_r$ follows for $y \in \mathscr{B}_r' \cap \mathscr{B}_+'$. Thus $\mathscr{B}_r' \cap \mathscr{B}_+'$ is the positive cone in $\mathscr{B}_r'$ dual to $\mathscr{B}_{r+}$.

At once we get $L_r \subset \mathscr{B}_r'$ with $L_r = [\mathbf{0}, r]$. From $y \in \mathscr{B}_r'$ and $\mathbf{0} \leqq y \leqq \mathbf{1}$ follows $\mathbf{0} \leqq y \leqq r$ and hence $L_r = \mathscr{B}_r' \cap [\mathbf{0}, \mathbf{1}] = \mathscr{B}_r' \cap L$. From VI T3.1 then follows $[\mathbf{0}, r] = L_0 K_0(r)$. □

From T5.2.8 follows that the norm of an element in $\mathscr{B}_r$ is the same relative to the basis K of $\mathscr{B}$ as to the basis K_r of $\mathscr{B}_r$. Correspondingly, the norm of an element of $\mathscr{B}_r'$ is the same relative to $[-r, r]$ as the unit ball of $\mathscr{B}_r$, as it is relative to $[-\mathbf{1}, \mathbf{1}]$ as the unit ball of $\mathscr{B}'$.

Hence the decomposition (5.2.3) with (5.2.4) yields

$$\begin{aligned} \| x \| &= \sup_{y \in [-\mathbf{1}, \mathbf{1}]} | \mu(x, y) | = \sup_{y_\nu \in [-r_\nu, r_\nu]} \Big| \sum_\nu \mu(x_\nu, y_\nu) \Big| \\ &= \sum_\nu \sup_{y_\nu \in [-r_\nu, r_\nu]} | \mu(x_\nu, y_\nu) | = \sum_\nu \| x_\nu \| \end{aligned} \tag{5.2.6}$$

and

$$\| y \| = \sup_{w \in K} | \mu(w, y) | = \sup \left\{ \sum_\nu \mu(w_\nu, y_\nu) \mid w_\nu \in \check{K}_\nu, \sum_\nu \mu(w_\nu, \mathbf{1}) = 1 \right\}.$$

For $\lambda_\nu = \mu(w_\nu, \mathbf{1}) \geqq 0$, we have $\sum_\nu \lambda_\nu = 1$ and

$$\| y \| = \sup \left\{ \sum_\nu \lambda_\nu \mu(\tilde{w}_\nu, y_\nu) \mid \tilde{w}_\nu \in K_r, \lambda_\nu \geqq 0, \sum_\nu \lambda_\nu = 1 \right\},$$

whence we obtain

$$\| y \| = \sup_\nu \| y_\nu \|. \tag{5.2.7}$$

The T5.2.8 also asserts that AV1.2s with $L_r = \mathscr{B}_r' \cap L$ is satisfied for $\mathscr{B}_r, \mathscr{B}_r'$. But AV1.1 and AVid also hold for $\mathscr{B}_r, \mathscr{B}_r'$, as proved in the theorems to follow.

T5.2.9 The sets $L_{r0}(k_r) = \{ g \mid g \in L_r$ and $\mu(w, g) = 0$ for all $w \in k_r \subset K_r \}$ are upward directed with a largest element $e_r L_{r0}(k_r) \in G \cap L_r$. Moreover, $e_r L_{r0}(k_r) = r \wedge e L_0(k_r) = T_r e L_0(k_r)$.

Proof. We have $r \wedge e L_0(k_r) \in G \cap L_r$ and $r \wedge e L_0(k_r) \in L_{r0}(k_r)$. For each $g \in L_{r0}(k_r)$, we get $g \in L_0(k_r)$ and $g \in L_r$, i.e. $g \leqq e L_0(k_r)$ and $g \leqq r$, so that $g \leqq r \wedge e L_0(k_r)$. From (5.2.2) follows $r \wedge e L_0(k_r) = T_r e L_0(k_r)$. □

This T 5.2.9 shows that AV 1.1 is satisfied. We denote the set of all $e_r\, L_r(k_r)$ by G_r. For this set holds

T 5.2.10 $G_r = G \cap L_r = \{e \mid e \in G,\, e \leqq r\} = \{e \mid e = T_r \tilde{e},\, \tilde{e} \in G\}$.

Proof. Obvious is $G \cap L_r = \{e \mid e \in G,\, e \leqq r\}$ since $L_r = [\mathbf{0}, r]$. For each subset k_r of K_r, we get $r \wedge e\, L_0(k_r) \leqq r$. Conversely, $e_1 \in G$ and $e_1 \leqq r$ yield $K_0(e_1) \cap K_r = S_r K_0(e_1)$, because $w' \in K_0(e_1)$ gives $0 = \mu(w', e_1) = \mu(w', S'_r e_1) = \mu(S_r w', e_1)$. For $w \in K_0(e_1) \cap K_r$ we have $0 = \mu(w, e_1)$ and $w \in S_r w$. From $\mu(w, g) = 0$ for all $w \in S_r K_0(e_1)$ follows $\mu(S_r w, g) = \mu(w, S'_r g) = 0$, i.e. $S'_r g \in L_0 K_0(e_1)$. Thus we get $T_r g \leqq e_1$, i.e. $e\, L_0(S_r K_0(e_1)) = \{g \mid T_r g \leqq e_1\}$. From $g = T_r g + T_{\mathbf{1}-r} g$ and $T_r g \leqq e_1$ follows $e\, L_0(S_r K_0)\, e_1)) = e_1 + (\mathbf{1} - r)$, hence $e_1 = r \wedge e\, L_0(S_r K_0(e_1))$. □

For $K_{r1}(e_1) = \{w \mid w \in K_r$ and $\mu(w, e_1) = 1\}$ with an $e_1 \leqq r$, we obtain $K_{r1}(e_1) = K_1(e_1)$ because $\mu(w, e_1) = 1$, $T_r e_1 = e_1$, and $0 \leqq \mu(w, e_1 + (\mathbf{1} - r)) = \mu(w, e_1) + \mu(w, \mathbf{1} - r) \leqq 1$ yield $\mu(w, \mathbf{1} - r) = 0$, i.e. $w \in K_r$. We have thus proved

T 5.2.11 For $e \in G_r$ we have $K_{r1}(e) = K_1(e)$; hence AVid holds in $\mathscr{B}_r, \mathscr{B}'_r$. □

$K_{r1}(e) = K_1(e)$ means that the isomorphic mapping K_{r1} of G_r onto W_{L_r} in $(\mathscr{B}_r, \mathscr{B}'_r)$ is only the restriction of the isomorphic mapping $G \xrightarrow{K_1} \mathscr{W}_L$ to G_r.

With Δ as in VI (6.1.11), for $e_1 \leqq r$, $e_2 \leqq r$ we get

$$\Delta(e_1, e_2) = \max\left\{\inf_{w \in K_{r1}(e_2)} \mu(w, \mathbf{1} - e_1),\ \inf_{w \in K_{r1}(e_1)} \mu(w, \mathbf{1} - e_2)\right\}.$$

Because $\mu(w, \mathbf{1} - r) = 0$ for $w \in K_r$, we finally have

$$\Delta(e_1, e_2) = \max\left\{\inf_{w \in K_{r1}(e_2)} \mu(w, r - e_1),\ \inf_{w \in K_{r1}(e_1)} \mu(w, r - e_2)\right\}.$$

Since K_{r_1} and K_1 are the same isomorphic mappings of G_r onto $\mathscr{W}_{L_r}$, we obtain

T 5.2.12 If AV 3 holds for $\mathscr{B}, \mathscr{B}'$, it also holds for $\mathscr{B}_r, \mathscr{B}'_r$.

Since $K_{r_1}(e) = K_1(e)$ for $e \leqq r$, we conclude

T 5.2.13 If AV 4 resp. AV 4s holds for $\mathscr{B}, \mathscr{B}'$, it also holds for $\mathscr{B}_r, \mathscr{B}'_r$.

If F is a closed face of $K_r = K_1(r)$, it is also a closed face of K. If this face F due to AV 2s or AV 2f is exposed, $F = K_1(e)$ holds with $e \in G$. Because $F \subset K_1(r)$, we have $e \leqq r$ and hence also $F = K_{r_1}(e)$, i.e. F is also an exposed face of K_r in $\mathscr{B}_r$. This also proves

T 5.2.14 If AV 2s resp. AF 2f holds for $\mathscr{B}, \mathscr{B}'$, it also holds for $\mathscr{B}_r, \mathscr{B}'_r$.

T 5.2.15 The center Z_r of G_r equals $G_r \cap Z = \{e \mid e \in Z$ and $e \leqq r\}$.

Proof. Z_r is the set of those $\tilde{r} \in G$ with $\tilde{r} \leqq r$ which are commensurable with all $e \in G$ for which $e \leqq r$. Having r commensurable with $e \in G_r$ in G_r, however, is equivalent to $e = e_1 + e_2$ with $e_1 \leqq \tilde{r}$, $e_2 \leqq r - \tilde{r}$.

From this follows $e_1 \leqq \tilde{r}$, $e_2 \leqq \mathbf{1} - \tilde{r}$; thus $\tilde{r}$ is commensurable with e in G. Conversely, if $\tilde{r} \leqq r$ and $\tilde{r} \in Z$, then $\tilde{r}$ is commensurable with all $e \in G_r$ in G; hence $e = e_1' + e_2'$ holds with $e_1' \leqq \tilde{r}$ and $e_2' \leqq \mathbf{1} - \tilde{r}$. From $\tilde{r} \leqq r$ we conclude $e_2' \leqq r - \tilde{r} + (\mathbf{1} - r)$. Because of $e_1' \leqq \tilde{r} \leqq r$, we have $T_r e_1' = e_1'$. From $T_r e = e$ then follows $T_r e_2' = e_2'$ and hence $T_r e_2' \leqq T_r(r - \tilde{r}) + T_r(\mathbf{1} - r) = r - \tilde{r}$. □

Thus we can conclude the structure analysis of $(\mathscr{B}_r, \mathscr{B}_r')$, emphasizing that we only needed to presume AV 1.1, AV 1.2s, AVid.

§ 5.3 Classical Systems

To characterize the "classical" systems as effect carriers, in VI § 7.1 we introduced the relation AV kl.

Due to D 1.2, this is equivalent to $Z = G$ and the additional requirement that each face of K be infinite-dimensional. But first let us in generality investigate the condition $Z = G$, of course without assuming AV 4. Inasmuch as $Z = G$, Axiom AV 3 is due to VI § 6.2 without meaning, since it follows as a theorem from AV 1.1, AV 1.2s, AVid and $Z = G$ (see just before VI T 6.2.3).

Due to T 1.2, we find $Z = G$ equivalent to the lattice G being a Boolean ring. With $\Sigma = G$, we can by V § 3.1 construct the Banach spaces $\mathscr{B}(G)$ and $\mathscr{B}'(G)$. Then we find the important theorem that for $Z = G$ the Banach spaces $\mathscr{B}$ and $\mathscr{B}'$ can be identified with $\mathscr{B}(G)$ resp. $\mathscr{B}'(G)$.

T 5.3.1 Let $Z = G$. There is an isomorphic mapping $\mathscr{B} \xrightarrow{S} \mathscr{B}(G)$ which maps K onto $K(G)$. Then S' maps $\mathscr{B}'(G)$ isomorphically onto $\mathscr{B}'$.

Proof. The identity mapping $\Sigma = G \xrightarrow{F} G$ is a decision observable. By T 3.1, for the mapping $\mathscr{B} \xrightarrow{S} \mathscr{B}(\Sigma) = \mathscr{B}(G)$ of V § 3.2 (which corresponds to an observable) we find that $\mathscr{B}'(\Sigma) = \mathscr{B}'(G) \xrightarrow{S'} \mathscr{B}'$ is a norm isomorphism of $\mathscr{B}'(G)$ onto the $\sigma(\mathscr{B}', \mathscr{B})$-closed subspace $S'\mathscr{B}'(G)$.

Here the restriction of S' to G is identical with the mapping F of the observables, i.e. to the identity mapping $G \to G$. Since VI T 3.1 imples $[\mathbf{0}, \mathbf{1}] = L = \overline{\mathrm{co}}^{\sigma} G$ in $\mathscr{B}'$, we have $S'\mathscr{B}'(G) = \mathscr{B}'$. Together with V T 2.3 this proves the assertion. □

We must still show in generality that $\mathscr{B}(\Sigma)$, $K(\Sigma)$, $\mathscr{B}'(\Sigma)$, $L(\Sigma)$ really satisfy AV 1.1, AV 1.2s, AVid and $G = \Sigma$. We have satisfied AV 1.2s by fixing $L(\Sigma)$ as the order interval $[\mathbf{0}, \mathbf{1}]$ from $\mathscr{B}'(\Sigma)$.

By VI T 1.3.1, relation AV 1.1 is equivalent to the fact that each set $L_0(k)$ with $k \subset K$ has a largest element. Since we assume Σ separable, it suffices to replace k by elements $m \in K(\Sigma)$ and to show that each $L_0(m)$ has a largest element.

Following [2] IV D 2.1.2, let σ_s be the support of m. Applying the spectral theorem V (3.5.1) to an element of $L(\Sigma)$, we find that $L_0(m)$ is the set of all $g \in L(\Sigma)$ with $\sigma(g \leqq 0) > \sigma_s$. Because $g \leqq \mathbf{1}$, the spectral theorem then yields $g \leqq \sigma_s^*$. On the other hand we have $\sigma_s^* \in L_0(m)$; thus AV 1.1 is proven.

In case $K = K(\Sigma)$, each closed face F of K is exposed. Since Σ, and hence $K(\Sigma)$, is separable, there is an $m \in K$ with $F = C(m)$. With σ_s as the support of m, this makes $F \subset K_0(\sigma_s^*)$. It remains to show $F \supset K_0(\sigma_s^*)$. With $m_1 \in K_0(\sigma_s^*)$, the relation $\sigma_{s1} \subset \sigma_s$ follows for the support σ_{s1} of m_1; and we must show $m_1 \in C(m)$. Applying [2] IV Th 2.1.11 to the subring $[0, \sigma_s]$ (rather than to Σ) and to the m given by $F = C(m)$, we find $m_1 \in F$.

Moreover we have shown above that $e L_0(m) = \sigma_s^*$. Since an m_{σ_1} with the support σ_1 is from an effective $m \in K(\Sigma)$ obtained by $m_{\sigma_1}(\sigma) = m(\sigma_1)^{-1} m(\sigma \wedge \sigma_1)$, the set G of all $e L_0(m)$ equals Σ; thus we have also proven AVid.

Due to AVkl, classical systems thus are effect carriers making $K = K(\Sigma)$ for a Boolean ring Σ; and this Σ has no atom. For, if K has finite-dimensional faces then it also has atoms. If there is an atom σ_0 in Σ, the face $K_1(\sigma_0)$ belongs to it. Then $K_1(\sigma_0)$ is the set of all measures m with supports $\sigma_s \subset \sigma_0$. Since σ_0 is an atom, only one such m therefore exists. Thus $K_1(\sigma_0)$ is an extreme point of K and hence finite-dimensional.

Including considerations from V § 10, we recognize that physical objects are "classical systems" provided they have no atomic, objective properties.

In closing this section let us prove the following two theorems without presuming AV 1.1 and AVid. We merely assume AV 1.2s, i.e. $L = [0, \mathbf{1}]$.

T 5.3.2 $\mathscr{B} = \mathscr{B}(\Sigma)$ (with a Boolean ring Σ) is equivalent to the case that every pair of effects is coexistent.

Proof. From $g_1 = g_{12} + g'$, $g_2 = g_{12} + g''$, $h = g_{12} + g' + g'' \in L$ (see V § 1.2) and $g_1, g_2 \in L_0(w)$ follows $h \in L_0(w)$, i.e. AV 1.1. Therefore the set G of decision effects exists.

It remains to prove AVid. Then T 5.3.1 shows $\mathscr{B} = \mathscr{B}(G)$ where G is a Boolean ring.

With $e \in G$ and $g \in L$ we have $g = \bar{g} + h$, $e = \bar{g} + k$ and $\bar{g} + h + k \leqq \mathbf{1}$, i.e. $h \leqq \mathbf{1} - e$.

Therefore e reduces g in the sense of D 5.1.1. According to T 5.1.1, $\bar{g}$ is uniquely determined (the proof of T 5.1.1 does not use AVid!). Therefore a mapping $L \xrightarrow{T_e} L$ is defined by $\bar{g} = T_e g$. Then $h = (\mathbf{1} - T_e)\, g$. Let us prove that T_e is affine on L.

With $g_1, g_2, g_1 + g_2 \in L$ we have $g_1 = \bar{g}_1 + h_1$, $g_2 = \bar{g}_2 + h_2$ and $\bar{g}_1 + \bar{g}_2 \in L$. Hence $\bar{g}_1 \leqq e$ and $\bar{g}_2 \leqq e$ imply $\bar{g}_1 + \bar{g}_2 \leqq e$.

With $y = \bar{g}_1 + \bar{g}_2 - T_e(g_1 + g_2)$ we have $-e \leqq y \leqq 2e$.

From $g_1 + g_2 = \bar{g}_1 + \bar{g}_2 + h_1 + h_2 = T_e(g_1 + g_2) + (\mathbf{1} - T_e)(g_1 + g_2)$ follows

$$y = (\mathbf{1} - T_e)(g_1 + g_2) - h_1 - h_2 .$$

Because of $h_1, h_2 \leqq \mathbf{1} - e$, we have

$$-2(\mathbf{1} - e) \leqq y \leqq \mathbf{1} - e .$$

Hence $-e \leqq \frac{1}{2} y \leqq e$ and $-(\mathbf{1} - e) \leqq \frac{1}{2} y \leqq \mathbf{1} - e$.

According to the proof of T 5.1.1, this implies $y = 0$, i.e. $T_e(g_1 + g_2) = T_e g_1 + T_e g_2$.

With $0 \leqq \lambda \leqq 1$ from $g = \bar{g} + h$ and $\bar{g} \leqq e$, $h \leqq \mathbf{1} - e$ we get $\lambda g = \lambda \bar{g} + \lambda h$ and $\lambda \bar{g} \leqq e$, $\lambda h \leqq \mathbf{1} - e$; i.e. $T_e(\lambda g) = \lambda T_e g$.

Since T_e is affine, it can be extended as a linear map to all of $\mathscr{B}'$. This mapping is $\sigma(\mathscr{B}', \mathscr{B})$-continuous if it is so on the unit ball and therefore if it is so on L. Since the $\sigma(\mathscr{B}', \mathscr{B})$-topology is metrizable on L, it suffices to take convergent sequences $g_n \to g$. Since L is compact in the $\sigma(\mathscr{B}', \mathscr{B})$-topology we may choose such a subsequence g_{n_i} that $T_e g_{n_i} \to g_1 \leqq e$ and $(\mathbf{1} - T_e) g_{n_i} \to g_2 \leqq \mathbf{1} - e$. From $g_{n_i} = T_e g_{n_i} + (\mathbf{1} - T_e) g_{n_i}$ follows $g = g_1 + g_2$ and thus $g_1 = T_e g$, i.e. the continuity of T_e.

Therefore there exists the dual mapping T_e' of $\mathscr{B}$ into $\mathscr{B}$, where T_e' maps K into $\check{K}$. If $w \in K$ is effective we have $\mu(w, e) \neq 0$. With $\tilde{w} = \mu(w, e)^{-1} T_e' w \in K$ we get

$$\begin{aligned} \mu(\tilde{w}, e) &= \mu(w, e)^{-1} \mu(T_e' w, e) = \mu(w, e)^{-1} \mu(w, T_e e) \\ &= \mu(w, e)^{-1} \mu(w, e) = 1, \end{aligned}$$

i.e. AVid. □

T 5.3.3 $\mathscr{B} = \mathscr{B}(\Sigma)$ is equivalent to the case that every pair of demixings of an ensemble is coexistent; hence all demixings coexist.

Proof (for parts of the proof see also [32]).

We shall first show that $\mathscr{B}$ is a Riesz space if two demixings always coexist. $\mathscr{B}$ is called a Riesz space if $[\mathbf{0}, x_1] + [\mathbf{0}, x_2] = [\mathbf{0}, x_1 + x_2]$ holds for $x_1, x_2 \in \mathscr{B}_+$.

To show this property, it suffices to consider the case $\|x_1 + x_2\| = 1$. With $w = x_1 + x_2$, we then have $w \in K$, and $w = x_1 + x_2$ is a demixing of w. Each $v \in [\mathbf{0}, w]$ is a mixture component because $w = v + (w - v)$. Since both demixings coexist, there are a Boolean ring Σ and a measure $\Sigma \xrightarrow{W} \check{K}$ with $W(\varepsilon) = w$, $W(\sigma_1) = x_1$, $W(\sigma_2) = x_2$, $W(\sigma_3) = v$. From this follows

$$v = W(\sigma_3) = W(\sigma_3 \wedge \sigma_1) + W(\sigma_3 \wedge \sigma_1^*),$$

i.e. we have a demixing of v with

$$W(\sigma_3 \wedge \sigma_1) \leqq W(\sigma_1) = x_1$$

and $W(\sigma_3 \wedge \sigma_1^*) \leqq w - W(\sigma_1) = w - x_1 = x_2$.

Thus $\mathscr{B}$ is a Riesz space since $[\mathbf{0}, x_1] + [\mathbf{0}, x_2] \subset [\mathbf{0}, x_1 + x_2]$ holds trivially.

If $\mathscr{B}$ is a Riesz space, the Riesz theorem (see A II T 2) makes $\mathscr{B}'$ a vector lattice. Then (see A II T 1) every pair of elements $y_1, y_2 \in \mathscr{B}'$ fulfills

$$y_1 + y_2 = (y_1 \wedge y_2) + (y_1 \vee y_2).$$

This implies that every pair of effects $g_i (i = 1, 2)$ is coexistent: Since $g_1 \wedge g_2 \leqq g_1, g_2$ we have $g_1 = g_1 \wedge g_2 + g_1'$, $g_2 = g_1 \wedge g_2 + g_2'$ with $g_1', g_2' \in L$. $g_1 + g_2 = (g_1 \wedge g_2) + (g_1 \vee g_2)$ implies $g_1' + g_2' + g_1 \wedge g_2 = g_1 + g_2 - (g_1 \wedge g_2) = g_1 \vee g_2 \in L$, i.e. g_1, g_2 are coexistent according to V § 1.2. Then T 5.3.2 finishes the proof of $\mathscr{B} = \mathscr{B}(\Sigma)$.

Conversely, now let $\mathscr{B} = \mathscr{B}(\Sigma)$. The set of mixture components of an ensemble $m \in K(\Sigma)$ is the order interval $[\mathbf{0}, m]$ from $\mathscr{B}(\Sigma)$. Therefore all demixings of m coexist if there is a Boolean ring with additive measure in $K(\Sigma)$ whose range is just $[\mathbf{0}, m]$. To check this, it suffices (see the remarks after V D 3.3.1 and in [2] IV § 2.3, carried over to preparators) to specify a Boolean ring $\tilde{\Sigma}$ with a measure $\tilde{\Sigma} \xrightarrow{W} \check{K}(\Sigma)$ such that $[\mathbf{0}, m] = \overline{\text{co}}\,(W \tilde{\Sigma})$. We choose $\tilde{\Sigma}$ as the section $[0, \sigma_s]$ with σ_s the support of m. For a $\tilde{\sigma} \in \tilde{\Sigma}$ we set $W(\tilde{\sigma}) = m'$ with $m'(\sigma) = m(\tilde{\sigma} \wedge \sigma)$ for $\sigma \in \Sigma$. By a result analogous to [2] IV Th 2.1.11 we conclude $[\mathbf{0}, m] = \overline{\text{co}}\,\{W(\sigma) \mid \tilde{\sigma} \in \tilde{\Sigma} = [\mathbf{0}, \sigma_s]\}$. □

§ 5.4 Decomposition into Irreducible Parts

For microsystems, AV 4 provides the important theorem that G is atomic (VI T 7.2.1). Hence the decomposition of $(\mathscr{B}, \mathscr{B}')$ into irreducible parts shall for this case be described more precisely (though almost all theorems have been proven in § 5.2). Since AV 4 is not needed for this decomposition, let us also in this § 5.4 *only* presume AV 1.1, AV 1.2s and AVid.

Instead of assuming AV 4, we formulate the following theorems in the form: From G atomic follows ...; from Z atomic follows .... This enables us to apply the theorems when AV 4 does not hold, but G is atomic; or when G is not atomic but Z is. We shall first show AV 4 $\Rightarrow$ G atomic $\Rightarrow$ Z atomic. That AV 4 $\Rightarrow$ G atomic was already shown in VI T 7.2.1. That G atomic $\Rightarrow$ Z atomic is the content of the following important theorem.

T 5.4.1 If G is atomic, Z is also atomic.

Proof. If Z were not atomic, a $q \in Z$ would exist which contains no atom of Z. Then considering $G(q) = T_q G$, we find

$$G(q) = \{e \mid e \in G \text{ and } e \leqq q\}.$$

Since G is atomic, $G(q)$ must contain an atom p of G. We introduce the following subsets of Z:

$$Z(q;p) = \{\tilde{q} \mid \tilde{q} \in Z, \tilde{q} \leqq q, \tilde{q} \geqq p\},$$
$$Z(q) = \{\tilde{q} \mid \tilde{q} \in Z, \tilde{q} \leqq q\}.$$

Since $Z(q)$ is simply the order interval $[\mathbf{0}, q]$ in Z, it is a Boolean ring with q as unit element. The set $Z(q;p)$ is a subset of $Z(q)$; as we shall straightaway show, *inside* $Z(q)$ it satisfies

(α) $q_1 \in Z(q;p), q_2 \in Z(q;p) \Rightarrow q_1 \wedge q_2 \in Z(q;p)$,
(β) $q_1 \geqq q_2, q_2 \in Z(q;p) \Rightarrow q_1 \in Z(q;p)$,
(γ) $q_1 \notin Z(q;p) \Rightarrow q_1^{\perp} \wedge q \in Z(q;p)$,

where $q_1^{\perp} \wedge q$ is simply the complement of q_1 *in* $Z(q)$.

The relations (α)–(γ) are – to emphasize it again – meant to be relations in $Z(q)$, i.e. all occurring elements q_i belong to $Z(q)$. The relations (α) and (β) are obvious so that it only remains to prove (γ).

Because $q_1 \leqq q$, from $\mathbf{1} = q_1 + q_1^{\perp} = q_1 + q_1^{\perp} \wedge q + q_1^{\perp} \wedge q^{\perp}$ follows $\mathbf{1} = q_1 + q_1 \wedge q + q^{\perp}$. Hence (5.2.2) and T 5.2.5 for $p \in G$ yield

$$p = p \wedge q_1 + p \wedge q_1^{\perp} \wedge q + p \wedge q^{\perp}.$$

Due to $p \leqq q$ and $p = p \wedge q + p \wedge q^{\perp}$, we have $p \wedge q^{\perp} = 0$. Since p is an atom, we get either $p = p \wedge q_1$ or $p = p \wedge q_1^{\perp} \wedge q$. Because $q_1 \notin Z(q;p)$, we find $p \neq p \wedge q_1$ and hence $p = p \wedge q_1^{\perp} \wedge q$. Therefore $p \leqq q_1^{\perp} \wedge q$ and thus $q_1^{\perp} \wedge q \in Z(q;p)$, which proves ($\gamma$).

As a complete lattice, Z yields

$$\bar{q} = \bigwedge_{q \in Z(q;p)} q \in Z(q).$$

Because $p \leqq q$ for all $q \in Z(q;p)$, we also have $p \leqq \bar{q}$ and hence $\bar{q} \in Z(q;p)$. Since $Z(q)$ must not contain any atoms, there is a $\tilde{q} \in Z(q)$ with $\mathbf{0} \neq \tilde{q} \nleqq \bar{q}$. This gives $\tilde{q}^{\perp} \wedge \bar{q} \neq 0$ and $\tilde{q}^{\perp} \wedge \bar{q} \neq \bar{q}$. Because $\tilde{q} \nleqq \bar{q}$, we have $\tilde{q} \notin Z(q;p)$; hence (γ) makes $\tilde{q}^{\perp} \wedge q \in Z(q;p)$, so that (α) yields $\tilde{q}^{\perp} \wedge q \wedge \bar{q} = \tilde{q}^{\perp} \wedge \bar{q} \in Z(q;p)$. This contradicts $\mathbf{0} \neq \tilde{q}^{\perp} \wedge \bar{q} \neq \bar{q}$ and that $\bar{q}$ was the smallest element of $Z(q;p)$. □

T 5.4.2 If Z is atomic, the set of atoms in Z is at most countable. Writing this set $A_Z = \{q_\nu\}$, we have $\sum_\nu q_\nu = \mathbf{1}$ and to each $r \in Z$ there is uniquely associated a subset $A_r \subset A_Z$ which makes $\sum_{q \in A_r} q = r$. The mapping $r \to A_r$ is an isomorphism of Z with the lattice of subsets of A_Z.

Proof. Since the atoms q, q' of each pair with $q \neq q'$ are commensurable and have $q' \wedge q = \mathbf{0}$, from $q' = (q' \wedge q) \vee (q \wedge q^{\perp})$ follows that q' is orthogonal to q. Therefore, due to VI T 3.9 there are only countably many atoms q_ν, and $\sum_\nu q_\nu = \bigvee_\nu q_\nu \in Z$ holds. If we had $\sum_\nu q_\nu \neq 1$, there would be an atom $q \in Z$ with $q \leqq \mathbf{1} - \sum_\nu q_\nu$, contradicting the fact that $\{q_\nu\}$ was the set of all atoms. In analogous manner, for an $r \in Z$ with $A_r = \{q \mid q \in A_Z \text{ and } q \leqq r\}$ we obtain

$$r = \bigvee_{q \in A_r} q = \sum_{q \in A_r} q .$$

Conversely, each subset A of A_Z makes $\sum_{q \in A} q$ an element of Z. It is easy to show that the mapping $r \to A_r$ (with $r_1 \wedge r_2 \to A_{r_1} \cap A_{r_2}$, $r_1 \vee r_2 \to A_{r_1} \cup A_{r_2}$ and $\mathbf{1} - r \to A_Z \backslash A_r$) is an isomorphism. □

From § 5.2, especially from T 5.2.5 and T 5.2.7 and (5.2.2)–(5.2.7), in particular for $r_\nu = q_\nu$ we conclude: Each $x \in \mathscr{B}$ can uniquely be written

$$x = \sum_\nu x_\nu ,$$

where $x_\nu \in \mathscr{B}_\nu = T_{q_\nu} \mathscr{B}$. In particular, for $w \in K$ we have

$$w = \sum_\nu w_\nu$$

with $w_\nu \in K_\nu$, where $K_\nu = K_1(q_\nu)$.

Each $y \in \mathscr{B}'$ can uniquely be written

$$y = \sum_\nu y_\nu$$

with $y_\nu \in \mathscr{B}'_\nu = S_{q_\nu} \mathscr{B}'$. In particular, for $g \in L$ we have

$$g = \sum_\nu g_\nu$$

with $g \in L_\nu = \{g \mid g \in L \text{ and } g \leqq q_\nu\}$. Here, G_ν is the set of all $e \in G$ with $e \leqq q_\nu$ and G is the set of all $e = \sum_\nu e_\nu$ with $e_\nu \in G_\nu$.

We have

$$\mu(x, y) = \sum_\nu \mu(x_\nu, y_\nu)$$

and $\| x \| = \sum_\nu \| x \|_\nu$ and $\| y \| = \sup_\nu \| y_\nu \|$.

From T 5.2.15 follows:

Each Z_ν consists only of the zero and the unit element, i.e. each $(\mathscr{B}_\nu, \mathscr{B}'_\nu)$ is irreducible.

These results can be inverted:

T 5.4.3 If $\mathscr{B}_\nu$ is a sequence of at most countably many base-normed spaces, the real vector space $\mathscr{B}$ defined by the sequences $\{x_\nu\}$ with $x_\nu \in \mathscr{B}_\nu$ and finite $\sum_\nu \|x_\nu\|$ is a base-normed Banach space with the basis

$$K = \{w \mid w = \{w_\nu\},\ w_\nu \in \check{K}_\nu,\ \sum_\nu \|w_\nu\| = 1\}.$$

Proof. One can easily see that $\mathscr{B}$ is a Banach space since all the $\mathscr{B}_\nu$ are Banach spaces. $\mathscr{B}_+$ is given by $\{x_\nu\}$ with $x_\nu \in \mathscr{B}_{\nu+}$. The unit ball of $\mathscr{B}$ is formed by the $\{x_\nu\}$ with $\sum_\nu \|x_\nu\| \leqq 1$. Hence, x_ν belongs to the unit ball of $\mathscr{B}_\nu$, i.e. $x_\nu \in \overline{\text{co}}\,(K_\nu \cup (-K_\nu))$.

Therefore, with $\lambda_\nu \geqq 0$ and $\sum_\nu \lambda_\nu = 1$, the unit ball of $\mathscr{B}$ is formed by all $\{x_\nu\}$ with $x_\nu \in \overline{\text{co}}\,(\lambda_\nu K_\nu \cup (-\lambda_\nu K_\nu))$; hence it equals $\overline{\text{co}}\,(K \cup (-K))$ with the K defined above. □

T 5.4.4 With $\mathscr{B}$ from T 5.4.3, its dual $\mathscr{B}'$ can be identified with the space of all $y = \{y_\nu\}$ with $y_\nu \in \mathscr{B}'_\nu$ and $\sup_\nu \|y_\nu\| < \infty$. The order-unit in $\mathscr{B}'$ is given by $y_\nu = \mathbf{1}_\nu$ with $\mathbf{1}_\nu$ the order-unit in $\mathscr{B}'_\nu$; and we have $\|y\| = \sup_\nu \|y_\nu\|$.

Proof. It is easily seen that a norm-continuous linear form y on $\mathscr{B}$ uniquely determines norm-continuous linear forms $y_\nu \in \mathscr{B}'_\nu$ with $\|y_\nu\| \leqq \|y\|$. Conversely, each such sequence $\{y_\nu\}$ with $\sup_\nu \|y_\nu\| < \infty$ determines a norm-continuous linear form y with $\|y\| = \sup_\nu \|y_\nu\|$. □

T 5.4.5 If, for the $\mathscr{B}$ from T 5.4.3, the $\mathscr{B}_\nu, \mathscr{B}'_\nu$ satisfy AV 1.1, AV 1.2s, AV 2f, AVid, AV 3, AV 4, AV 4s, then the same holds for $\mathscr{B}, \mathscr{B}'$.

Proof. For a subset $k \subset K$, we find $L_0(k)$ given by the set of all $g = \{g_\nu\}$ with $g_\nu \in L_0(k_\nu)$, where k_ν is the set of the ν-components in the elements of k, i.e.

$$L_0(k) = \{g \mid g = \{g_\nu\} \text{ with } g_\nu \in L_0(k_\nu)\}.$$

Therefore, $L_0(k)$ has a largest element $e\, L_0(k) = \{e_\nu\, L_0(k_\nu)\}$, and thus AV 1.1 for $\mathscr{B}, \mathscr{B}'$ is a consequence of AV 1.1 for all $\mathscr{B}_\nu, \mathscr{B}'_\nu$.

AV 1.2 s now means $[\mathbf{0}, \mathbf{1}] = \{y \mid y = \{y_\nu\},\ y_\nu \in [\mathbf{0}_\nu, \mathbf{1}_\nu]\}$.

For an $e \in G$ with $e = \{e_\nu\}$ and $e_\nu \in G_\nu$, from AVid follows that there are elements $w_\nu \in K_\nu$ with $\mu(w_\nu, e_\nu) = 1$. With $\lambda_\nu \geqq 0$ and $\sum_\nu \lambda_\nu = 1$, for $w = \{\lambda_\nu w_\nu\}$ follows

$$\mu(w, e) = \sum_\nu \lambda_\nu\, \mu(w_\nu, e_\nu) = \sum_\nu \lambda_\nu = 1;$$

therefore AVid also holds for $\mathscr{B}$.

In order to show AV 3 for $\mathscr{B}, \mathscr{B}'$, let us use relation (iv) from VI T 6.2.3 and also (6.1.11). From

$$\mu(w, \mathbf{1} - e) = \sum_\nu \mu_\nu(w_\nu, \mathbf{1}_\nu - e_\nu)$$

and

$$K_1(e) = \left\{ w \mid w = \{w_\nu\} \in K, \frac{w_\nu}{\| w_\nu \|} \in K_1(e_\nu) \text{ for } w_\nu \neq 0 \right\}$$

then follows

$$\inf_{w \in K_1(e_1)} \mu(w, \mathbf{1} - e_2) = 0,$$

provided that at least a single ν obeys

$$\inf_{w_\nu \in K_1(e_{1\nu})} \mu_\nu(w_\nu, \mathbf{1} - e_{2\nu}) = 0.$$

Therefore, from $\Delta(e_1, e_3) \neq \emptyset$ follows $\Delta_\nu(e_{1\nu}, e_{3\nu}) \neq 0$ for all ν.

From $e_2 \leqq e_1 \leqq e_2 \vee e_3$ follows $e_{2\nu} \leqq e_{1\nu} \leqq e_{2\nu} \vee e_{3\nu}$ for all ν; hence AV 3, i.e. (ix) from VI T 6.2.3 yields $e_{1\nu} = e_{2\nu}$ for all ν. Thus we find $e_1 = e_2$, whereby AV 3 is shown for $\mathscr{B}, \mathscr{B}'$.

A closed face F of K is determined by a sequence $\{F_\nu\}$ of closed faces of K_ν:

$$F = \{w \mid w = \{\lambda_\nu w_\nu\} \text{ with } \lambda_\nu \geqq 0, \sum_\nu \lambda_\nu = 1, w_\nu \in F_\nu\}.$$

Therefore, the dimension of F equals $[-1 + \sum_\nu{}' (1 + \text{dimension of } F_\nu)$, with Σ_ν' taken only over those ν for which $F_\nu \neq \emptyset$. Thus, the dimension of F is finite if and only if that sum runs over only finitely many ν and the dimension of F_ν is finite for all ν. Hence if AV 2f holds for all ν then all the F_ν are exposed faces, i.e. $F_\nu = K_0(g_\nu)$. Then we get $F = K_0(g)$ with $g = \{g_\nu\}$, where one must set $g_\nu = 1_\nu$ for $F_\nu = \emptyset$. Therefore AV 2f holds for $\mathscr{B}, \mathscr{B}'$.

If AV 4 holds for all ν, then in an $F_{\nu_1} \neq \emptyset$ there is a finite-dimensional face $\tilde{F}_{\nu_1}$ of K_ν. Thus $\tilde{F}_1 = \{w \mid w = \{0, 0, \ldots, w_{\nu_1}, 0, 0, \ldots\}$ with $w_{\nu_1} \in \tilde{F}_{\nu_1}$ is a finite-dimensional face of K with $\tilde{F}_1 \subset F$. Hence AV 4 holds for $\mathscr{B}, \mathscr{B}'$.

If even AV 4s holds for all ν, then each ν makes

$$F_\nu = \bigvee_i F_{\nu i},$$

where the $F_{\nu i}$ are finite-dimensional and increasing. Then a finite-dimensional face $\tilde{F}_i$ of K is determined by the finite sequence $\{\tilde{F}_{\nu i}\}$ with $\tilde{F}_{\nu i} = F_{\nu i}$ for $\nu \leqq i$ and $\tilde{F}_{\nu i} = \emptyset$ for $\nu > i$. Then $F = \bigvee_i F_i$ and hence AV 4s holds for $\mathscr{B}, \mathscr{B}'$.

If Axiom AV 4 holds for a pair $\mathscr{B}, \mathscr{B}'$, we can perform the given decomposition of $\mathscr{B}, \mathscr{B}'$ into the irreducible parts $\mathscr{B}_\nu, \mathscr{B}_\nu'$. In the converse way, we can then by T 5.4.3 construct a pair $\tilde{\mathscr{B}}, \tilde{\mathscr{B}}'$ from the irreducible $\mathscr{B}_\nu, \mathscr{B}_\nu'$. We see immediately the isomorphism of $\tilde{\mathscr{B}}, \tilde{\mathscr{B}}'$ with $\mathscr{B}, \mathscr{B}'$. Thus, only the decisive problem of further investigating the structure of $\mathscr{B}, \mathscr{B}'$ for the *irreducible* case remains.

§ 6 System Types and Super Selection Rules

To the decomposition of $\mathscr{B}, \mathscr{B}'$ and hence of K, L into irreducible parts described in § 5.4, there correspond some important physical concepts which are not always uniformly named.

In V D 10.4, we defined the set $\mathscr{E}_m$ of objective properties. Let us show that the mapping $\mathscr{E}_m \xrightarrow{\chi} L$ from V § 10 yields $\mathscr{E}_m \xrightarrow{\chi} Z$, and that we can set $\chi(\mathscr{E}_m) = Z$ (as a certain hypothesis).

If $\chi(\mathscr{E}_m) = Z$, then $\mathscr{E}_m \xrightarrow{\chi} Z$ is a decision observable while T 2.2 makes χ an isomorphism between $\mathscr{E}_m$ and Z. Hence one uses to denote Z also as the set of objective properties of the microsystem. Thus we must first show $\chi(\mathscr{E}_m) \subset Z$. To this end, we proceed step by step.

T 6.1 By

$$\mu(w, \psi(b_0, b \cap p)) = \mu(T_p w, \psi(b_0, b))$$

a mapping $K \xrightarrow{T_p} \breve{K}$ uniquely corresponds to each p. While it obeys $T_{M \setminus p} = \mathbf{1} - T_p$ and $T_{p_1 \cap p_2} = T_{p_1} T_{p_2} = T_{p_2} T_{p_1}$, the dual T'_p makes $T'_p \mathbf{1} = \chi(p)$.

Proof. Use [2] III Th 4.1.6 or [3] § 12.3. In particular, from the defining equation for T_p follows

$$\begin{aligned}\mu(w, \chi(p)) &= \mu(w, \psi(b_0, b_0 \cap p)) \\ &= \mu(T_p w, \psi(b_0, b_0)) = \mu(T_p w, \mathbf{1}) \\ &= \mu(w, T'_p 1). \quad \square\end{aligned}$$

T 6.2 $\mathscr{E}_m \xrightarrow{\chi} Z$ holds.

Proof. See [2] IV § 8.1.

That $\mathscr{E}_m \xrightarrow{\chi} Z$ is surjective has (also in the extended theory) turned out as certain hypothesis: One can always "think" that for each registration procedure b_0 and prescribed $z \in Z$ one can ideally register those $b_0 \cap p$ which make $\chi(p) = \psi(b_0, b_0 \cap p) = z$.

Since Z is atomic, one can completely characterize the objective properties by the set A_Z of the atoms of Z. For a $q \in A_Z$, the set $p(q)$ with $\chi(p(q)) = q$ consists of the systems with the (atomic) objective property q.

D 6.1 For a $q \in A_Z$, we call $p(q)$ the set of systems of type q or briefly a "system type".

Thus we have $M = \bigcup_q p(q)$ and $p(q_1) \cap p(q_2) = 0$ for $q_1 \neq q_2$. Hence the set M can be decomposed into system types.

Examples of such type are, for instance, "one electron", "one helium atom", "a system composed of one helium atom and one hydrogen atom", and so forth. Of course, one must first go over from the theory presented so far to the standard extensions in the sense of [3] § 8, in order to give such "words" a meaning (see IX and [2]).

It is customary in physics to consider each system type separately, because only this is interesting. That is, one singles out a $q_\nu \in A_Z$ and takes only $M_\nu \overset{\text{def}}{=} p(q_\nu)$ as the "set of systems", K_ν as the set of ensembles, L_ν as the set of effects and G_ν as the set of decision effects. One considers in this way only one irreducible part of the decomposition described in § 5.4. In the course of such an investigation (e.g. of "hydrogen atoms"), one drops the index ν, since one has prescribed which system type is considered.

The fact that the element of A_Z are atomic objective properties, commensurable with all the decision effects, is frequently characterized by the concept of *super selection* rules. It is described in [2] IV § 8.1, how this concept was arrived at. In this sense, a $q \in A_Z$ determines a super selection rule. To be sure, there enters something from the fact that the transports introduced in II § 7 map each irreducible L_v into itself (see IX § 1).

VIII Representation of $\mathscr{B}$, $\mathscr{B}'$ by Banach Spaces of Operators in a Hilbert Space

In this chapter let us show that the irreducible parts $\mathscr{B}_\nu$, $\mathscr{B}'_\nu$ (see VII § 5.4) permit a representation by Hilbert space operators, i.e. for $\mathscr{B}_\nu$, $\mathscr{B}'_\nu$ there is a Hilbert space $\mathscr{H}_\nu$ over the field **R** of real numbers or over the **C** of complex numbers or over the **Q** of quaternions where $\mathscr{B}_\nu$ can be identified with the set of self-adjoint operators of the trace class and $\mathscr{B}'_\nu$ with the set of all bounded, self-adjoint operators, so that $\mu(x, y) = \operatorname{tr}(x\,y)$. Then K_ν is the set of all operators $w \in \mathscr{B}_\nu$ with $w \geqq 0$, $\operatorname{tr}(w) = 1$, while L_ν is the set of all operators $g \in \mathscr{B}'_\nu$ with $\mathbf{0} \leqq g \leqq \mathbf{1}$.

§ 1 The Finite Elements of *G*

We now presume AV 1.1, AV 1.2s, AVid, AV 3 and AV 4, without always mentioning it. Then VII § 5.2 makes them valid for each irreducible part. Likewise the validity of AV 2f and AV 4s carries over. Hence the theorems to be proved hold for all of G as well as for each irreducible part.

By VI T 7.2.1 the lattice G is atomic. According to VI T 6.3.5, for an $e \in G$ with finite dimensional $K_1(e)$, the order interval $[\mathbf{0}, e]$ is a complete orthocomplemented and modular lattice.

T 1.1 Each $e \in G$ is the sum of at most countably many pairwise orthogonal atoms.

Proof. The countability follows from VI T 3.6.

The set $A = \{(p_1, p_2, \ldots) \mid \text{the } p_\nu \text{ are pairwise orthogonal atoms with } p_\nu \leqq e\}$ satisfies the assumptions of Zorn's lemma. Hence, there is a maximal element. When $(p_1, p_2, \ldots)$ is such an element, with $e' = \sum_\nu p_\nu = \bigvee_\nu p_\nu$ we have $e' \leqq e$. If we had $e' \neq e$, hence $e - e' \neq 0$, there would be an atom $\tilde{p} \leqq e - e'$, i.e. an atom $\tilde{p}$ with $\tilde{p} \perp e'$ and hence $\tilde{p} \perp$ all p_ν, contradicting the fact that $(p_1, p_2, \ldots)$ was maximal. □

T 1.2 If $K_1(e)$ is finite dimensional, then e is the sum of finitely many atoms. From AV 4s follows: If e is the sum of finitely many atoms, then $K_1(e)$ is finite dimensional.

Proof. Let $K_1(e)$ be finite dimensional. By T 1.1 we have $e = \sum_\nu p_\nu$, where the p_ν are atoms. For $e_n = \sum_{\nu=1}^{n} p_\nu$ we find $K_1(e) \supset K_1(e_n) \supsetneqq K_1(e_{n-1})$, hence "dimension $K_1(e_n) \gneqq$ dimension $K_1(e_{n-1})$". Therefore the sequence of e_n must break off.

Let $e = \sum_{\nu=1}^{n} p_\nu$, where the p_ν are atoms (pairwise orthogonal; see VI T 3.9). Putting $\tilde{e}_k = \sum_{\nu=1}^{k} p_\nu$, we get

$$0 \leqq \tilde{e}_0 < \tilde{e}_1 < \tilde{e}_2 \ldots < \tilde{e}_{n-1} < \tilde{e}_n = e \, .$$

By the covering condition (VI T 7.2.3), between any two $\tilde{e}_\nu$, $\tilde{e}_{\nu+1}$ there does not exist any element of G different from $\tilde{e}_\nu$ or $\tilde{e}_{\nu+1}$.

By AV 4s there is a sequence e_i with

$$\mathbf{0} < e_0 < e_1 < e_2 \ldots < e \, , \qquad e = \bigvee_i e_i$$

and $K_1(e_i)$ finite dimensional for all i. We will show that there must be a finite number M such that $e_M = e$, so that $K_1(e) = K_1(e_M)$ is finite dimensional.

We define

$$c_{ij} = (\tilde{e}_i \wedge e_j) \vee \tilde{e}_{i-1} \, , \qquad d_{ji} = (e_j \wedge \tilde{e}_i) \vee e_{j-1} \, ,$$

where i ranges from 0 to n and j over all its (still unknown) values.

One finds immediately

$$\tilde{e}_{i-1} = c_{i0} \leqq c_{i1} \leqq c_{i2} \leqq \ldots \leqq \tilde{e}_i$$

and

$$e_{j-1} = d_{j0} \leqq d_{j1} \leqq \ldots \leqq d_{j,\, n-1} \leqq d_{jn} = e_j \, .$$

Since between $\tilde{e}_{i-1}$ and $\tilde{e}_i$ there cannot be any of these distinct elements of G, there must exist an m_i (possibly $m_i = 0$) with $c_{ik} = c_{im_i}$ for $k \geqq m_i$. With M as the largest of the numbers $m_1, m_2, \ldots m_n$, we thus get $c_{ik} = c_{iM}$ for $k \geqq M$ and for all i.

Let us show that $c_{ij} = c_{i,j-1}$ is equivalent to $d_{j,i-1} = d_{ji}$. Here we use the relations $M(\tilde{e}_i, e_j)$, $M(e_j, \tilde{e}_i)$ for all j and i, which follow from VI T 6.3.3, because $K_1(e_j)$ is finite dimensional. Because of these modular relations, we also have

$$c_{ij} = \tilde{e}_i \wedge (\tilde{e}_{i-1} \vee e_j) \, .$$

Hence $c_{ij} = c_{i,j-1}$ is equivalent to

$$\tilde{e}_i \wedge (\tilde{e}_{i-1} \vee e_{j-1}) = \tilde{e}_i \wedge (\tilde{e}_{i-1} \vee e_j) \, . \tag{1.1}$$

This implies

$$\tilde{e}_i \wedge (\tilde{e}_{i-1} \vee e_{j-1}) \wedge (\tilde{e}_{i-1} \vee e_j) = \tilde{e}_i \wedge (\tilde{e}_{i-1} \vee e_j) \, ,$$

which provides

$$\tilde{e}_i \wedge (\tilde{e}_{i-1} \vee e_j) \leqq \tilde{e}_{i-1} \vee e_{j-1} \, . \tag{1.2}$$

The latter gives

$$\tilde{e}_i \wedge (\tilde{e}_{i-1} \vee e_j) \leqq \tilde{e}_i \wedge (\tilde{e}_{i-1} \vee e_{j-1})$$

and by $e_j \geqq e_{j-1}$ also yields (1.1). Hence $c_{ij} = c_{i,j-1}$ is equivalent to (1.2). Because of the modular relations, one can also replace (1.2) by

$$\tilde{e}_{i-1} \vee (\tilde{e}_i \wedge e_j) \leqq \tilde{e}_{i-1} \vee e_{j-1} \, .$$

Thus (1.2) is equivalent to

$$\tilde{e}_{i-1} \leqq \tilde{e}_{i-1} \vee e_{j-1} \quad \text{and} \quad \tilde{e}_i \wedge e_j \leqq \tilde{e}_{i-1} \vee e_{j-1} \, ,$$

hence to

$$\tilde{e}_i \wedge e_j \leqq \tilde{e}_{i-1} \vee e_{j-1} . \tag{1.3}$$

Thus $c_{ij} = c_{i,j-1}$ is equivalent to (1.3). Therefore, from $c_{i,k+1} = c_{ik}$ for $k \geqq M$ and for all i follows $d_{k+1,i-1} = d_{k+1,i}$ for $k \geqq M$ and all i.

From this follows

$$e_k = d_{k+1,0} = d_{k+1,1} = \ldots = d_{k+1,n-1} = d_{k+1,n} = e_{k+1}$$

for all $k \geqq M$ and hence $e_k = e_M$ for all $k \geqq M$. Thus we get $\bigvee_i e_i = e_M$ and hence $e = e_M$. □

T 1.3 AV 4s implies the following: If e is the sum of finitely many atoms, then $[\mathbf{0}, e]$ is an atomic, complete, orthocomplemented and modular lattice.

Proof. This follows immediately from T 1.2 and VI T 6.3.5. □

The method of proving T 1.2 contains practically also the proof of the well-known theorem

T 1.4 If V is an atomic, orthocomplemented lattice, whose unit element is the union of finitely many atoms, each sequence

$$\mathbf{0} = a_0 \leqq a_1 \leqq a_2 \ldots a_n = a \in V ,$$

where $a_{\nu+1} \neq a_\nu$ and $a_{\nu+1}$ covers a_ν (i.e. $b \in V$ and $a_\nu < b < a_{\nu+1}$ imply $b = a_\nu$ or $b = a_{\nu+1}$), has the *same* finite length n. The atoms $p_\nu = a_{\nu+1} \wedge a_\nu^\perp$ then are pairwise orthogonal and make $a = \bigvee_{\nu=1}^{n} p_\nu$. Hence each set of pairwise orthogonal atoms, whose union equals a, has the same cardinality $d(a)$. To each a, this cardinality thus assigns a positive integer $d(a)$ called "dimension of a".

Proof. Consider two distinct sequences,

$$\mathbf{0} = a_0 < a_1 < a_2 < \ldots < a_n = a ,$$
$$\mathbf{0} = \tilde{a}_0 < \tilde{a}_1 < \tilde{a}_2 < \ldots < \tilde{a}_n = a ,$$

without the assumption that $a_{\nu+1}$ covers a_ν and $\tilde{a}_{\nu+1}$ covers $\tilde{a}_\nu$. To these sequences one applies the procedure from the proof of T 1.2. For each of the two sequences one thus obtains a refinement of the *same* length.

If $a_{\nu+1}$ covers a_ν and $\tilde{a}_{\nu+1}$ covers $\tilde{a}_\nu$, then no proper refinement exists; hence all such sequences have the same length.

The remainder of the proof follows easily. □

§ 2 The General Representation Theorem for Irreducible G

Here let us show that AV 1.1, AV 1.2s, AVid, AV 3 and AV 4s make an irreducible G isomorphic to the lattice of the closed subspaces of a Hilbert space.

Let D be a field. Let V be a vector space over D. As antiautomorphism Θ of D one denotes a mapping $D \xrightarrow{\Theta} D$ with $\Theta\,(a + b) = \Theta\, a + \Theta\, b$, $\Theta\,(a\, b) = (\Theta\, b)\,(\Theta\, a)$.

A Θ-bilinear form over V is understood as a mapping $V \times V \to D$, written $\langle x, y \rangle$ with

$$\langle x_1 + x_2, y \rangle = \langle x_1, y \rangle + \langle x_2, y \rangle, \quad \langle x, y_1 + y_2 \rangle = \langle x, y_1 \rangle + \langle x, y_2 \rangle,$$
$$\langle a\, x, b\, y \rangle = (\Theta\, a) \langle x, y \rangle\, b .$$

This bilinear form is called symmetric if

$$\langle x, y \rangle = \Theta\, (\langle y, x \rangle) .$$

Then Θ is involutory, i.e. $\Theta^2 = 1$; here we assume that $\langle \ldots, \ldots \rangle$ is not identically zero.

Proof. Let $\langle x, y \rangle \neq 0$. From $\langle x, a\, y \rangle = \langle x, y \rangle\, a$ follows $\{\langle x, a\, y \rangle \mid a \in D\} = D$. Therefore, *a fortiori* $\{\langle x, y \rangle \mid x, y \in V\} = D$. From this follows $\Theta^2 \langle x, y \rangle = \Theta \langle y, x \rangle = \langle x, y \rangle$ and hence $\Theta^2 = 1$. □

A symmetric Θ-bilinear form is called *definite* if $\langle x, x \rangle = 0$ implies $x = 0$.

In a vector space V over D, let Θ be an involutory antiautomorphism and $\langle \ldots, \ldots \rangle$ a symmetric definite Θ-bilinear form. For a subset $M \subset V$, we define

$$M^{\perp} = \{y \mid y \in M, \langle y, x \rangle = 0 \text{ for all } x \in M\} .$$

This is obviously a linear subspace of V, which obeys

$$M \cap M^{\perp} = \{0\} \quad \text{and} \quad M \subset M^{\perp\perp} .$$

A linear subspace M of V is called $\langle \ldots, \ldots \rangle$-closed if $M = M^{\perp\perp}$. Because $M^{\perp\perp\perp} = M^{\perp}$, we find $M^{\perp}$ always closed.

V is called "Hilbert space" if each closed subspace M satisfies $V = M + M^{\perp}$. This means that each $x \in V$ can be written $x = y + z$ with $y \in M$, $z \in M^{\perp}$; as one easily sees, y, z are uniquely determined by x and M.

Presuming AV 1.1, AV 1.2s, AVid, AV 3 and AV 4s, we obtain:

T 2.1 An irreducible G, which has at least four orthogonal atoms, is isomorphic to the lattice of closed subspaces of a Hilbert space V.

Proof. If AV 1.1, AV 1.2s, AVid, AV 3 and AV 4s hold for $\mathscr{B}, \mathscr{B}'$, by VII § 5.2 they hold for each irreducible part.

For the proof we can use that of Theorem 7.40 in [9], if we show that its presumptions from [9] are satisfied.

According to VI T 1.3.3 the lattice G is complete. The assumptions (96) of page 176 in [9] are satisfied, for:

(i) G is atomic by VI T 7.2.1.
(ii) is satisfied by T 1.3.
(iii) is satisfied by VI T 7.2.4.
(iv) is satisfied by the assumption of T 2.1.

The assumptions from (2) of page 105 in [9] are satisfied, since G is complete and because $e = e_2 - e_1$ with $e_1 < e_2$ fulfills $e_2 = e_1 + e = e_1 \vee e$ and $e \perp e_1$. □

It is unknown whether *each* lattice G_V of the closed subspaces of a Hilbert space V can be embedded in a dual pair $\mathscr{B}, \mathscr{B}'$ so that the presumptions AV 1.1, AV 1.2s, AVid, AV 3 and AV 4s hold for $\mathscr{B}, \mathscr{B}'$ and so that G_V coincides with the G defined by $\mathscr{B}, \mathscr{B}'$. This question would be interesting as follows. When axiom AV 2f is adjoined, it turns out that D *must* be one of the fields **R**, **C**, **Q** whereas $\mathscr{H}$ is a Hilbert space (i.e. a norm-complete space) over one of these three fields. Thus, a field other than **R**, **C**, **Q** could appear only if there exist finite-dimensional faces of K, which are not exposed!

§ 3 Some Topological Properties of G

We now presume AV 1.1, AV 1.2s, AVid, AV 3 and AV 4s without mentioning it again. Under these presumptions, G is an atomic lattice. Let the set of atoms of G be denoted by $A(G)$. From AV 2f and VI T 7.2.2 with the isomorphism K_1 of G onto $\mathscr{W}_L$ follows that a bijective mapping of $A(G)$ onto $\partial_e K$ is given by K_1. Let $A(e)$ denote the set of all atoms $p \leqq e \in G$. Due to T 1.2, the relation "$K_1(e)$ is finite dimensional" is equivalent to "e is the sum of finitely many atoms". In this case we say briefly: e is finite.

We furthermore assume AV 2f.

T 3.1 $A(e)$ with finite e is a $\sigma(\mathscr{B}', \mathscr{B})$-compact subset of $\mathscr{B}'$, of course obeying $A(e) \subset G \subset L \subset \mathscr{B}'$.

Proof. Since the $\sigma(\mathscr{B}', \mathscr{B})$-topology is metrizable on L, it suffices to consider a convergent sequence $p_\nu \in A(e)$ and to prove $p_\nu \to g \in A(e)$. Because $p_\nu \leqq e$, we have $g \leqq e$ and hence $g \in L_0 K_0(e)$. For each p_ν, one can find atoms $p_\nu^1, p_\nu^2, \ldots, p_\nu^{n-1}$, with $e = \sum_{\alpha=0}^{n-1} p_\nu^\alpha$ and $p_\nu^0 = p_\nu$ (by T 1.4 the number n is the same for each ν!), since $e - p_\nu$ due to T 1.1 need only be represented as a sum of orthogonal atoms. From $p_\nu^\alpha \leqq e$ follows $K_1(p_\nu^\alpha) \subset K_1(e)$; hence the $K_1(p_\nu^\alpha)$ are elements from $\partial_e K \cap K_1(e) = \partial_e K_1(e)$. We briefly write $w_\nu^\alpha = K_1(p_\nu^\alpha)$. Since $L_0 K_0(e)$ is $\sigma(\mathscr{B}', \mathscr{B})$-compact and $K_1(e)$ is likewise compact (being a finite dimensional face) one can choose a subsequence of the ν (again writing ν) so that for each α the sequences p_ν^α, w_ν^α are convergent. Then we get $p_\nu^\alpha \underset{\nu}{\to} g^\alpha \in L_0 K_0(e)$ (with $g^0 = g$) and $w_\nu^\alpha \to w^\alpha \in K_1(e)$.

Using

$$\begin{aligned}
|\mu(w^\alpha, g^\alpha) - 1| &= |\mu(w^\alpha, g^\alpha) - \mu(w_\nu^\alpha, p_\nu^\alpha)| \\
&\leqq |\mu(w^\alpha, g^\alpha) - \mu(w^\alpha, p_\nu^\alpha)| + |\mu(w^\alpha - w_\nu^\alpha, p_\nu^\alpha)| \\
&\leqq |\mu(w^\alpha, g^\alpha - p_\nu^\alpha)| + \| w^\alpha - w_\nu^\alpha \|
\end{aligned}$$

with $g^\alpha - p_\nu^\alpha \to 0$ (in the $\sigma(\mathscr{B}', \mathscr{B})$-topology) and $\| w^\alpha - w_\nu^\alpha \| \to 0$, from $\mu(w_\nu^\alpha, p_\nu^\alpha) = 1$ we get $\mu(w^\alpha, g^\alpha) = 1$. From $\sum_{\alpha=0}^{n-1} p_\nu^\alpha = e$ follows $\sum_{\alpha=0}^{n-1} g^\alpha = e$. Because $w^\alpha \in K_1(g^\alpha)$, we have $g^\alpha \neq 0$. With $e^\alpha \in G$ and $K_1(e^\alpha) = K_1(g^\alpha)$ follows $g^\alpha \geqq e^\alpha \neq 0$; hence there is an atom $q^\alpha \leqq e^\alpha \leqq g^\alpha$. From $\sum_{\alpha=0}^{n-1} q^\alpha \leqq \sum_{\alpha=0}^{n-1} g^\alpha = e$ follows that the q^α are pairwise orthogonal. Since there are n pairwise orthogonal q^α, we obtain $\sum_{\alpha=0}^{n-1} q^\alpha = e$ and hence $\sum_{\alpha=0}^{n-1} (g^\alpha - q^\alpha) = 0$. From $g^\alpha - q^\alpha \geqq 0$ follows $g^\alpha = q^\alpha$ for all α, also for $\alpha = 0$; hence $g = g^0 \in A(e)$. □

A finite-dimensional convex set C is called *strictly convex* if its boundary consists solely of extreme points. Then C is homeomorphic to a ball.

T 3.2 If p_1, p_2 are two distinct elements of $A(G)$, then $e = p_1 \vee p_2$ is finite; if G is irreducible, then $K_1(e)$ is strictly convex and hence homeomorphic to a ball of at least dimension two.

Proof. According to the covering theorem, $q = e - p_1$ must be an atom, so that $e = p_1 + q$ is two dimensional and hence finite. Therefore, $K_1(e)$ is finite dimensional.

If G is irreducible, there must be a third atom $p_3 \leqq e$ distinct from p_1 and p_2 (see AI). Since $K_1(p_3) = w_3 \in K_1(e)$ and w_3 is an extreme point, w_3 cannot lie on the segment between the extreme points $w_1 = K_1(p_1)$ and $w_2 = K_1(p_2)$. Therefore the whole triangle generated by w_1, w_2, w_3 lies in $K_1(e)$, so that $K_1(e)$ is at least two dimensional.

If F is a face of K with $F \subset K_1(e)$, then F is finite dimensional and (by AV 2f) an element of $\mathscr{W}_L$. If F is not an extreme point and $F \neq K_1(e)$, with w_1 as an extreme point of F one obtains the following chain of properly increasing faces of $K_1(e)$:

$$\emptyset,\ w_1,\ F,\ K_1(e)\,.$$

But T 1.4 shows that each chain due to $e = p_1 \vee p_2$ has at most the length 2. Hence no faces $F \neq K_1(e)$ exist which are not extreme points.

Therefore $K_1(p_1 \vee p_2)$ is homeomorphic to a ball having at least the dimension two. □

T 3.3 If e is finite, K_1 is a homeomorphic mapping of $A(e)$ onto $\partial_e K_1(e)$ with the norm topology in $\mathscr{B}$ and the $\sigma(\mathscr{B}', \mathscr{B})$-topology in $\mathscr{B}'$.

Proof. Since K_1 is a bijective mapping of $A(e)$ onto $\partial_e K_1(e)$ and $A(e)$ is compact, by T 3.1 it suffices to show that the mapping K_1 is continuous.

Let $p_\nu \in A(e)$ be a convergent sequence of atoms, so that T 3.1 makes $p_\nu \to p \in A(e)$. It must be shown that $w_\nu \to w$ holds with $w_\nu = K_1(p_\nu)$ and $w = K_1(p)$. Since $K_1(e)$, as a finite-dimensional face, is compact in the norm topology, the sequence w_ν has an accumulation point $\tilde{w}$; then a subsequence exists with $w_{\nu_i} \to \tilde{w}$. From

$$\begin{aligned} |\mu(\tilde{w}, p) - 1| &\leqq |\mu(\tilde{w}, p) - \mu(\tilde{w}, p_{\nu_i})| + |\mu(\tilde{w}, p_{\nu_i}) - \mu(w_{\nu_i}, p_{\nu_i})| \\ &\leqq |\mu(\tilde{w}, p - p_{\nu_i})| + \|\tilde{w} - w_{\nu_i}\| \end{aligned}$$

follows $\mu(w, p) = 1$ and hence $\tilde{w} \in K_1(p) = w$. Thus, w is the only accumulation point and hence $w_\nu \to w$. □

T 3.4 For each $e \in G$, the set $A(e)$ is a connected set in the $\sigma(\mathscr{B}', \mathscr{B})$-topology.

Proof. It will suffice to show that any two atoms p_1 and p_2 can in $A(p_1 \vee p_2)$ be connected by a continuous path. But T 3.2 and T 3.3 make $A(p_1 \vee p_2)$ homeomorphic to the surface of a ball having at least the dimension two, from which the assertion follows.

T 3.5 If e is finite, $[\mathbf{0}, e]$ is $\sigma(\mathscr{B}', \mathscr{B})$-compact (as a subset of G).

Proof. We have only to show that $e_\nu \in G$, $e_\nu \leqq e$ and $e_\nu \to g$ imply $g \in G$. For $e_\nu \leqq e$, T 1.3 and T 1.4 yield

$$e_\nu = \sum_{i=1}^{d(e_\nu)} p_\nu^i \tag{3.1}$$

with atoms p_ν^i and $d(e_\nu) \leqq d(e)$ the dimension of e_ν. One can complete the p_ν^i by zero-elements and instead of (3.1) write

$$e_\nu = \sum_{i=1}^{d(e)} p_\nu^i \tag{3.2}$$

with p_ν^i an atom or $p_\nu^i = \mathbf{0}$.

Then we can choose a subsequence of the ν (again written ν) so that $p_\nu^i \to g^i$ for each $i = 1, \ldots, d(e)$. By T 3.1 we have $g^i \in A(e)$ or $g^i = 0$, so that

$$g = \sum_{i=1}^{d(e)} g^i \leqq \mathbf{1}.$$

Hence the g^i are pairwise orthogonal and thus $g \in [\mathbf{0}, e]$. □

From this proof follows easily

T 3.6 If e is finite, the set $[\mathbf{0}, e]_d$ of all $e' \in [\mathbf{0}, e]$ with $d(e') = d$ is a $\sigma(\mathscr{B}', \mathscr{B})$-compact set.

The last theorems permit us to formulate the continuity properties from VI T 3.11 especially simply for sequences $e_\nu^{(1)}, e_\mu^{(2)}$ from an order interval $[\mathbf{0}, e]$ with finite e:

T 3.7 Let $e_\nu^{(1)} \leqq e$ and $e_\nu^{(2)} \leqq e$ for a finite e. From $e_\nu^{(1)} \to e^{(1)}$ and $e_\nu^{(2)} \to e^{(2)}$ in the $\sigma(\mathscr{B}', \mathscr{B})$-topology follows

(i) $e_\nu^{(1)\perp} \wedge e \to e^{(1)\perp} \wedge e$.
(ii) $e_\nu^{(1)} \vee e_\mu^{(2)} \to e^{(1)} \vee e^{(2)}$ if $d(e^{(1)} \vee e^{(2)}) = \{\min d(e), d(e^{(1)}) + d(e^{(2)})\}$.
(iii) $e_\nu^{(1)} \wedge e_\mu^{(2)} \to e^{(1)} \wedge e^{(2)}$, if $d(e^{(1)} \wedge e^{(2)}) = \max\{0, d(e^{(1)}) + d(e^{(2)}) - d(e)\}$.

Proof. From $e_\nu^{(1)\perp} \wedge e = e - e_\nu^{(1)}$ and $e_\nu^{(1)} \to e^{(1)}$ follows $e^{(1)} \leqq e$ and $e_\nu^{(1)\perp} \wedge e \to e - e^{(1)} = e^{(1)\perp} \wedge e$, i.e. we obtain (i).

As in the proof of VI T 3.11, for an accumulation point g of $e_\nu^{(1)} \vee e_\mu^{(2)}$ follows $g \geqq e^{(1)} \vee e^{(2)}$. By T 3.5 follows $g \in G$ with $g \leqq e$, hence $g = e^{(1)} \vee e^{(2)}$ if $d(g) = d(e^{(1)} \vee e^{(2)})$. By T 3.6 one obtains an accumulation point g with the largest possible $d(g)$ when $d(g) = \overline{\lim}\, d(e_\nu^{(1)} \vee e^{(2)})$. Thus $g = e^{(1)} \vee e^{(2)}$ is the only accumulation point if $\overline{\lim}\, d(e_\nu^{(1)} \vee e_\mu^{(2)}) \leqq d(e^{(1)} \vee e^{(2)})$. Because $e_\nu^{(1)} \to e^{(1)}$, by T 3.6 we get $d(e_\nu^{(1)}) \to d(e^{(1)})$; likewise follows $d(e_\mu^{(2)}) \to d(e^{(2)})$. From $d(e_\nu^{(1)} \vee e_\mu^{(2)}) \leqq \min\{d(e), d(e_\nu^{(1)}) + d(e_\mu^{(2)})\}$ results $\overline{\lim}\, d(e_\nu^{(1)} \vee e_\mu^{(2)}) \leqq \min\{d(e), d(e^{(1)}) + d(d(e^{(2)})\}$; thus (ii) follows.

(iii) follows from (ii) by means of (i). □

§ 4 The Representation Theorem for K, L

We now presume AV 1.1, AV 1.2s, AV 2f, AVid, AV 3 and AV 4s. These axioms suffice to prove the important representation theorem T 4.1.5. Since in VII § 5.3 we have shown the decomposition into irreducible parts, we can now assume G irreducible.

§ 4.1 The Representation Theorem for G

In § 2 we proved the general representation theorem T 2.1 for G, not assuming AV 2f. Let us now show that under AV 2f the field D can only be one of **R**, **C**, **Q**. The involutory antiautomorphism Θ is then uniquely determined by **R**, **C**, **Q** (the field of real or complex numbers or of quaternions).

As we saw in IV §§ 2 and 3, the topology of the physical imprecision on L is given by $\sigma(L, \mathscr{K})$, where L is compact and $\sigma(L, \mathscr{K})$ metrizable. Moreover, in IV § 3 we saw that on L the $\sigma(L, \mathscr{K})$-topology coincides with the $\sigma(\mathscr{B}', \mathscr{B})$-topology. From this follows that $\sigma(L, \mathscr{K}_1)$ and $\sigma(L, \mathscr{K}_2)$ coincide on L for any two countable subsets $\mathscr{K}_1, \mathscr{K}_2$ norm-dense in K. For many practical purposes, it is convenient in $\mathscr{B}'$ to use a norm whose topology on L coincides with the $\sigma(\mathscr{B}' \mathscr{B})$-topology.

T 4.1.1 The $\sigma(\mathscr{B}', \mathscr{B})$-topology is normable on L.

Proof. For a countable subset $\tilde{\mathscr{K}} = \{w_\nu\}$ norm-dense in K, and a sequence λ_ν (with $\lambda_\nu > 0$, $\sum_\nu \lambda_\nu = 1$), define

$$\| y \|_\sigma = \sum_\nu \lambda_\nu \, | \mu(w_\nu, y) | , \tag{4.1.1}$$

which is easily seen to be a norm.

From (4.1.1) follows

$$\lambda_\nu \, | \mu(w_\nu, y) | \leqq \| y \|_\sigma ;$$

hence the $\sigma(L, \tilde{\mathscr{K}})$-topology is weaker than the $\| \dots \|_\sigma$-topology.

For $g_1, g_2 \in L$ from (4.1.1) we due to $\mu(w, g) \leqq 1$ conclude

$$\| g_1 - g_2 \|_\sigma \leqq \sum_{\nu=1}^{N} \lambda_\nu \, | \mu(w_\nu, g_1 - g_2) | + \sum_{\nu=N+1}^{\infty} \lambda_\nu$$

and from this in turn that the $\sigma(K, \tilde{\mathscr{K}})$-topology is stronger than the $\| \dots \|_\sigma$-topology. □

From T 3.7 resp. VI T 3.11 thus follows that the lattice operations relative to the $\| \dots \|_\sigma$-topology are continuous in G. One can use this fact to prove:

T 4.1.2 The field D (of the representing Hilbert space) is a topological field and locally compact.

Proof. Since in this book we have not presented the introduction of coordinates, but have adopted it from other presentations of orthocomplementary, modular and atomic lattices available in the literature, also in this proof we must refer to other literature. Coordinatization is presented in many books; in T 2.1 we have referred to the presentation in [9]. In order to transfer the $\| \dots \|_\sigma$-topology from G to D, in the introduction of coordinates we must refer to a paper of Kolmogorov [12] (also see [13], [15], [16] and [9] p. 182). In [12] the proof is presented for T 4.1.2, i.e. that the field operations are continuous in the introduced topology and that D is locally compact. In our case, the local compactness follows quickly from T 3.1. □

T 4.1.3 D is one of the three fields **R**, **C**, **Q**.

Proof. The proof of T 4.1.2 given in [12] shows that the field D is homeomorphic to the set $A(p_1 \vee p_2)$ from T 3.2 if from $A(p_1 \vee p_2)$ one removes an element as the "point at infinity". This by T 3.2 and T 3.4 implies that D is connected. But each connected, locally compact topological field is one of **R**, **C**, **Q** (see [18] Satz 21, or [17]). □

T 4.1.4 The anti-automorphism Θ (which exists due to T 2.1) is continuous.

Proof. To prove this, we must again refer to the literature, since we have not presented an introduction of the Θ from T 2.1. The proof of T 4.1.4 is in [14] shown on the basis of [19]. Concerning [19], also see Theorem 4.6 in [9]. □

Since Θ is continuous, for $D = \mathbf{C}$ we get $\Theta a = \bar{a}$, where $\bar{a}$ is the complex number conjugate to a (see [14] and also [19] p. 181). This finally implies the important representation theorem of G:

T 4.1.5 An irreducible G, which has at least four orthogonal atoms, is isomorphic to the lattice of the (topologically) closed subspaces of a Hilbert space $\mathscr{H}$ over one of the fields **R**, **C**, **Q**, where $\perp$ is defined by the inner product in $\mathscr{H}$.

Proof. This follows from T 2.1, T 4.1.3, T 4.1.4 and Lemma 7.42 in [9], resp. from the Piron theorem (see [20]) which is formulated in [9] as Theorem 7.44. □

In T 4.1.5, the assumption enters that G contains at least four orthogonal atoms. To be sure, this condition is not aggravating for physics, because (for representing the Galileo resp. Poincaré group, see IX § 1), the irreducible parts must be either one- or infinite-dimensional. Without anticipating IX § 1 as an axiom we therefore formulate (with G a whole lattice, not only one of its irreducible parts):

AV 4a For each atom q of the center Z which is not an atom of G, the face $K_1(q)$ is not finite dimensional.

For each irreducible part G of the lattice G, this makes the assumptions of T 4.1.5 satisfied. Hence G_ν is isomorphic to the lattice of the closed subspaces of a Hilbert space $\mathscr{H}_\nu$ over one of the fields **R**, **C**, **Q**.

Only **C** turns out to be "realized" for the physics of the microsystems. There are various arguments which prefer **C** due to several indications from experience, e.g. the following: A two-dimensional representation of the rotation group (as a subgroup of the Galileo resp. Poincaré group), describing the spin of many microsystems, is not possible for $D = \mathbf{R}$. The production of "uncorrelated" ensembles for scattering processes (see IV § 2) suggests to exclude $D = \mathbf{Q}$. The consideration of "composite systems" in the way described in IX § 2 is not possible for $D = \mathbf{R}$ and $D = \mathbf{Q}$. But so far one has not succeeded, in a physically coincise way to formulate an axiom that excludes $D = \mathbf{R}$ and $D = \mathbf{Q}$. Therefore let us choose a mathematically simple formulation, which is at least based on a physically verifiable assumption, namely the dimension of $K_1(p_1 \vee p_2)$ with $p_1, p_2 \in A(G)$. This formulation rests on the following theorem, whose proof results from the considerations of § 4.2.

T 4.1.6 If G is the lattice of the closed subspaces of a Hilbert space over the field R resp. Q, then two atoms p_1, p_2 with $p_1 \perp p_2$ yield a face $K_1(p_1 \vee p_2)$ of the dimension 2 resp. 5.

As an axiom we therefore formulate (G is again the whole lattice):

AV 4b For two atoms p_1, p_2 of G with $p_1 \perp p_2$, the face $K_1(p_1 \vee p_2)$ is neither 2-dimensional nor 5-dimensional.

Due to T 4.1.6 this AV 4b can physically be checked as follows: For an irreducible part G, one picks *any two* atoms p_1, p_2 with $p_1 \perp p_2$, and for these verifies that $K_1(p_1 \vee p_2)$ is neither 2-dimensional nor 5-dimensional. In this sense, one could "physically" express AV 4b as follows: So far no system types (see § 6) have been found for which $D = \mathbf{R}$ or $D = \mathbf{Q}$; for this reason we reject the two cases $D = \mathbf{R}$ and $D = \mathbf{Q}$ by the axiom AV 4b.

§ 4.2 The Ensembles and Effects

Since the decomposition into irreducible parts is shown in VII § 5.4, let us again assume G irreducible. Henceforth let us consider only the case $D = \mathbf{C}$. If one wishes to prove a theorem analogous to T 4.2.5, one must perform the considerations of this § 4.2 analogously for $D = \mathbf{R}$ and $D = \mathbf{Q}$; for brevity let us not do so.

The following is shown in [2] A IV: The space $\mathscr{B}(\mathscr{H})$ of the self-adjoint operators in $\mathscr{H}$ of trace class is a base-normed Banach space with the basis $K(\mathscr{H})$ of all self-adjoint operators $W \geqq 0$ with $\operatorname{tr}(W) = 1$.

For $X \in \mathscr{B}(\mathscr{H})$, we have $\|X\| = \operatorname{tr}(\sqrt{X^2}) = \operatorname{tr}(X_+) + \operatorname{tr}(X_-)$ with $X_+ = \frac{1}{2}(\sqrt{X^2} + X)$ the positive part and $X_- = \frac{1}{2}(\sqrt{X^2} - X)$ the negative part of the operator $X = X_+ - X_-$. Thus $\mathscr{B}(\mathscr{H})$ satisfies the minimal decomposition property (see IV § 4). The Banach space $\mathscr{B}'(\mathscr{H})$ dual to $\mathscr{B}(\mathscr{H})$ can be identified with the space of all self-adjoint, bounded operators, if one writes the canonical bilinear form of $\mathscr{B}(\mathscr{H})$, $\mathscr{B}'(\mathscr{H})$ as $\operatorname{tr}(XY)$ with $X \in \mathscr{B}(\mathscr{H})$, $Y \in \mathscr{B}'(\mathscr{H})$. Here, $\mathscr{B}'(\mathscr{H})$ is an order unit space with the unit operator as the order unit. The norm in $\mathscr{B}'(\mathscr{H})$ is the operator norm of the operators in $\mathscr{H}$.

One can assign the closed subspaces of $\mathscr{H}$ bijectively to the projectors in $\mathscr{B}'(\mathscr{H})$. Let $G(\mathscr{H})$ denote the set of projectors, which is an orthocomplemented lattice, isomorphic to that of the closed subspaces in $\mathscr{H}$. Therefore, we can write the isomorphism given by T 4.1.5 as an isomorphism $G \xrightarrow{\alpha} G(\mathscr{H})$. Proceeding from this isomorphism α, let us draw further conclusions about K, L.

T 4.2.1 There is an injective mapping $K \xrightarrow{\beta} K(\mathscr{H})$ with $\mu(w, e) = \operatorname{tr}((\beta w)(\alpha e))$.

Proof. For pairwise orthogonal e_ν, from VI T 3.9 follows $\mu(w, \sum_\nu e_\nu) = \sum_\nu \mu(w, e_\nu)$. As an isomorphism, α gives $\alpha(\sum_\nu e_\nu) = \alpha(\bigvee_\nu e_\nu) = \bigvee_\nu (\alpha e_\nu) = \sum_\nu \alpha e_\nu$ (as sum of the operators αe_ν). Hence, $m_w(E) = \mu(w, \alpha^{-1} E)$ with $E \in G(\mathscr{H})$ defines a function $G(\mathscr{H}) \xrightarrow{m_w} [0, 1]$, with $m_w(\sum_\nu E_\nu) = \sum_\nu m_w(E_\nu)$ for pairwise orthogonal E_ν. By Gleason's theorem (see [2] A IV § 12) a $W \in K(\mathscr{H})$ therefore exists with $m_w(E) = \operatorname{tr}(WE)$. One can easily see that W is uniquely determined, since $\operatorname{tr}(W_1 - W_2) E) = 0$ for all $E \in G(\mathscr{H})$ implies $W_1 = W_2$. Hence, there is a mapping $K \xrightarrow{\beta} K(\mathscr{H})$ with $\mu(w, e) = \operatorname{tr}((\beta w)(\alpha e))$. Since $\mu(w_1, e) = \mu(w_2, e)$ for all $e \in G$ likewise implies $w_1 = w_2$ (the $\sigma(\mathscr{B}', \mathscr{B})$-closed space spanned by G is all of $\mathscr{B}'$; see VI T 3.1) β is injective. □

T 4.2.2 The mapping β in T 4.2.1 is affine.

Proof. Since β is injective, from $w = \lambda w_1 + (1 - \lambda) w_2$ follows

$$\begin{aligned} \operatorname{tr}((\beta w)(\alpha e)) &= \mu(w, e) = \lambda \mu(w_1, e) + (1 - \lambda) \mu(w_2, e) \\ &= \operatorname{tr}((\beta w_1)(\alpha e)) + (1 - \lambda) \operatorname{tr}((\beta w_2)(\alpha e)) \\ &= \operatorname{tr}([\lambda(\beta w_1) + (1 - \lambda)(\beta w_2)](\alpha e)) \end{aligned}$$

and hence

$$\beta w = \lambda \beta w_1 + (1 - \lambda) \beta w_2 . \quad \square$$

T 4.2.3 The mapping β in T 4.2.1 is surjective and hence bijective.

Proof. With p an atom of G, we have $\alpha p = P_\varphi$, with P_φ the projector on a vector $\varphi \in \mathscr{H}$ ($\|\varphi\| = 1$), i.e. projecting on a one-dimensional subspace of $\mathscr{H}$. The face $K_1(p)$ is an extreme point w_p of K. From $\mu(w_p, p) = 1 = \operatorname{tr}((\beta w_p) P_\varphi)$ follows $\beta w_p = P_\varphi$. Therefore all P_φ lie in βK.

Each $W \in K(\mathscr{H})$ can be written $W = \sum_\nu \lambda_\nu P_{\varphi_\nu}$ with $\lambda_\nu \geqq 0$ and $\sum_\nu \lambda_\nu = 1$. Hence a $w_\nu \in K$ with $\beta w_\nu = P_{\varphi_\nu}$ exists for each P_{φ_ν}. The conditions on the λ_ν make $w = \sum_\nu \lambda_\nu w_\nu \in K$, since K is norm closed. With this w, we then have

$$\mu(w, e) = \underbrace{\sum_\nu \lambda_\nu \mu(w, e)}_{= \operatorname{tr}(W(\alpha e))} = \sum_\nu \lambda_\nu \operatorname{tr}(P_{\varphi_\nu}(\alpha e)) ,$$

and hence $W = \beta w$. □

T 4.2.4 The bijective and affine mapping $K \xrightarrow{\beta} K(\mathscr{H})$ can be extended uniquely as an isomorphic mapping $\mathscr{B} \xrightarrow{\beta} \mathscr{B}(\mathscr{H})$ of Banach spaces.

Proof. This follows from the previous theorems and the fact that $\mathscr{B}$ as well as $\mathscr{B}(\mathscr{H})$ are linearly spanned by K resp. $K(\mathscr{H})$. □

From T 4.2.4 follows that the dual mapping β' is an isomorphic mapping $\mathscr{B}'(\mathscr{H}) \xrightarrow{\beta'} \mathscr{B}'$ of Banach spaces. From $\mu(w, e) = \operatorname{tr}((\beta w)(\alpha e))$ for $e \in G$ follows $\alpha = \beta'^{-1}$ on G.

Therefore, one can extend α as an isomorphic mapping $\mathscr{B}' \xrightarrow{\alpha} \mathscr{B}'(\mathscr{H})$ with $\alpha = \beta'^{-1}$.

Thus $L \xrightarrow{\alpha} L(\mathscr{H})$ is a bijective mapping with $L(\mathscr{H})$ as the order interval $[\mathbf{0}, \mathbf{1}]$ in $\mathscr{B}'(\mathscr{H})$, while

$$\mu(w, g) = \operatorname{tr}((\beta w)(\alpha g))$$

holds for all $w \in K$, $g \in L$. Hence one can identify K, L and μ with $K(\mathscr{H})$, $L(\mathscr{H})$ and tr.

Conversely, for $K = K(\mathscr{H})$ and $L = L(\mathscr{H})$ it remains to be shown that AV 1.1, AV 1.2s, AV 2f, AVid, AV 3 and AV 4s hold as theorems. AV 1.2s holds by the definition of $L(\mathscr{H})$.

In order to prove AV 1.1, we consider the set $L_0(W)$ for a $W \in K(\mathscr{H})$, i.e. the set of all F with $\mathbf{0} \leqq F \leqq \mathbf{1}$ and $\operatorname{tr}(WF) = 0$. Using $W = \sum_\nu \lambda_\nu P_{\varphi_\nu}$ (with $\lambda_\nu > 0$ and φ_ν pairwise orthogonal) we get $\langle \varphi_\nu, F \varphi_\nu \rangle = \| F^{1/2} \varphi_\nu \|^2 = 0$, i.e. $F^{1/2} \varphi_\nu = 0$ and hence $F \varphi_\nu = 0$. The projector $\sum_\nu P_{\varphi_\nu} = E$ (here $\lambda_\nu > 0$!) is called the support of W. Therefore, $FE = 0$ holds which easily gives $\operatorname{tr}(WF) = 0$. From $FE = 0$ also follows $FE^\perp = F(\mathbf{1} - E) = F$, hence $E^\perp F = F$ and thus $E^\perp F E^\perp = F$. From $F \leqq \mathbf{1}$ follows $E^\perp F E^\perp \leqq E^\perp$ and hence $F \leqq E^\perp$. Therefore $E^\perp$ is the largest element of $L_0(W)$, so that AV 1.1 holds.

As is easily seen, for each projector E there is a W with support E; therefore $G(\mathscr{H}) = \{e\, L_0(W)\}$ is equal to the set of all projectors. As a trivial result this provides AVid as a theorem.

In order to show AV 2f and AV 4s, let us determine the closed faces of $K(\mathscr{H})$. Each such face has the form $C(W)$. Using $W = \sum_\nu \lambda_\nu P_{\varphi_\nu}$ (with $\lambda_\nu > 0$ and pairwise orthogonal φ_ν), we obtain $P_{\varphi_\nu} \in C(W)$, hence $\frac{1}{n} \sum_{i=1}^{n} P_{\varphi_{\nu_i}} \in C(W)$ for each finite subset $\{\nu_i\}$ of $\{\nu\}$. Since one can write the projector $\sum_{i=1}^{n} P_{\varphi_{\nu_i}} = \sum_{i=1}^{n} P_{\psi_i}$ with any (!) other complete orthogonal system ψ_i in the subspace spanned by the φ_{ν_i}, we also have $P_{\psi_i} \in C(W)$. Thus $P_\psi \in C(W)$ holds for any linear combination ψ of finitely many φ_ν. For $\psi = \sum_\nu \alpha_\nu \varphi_\nu$ with $\sum_\nu |\alpha_\nu|^2 = 1$ and $\psi_N = \left(\sum_{\nu=1}^{N} |\alpha_\nu|^2 \right)^{-1/2} \sum_{\nu=1}^{N} \alpha_\nu \varphi_\nu$ we find $P_{\psi_N} \in C(W)$ and $\| P_{\psi_N} - P_\psi \| \to 0$ (with the norm in $\mathscr{B}(\mathscr{H})$, i.e. the trace norm). Since $C(W)$ is norm closed, we also have $P_\psi \in C(W)$ for all ψ in the projection space of the support E of W. From this in turn follows $W' \in C(W)$ for all W' with a support $E' \leqq E$. Since $E' \leqq E$ is equivalent to $\operatorname{tr}(W' E^\perp) = 0$, we have

$C(W) \supset K_0(E^\perp)$. Since $W \in K_0(E^\perp)$ always makes $C(W) \supset K_0(E^\perp)$, we conclude $C(W) = K_0(E^\perp)$. Hence each closed face of $K(\mathscr{H})$ is exposed, i.e. AV 2s and thus *a fortiori* AV 2f hold as theorems.

With $C(W) = K_0(E^\perp) = K_1(E)$ and an orthogonal system φ_ν with $\sum\limits_\nu P_{\varphi_\nu} = E$, the $K_1(E_n)$ with $E_n = \sum\limits_{\nu=1}^{n} P_{\varphi_\nu}$ are finite dimensional and give $K_1(E) = K_1(\bigvee\limits_n E_n) = \bigvee\limits_n K_1(E_n)$, whereby AV 4s is proven.

Therefore it remains to show AV 3. To this end, let us prove (iii) from VI T 6.2.3, first considering $\Delta(E_1, E_4)$ from VI (6.1.11). One obtains

$$\inf_{W \in K_1(E_1)} \operatorname{tr}(W(\mathbf{1} - E_1)) = \inf_{E_4 \varphi = \varphi} \operatorname{tr}(P_\varphi(\mathbf{1} - E_1)) = \inf_{\substack{E_4 \varphi = \varphi \\ \|\varphi\| = 1}} \|(\mathbf{1} - E_1)\varphi\|^2$$

and correspondingly

$$\inf_{W \in K_1(E_1)} \operatorname{tr}(W(\mathbf{1} - E_4)) = \inf_{\substack{E_1 \varphi = \varphi \\ \|\varphi\| = 1}} \|(\mathbf{1} - E_4)\varphi\|^2 .$$

Making $\Delta(E_1, E_4) \neq 0$ is equivalent to fulfilling at least one of the inequalities

$$\inf_{\substack{E_1 \varphi = \varphi \\ \|\varphi\| = 1}} \|(1 - E_4)\varphi\| \neq 0, \tag{4.2.1}$$

$$\inf_{\substack{E_4 \varphi = \varphi \\ \|\varphi\| = 1}} \|(1 - E_1)\varphi\| \neq 0. \tag{4.2.2}$$

Let us first show: If (4.2.1) or (4.2.2) holds, the set of all vectors $\chi + \eta$ with $E_1 \chi = \chi$ and $E_4 \eta = \eta$ is a closed subspace; hence it is the space on which $E_1 \vee E_4$ projects. Since the set of all $\chi + \eta$ trivially forms a subspace, we need only show that it is closed.

Let $\chi_\nu + \eta_\nu$ be a convergent sequence of vectors, i.e. $\|\chi_\nu + \eta_\nu - \chi_\mu - \eta_\mu\| < \varepsilon$ for $\nu, \mu > N$. Then $\varrho_{\nu\mu} = \chi_\nu - \chi_\mu$ and $\sigma_{\nu\mu} = \eta_\nu - \eta_\mu$ yield $\|\varrho_{\nu\mu} + \sigma_{\nu\mu}\| < \varepsilon$. Let us show $\varrho_{\nu\mu} \to 0$ and $\sigma_{\nu\mu} \to 0$. From the identity

$$\sigma_{\nu\mu} = \frac{\varrho_{\nu\mu}}{\|\varrho_{\nu\mu}\|} \left\langle \frac{\varrho_{\nu\mu}}{\|\varrho_{\nu\mu}\|}, \sigma_{\nu\mu} \right\rangle + \tau_{\nu\mu}$$

follows $\tau_{\nu\mu} \perp \varrho_{\nu\mu}$; hence

$$\varrho_{\nu\mu} + \sigma_{\nu\mu} = \ldots \varrho_{\nu\mu} + \tau_{\nu\mu}$$

yields the estimate $\|\varrho_{\nu\mu} + \sigma_{\nu\mu}\| \geqq \tau_{\nu\mu}$.

Since $(\mathbf{1} - E_1)\sigma_{\nu\mu}$ is the vector of least distance from $\sigma_{\nu\mu}$ to the subspace $E_1 \mathscr{H}$, we obtain

$$\|\tau_{\nu\mu}\| \geqq \|(\mathbf{1} - E_1)\sigma_{\nu\mu}\|$$

and hence

$$\|\tau_{\nu\mu}\| \geqq \|\sigma_{\nu\mu}\| \inf_{\substack{E_4 \varphi = \varphi \\ \|\varphi\| = 1}} \|(\mathbf{1} - E_1)\varphi\| .$$

If (4.2.2) holds, then $\|\varrho_{\nu\mu} + \sigma_{\nu\mu}\| < \varepsilon$ for $\nu, \mu > N$ implies that $\|\sigma_{\nu\mu}\|$ can be chosen arbitrarily small. Hence η_ν is a convergent sequence, and χ_ν also converges.

Conversely, if (4.2.1) is satisfied, one interchanges $\sigma_{\nu\mu}$ with $\varrho_{\nu\mu}$. Since $\eta_\nu \to \eta$ and $\chi_\nu \to \chi$ imply $\chi_\nu + \eta_\nu \to \chi + \eta$, by $\Delta(E_1, E_4) \neq 0$ the projection space given by $E_1 \vee E_4$ equals the set of vectors $\chi + \eta$ with $E_1 \chi = \chi$, $E_4 \eta = \eta$. The representation $\chi + \eta$ of such as vector is unique since $\Delta(E_1, E_4) \neq 0$ implies $E_1 \wedge E_4 = 0$.

Because $E_2 \leqq E_4$ in (iii), from VI T 6.2.3 follows $\Delta(E_1, E_2) \geqq \Delta(E_1, E_4)$, so that $\Delta(E_1, E_2) \neq 0$. Therefore, the subspace on which $E_1 \vee E_2$ projects is the set of all vectors $\tau + \varrho$ with $E_1 \tau = \tau$, $E_2 \varrho = \varrho$.

Therefore, $E_1 \vee E_2 = E_1 \vee E_4$ for a vector χ from $E_4 \mathscr{H}$ implies $\chi = \tau + \varrho$ with $E_1 \tau = \tau$ and $E_2 \varrho = \varrho$. Then $E_2 \leqq E_4$ also makes $E_4 \varrho = \varrho$ and hence $\chi = \tau + \varrho$ with $E_1 \tau = \tau$ and $E_4 \varrho = \varrho$. Hence $E_1 \wedge E_4 = 0$ and $\chi \in E_4 \mathscr{H}$ yield $\tau = 0$ and $\varrho = \chi$, i.e. $\chi = \varrho \in E_2 \mathscr{H}$. Thus we find $E_4 \leqq E_2$ and hence $E_4 = E_2$, whereby (iii) in VI T 6.2.3 is proven.

We have thus proven the important representation theorem

T 4.2.5 From the axioms AV 1.1, AV 1.2s, AV 2f, AVid, AV 3, AV 4s, AV 4a, AV 4b follows that K, L can be decomposed into irreducible parts in the way of VII § 5.4. Unless an irreducible part (K_ν, L_ν) is "trivial" (K_ν has only one element and L_ν is isomorphic to the interval $[0, 1]$ of real numbers), (K_ν, L_ν) can be identified with the sets $K(\mathscr{H}_\nu)$, $L(\mathscr{H}_\nu)$, with $\mathscr{H}_\nu$ a separable infinite-dimensional Hilbert space over the field $\mathbf{C}$. The probability function μ_ν for K_ν, L_ν is then the trace in $\mathscr{H}_\nu$.

By T 4.2.5 we have linked up with the representation of quantum mechanics in [2] (see [2] III § 3). This representation intersects in many points with that given here, since we could here show that not all the theorems derived in [2] require the complete system of the axioms listed in T 4.2.5.

It would be interesting to know whether many other theorems about the elements of K and L, which one uses to derive from the algebraic properties of the operators in $K(\mathscr{H})$ and $L(\mathscr{H})$, also follow without these properties, solely in $\mathscr{B}$ and $\mathscr{B}'$. In § 5 we shall face some of these problems, without being able to solve them in generality.

By the representation theorem T 4.2.5, the algebraic properties of the operators in $\mathscr{B}(\mathscr{H})$ and $\mathscr{B}'(\mathscr{H})$ follow as mathematical consequences of the above axioms for K and L. Whereas these axioms are transparent in their physical meaning, the algebraic structure of the operators is far removed from the original physically interpretable structures of K and L. For instance, the physical meaning in multiplying operators is not recognizable. The historic "correspondence principle" by means of which one inferred quantum mechanics, appears to speak against this view on algebraic structures. But the correspondence principle just consisted in mapping classical, physically interpretable quantities, such as the product of two quantities, in quantum mechanics on mathematical expressions, for which this "correspondence" by no means gives a physically interpretation.

And yet certain algebraic properties of operators often typically characterize physically meaningful relations. Therefore let us yet consider such relations.

§ 4.3 Coexistence, Commensurability, Uncertainty Relations, and Commutability of Operators

Due to VII § 5.4 one can represent the elements of $\mathscr{B}$ and $\mathscr{B}'$ as sequences $x = (x_1, x_2, \ldots)$, where the x_ν are operators in the Hilbert spaces $\mathscr{H}_\nu$. One can then regard these elements of $\mathscr{B}$ and $\mathscr{B}'$ themselves as forming an algebra of operators with the multiplication

$$(x_1, x_2, \ldots)(y_1, y_2, \ldots) = (x_1 y_1, x_2 y_2, \ldots).$$

It is interesting that mathematically equivalent algebraic relations can be given for many physical concepts. Examples are the theorems to follow:

T 4.3.1 For $g \in L$ and $e \in G$, the coexistence of $\{g, e\}$ is equivalent to $g\, e = e\, g$ (the commutability of the operators g and e).

Proof. See [2] IV Th 1.3.1. □

For $g_1 g_2 = g_2 g_1$ we get $g_1 g_2 = g_{12} \in L$ and $g_{12} \leqq g_1$, $g_{12} \leqq g_2$. This gives $g_1 = g_{12} + g_1'$ and $g_2 = g_{12} + g_2'$ with $g_1' = g_1(\mathbf{1} - g_2)$ and $g_2' = g_2(\mathbf{1} - g_1)$, hence $g_{12} + g_1 = g_1 + g_2(\mathbf{1} - g_1) \leqq g_1 + (\mathbf{1} - g_1) = \mathbf{1}$. Thus $g_1 g_2 = g_2 g_1$ makes g_1 and g_2 coexist (V T 1.2.3). But let it be emphasized that two effects g_1, g_2 can coexist even when $g_1 g_2 \neq g_2 g_1$!

T 4.3.2 Two decision effects $e_1, e_2 \in G$ are commensurable if and only if $e_1 e_2 = e_2 e_1$.

Proof. This follows from T 4.3.1 and VI T 4.4.

T 4.3.3 Two decision effects $e_1, e_2 \in G$ are commensurable if and only if $e_1 e_2 = e_1 \wedge e_2$.

Proof. See [2] IV Th 1.3.4.

T 4.3.4 A scale observable (V D 3.1) is uniquely determined by an $a \in \mathscr{B}'$. The corresponding Boolean ring is generated by the spectral family $e(\lambda)$ of a, so that

$$a = \int \lambda \, de(\lambda).$$

Proof. See [2] IV Th 2.5.9.

T 4.3.5 Two scale observables a_1, a_2 are commensurable if and only if $a_1 a_2 = a_2 a_1$.

Proof. See [2] IV Th 3.2.

The mixture components of an ensemble $w \in K$ are just all the elements of $[\mathbf{0}, w]$; they obey

T 4.3.6 Each $\tilde{w} \in [\mathbf{0}, w]$ can be written $\tilde{w} = w^{1/2} g\, w^{1/2}$ with some $g \in L$, while any $g \in L$ makes $w^{1/2} g\, w^{1/2} \in [\mathbf{0}, w]$. The mapping $g \to w^{1/2} g\, w^{1/2}$ is an order isomorphism of $[\mathbf{0}, e]$ onto $[\mathbf{0}, w]$ where e is a decision effect with $K_1(e) = C(w)$.

Proof. See [2] V Th 5.3.

If a is a scale observable while $w \in K$, the expression $\mathrm{Str}(a) = \mu(w, a'^2)$, with $a' = a - \mathbf{1}\,\mu(w, a)$, is the expectation value for the quadratic deviation of a in the ensemble w. Hence one defines

$$\Delta a = [\mathrm{Str}(a)]^{1/2}.$$

Then any two scale observables a, b obey the famous uncertainty relation

$$\Delta a\, \Delta b \geqq \tfrac{1}{2}\,|\mu(w, c)| \quad \text{with} \quad c = i(a\,b - b\,a). \tag{4.3.1}$$

Proof. See [2] IV (8.3.17).

It is recommended that the reader review the experimental meaning of (4.3.1) (see [2] IV § 8.3).

§ 5 Some Theorems for Finite-dimensional and Irreducible $\mathscr{B}$

The contents of this section is in principle not needed. We shall only point out mathematical problems which unfortunately have not been solved except for a finite-dimensional Banach space $\mathscr{B}$. We again assume $\mathscr{B}$ irreducible, since the general reduction has been performed in VII § 5.4.

If $\mathscr{B}$ is finite dimensional, AVid is satisfied as a theorem since K is compact and thus $\mu(w, e)$ must for $e \in G$ attain its supremum on K. Also AV 4s is satisfied since all faces of K are finite dimensional. Thus we need only presume AV 1.1, AV 1.2s, AV 2f and AV 3 where AV 2f just says that all faces of K are exposed.

It is remarkable that many well-known theorems for operators in finite-dimensional Hilbert spaces can already be proved without AV 3 and some even without AV 2f. Therefore we now assume AV 1.1 and AV 1.2s without mentioning this again. Whether AV 2f or AV 3 or both axioms are assumed, will on occasion be stated.

In the finite-dimensional case, solely AV 1.1 and AV 1.2 s are therefore needed, when by means of VI T 3.3 and VI T 3.4 one proves

T 5.1 The following are valid:

(i) $\mathbf{1} - e \in G$ for all $e \in G$;
(ii) $\mathbf{1} - G = G$;
(iii) $e \in G, e \neq 0 \Rightarrow K_0(\mathbf{1} - e) \neq \emptyset$;
(iv) $e_1, e_2 \in G$ and $e_1 \leqq e_2 \Rightarrow e_2 - e_1 \in G$;
(v) all elements of G are exposed points of $L = [\mathbf{0}, \mathbf{1}]$.

Without reformulating them as further theorems, we thus find that VI T 3.5 through VI T 3.11 also hold in this context. One must only specialize that the $\sigma(\mathscr{B}', \mathscr{B})$-topology coincides with the uniquely determined topology for finite-dimensional vector spaces, and that only *finitely* many $e_\nu \in G$ can be pairwise orthogonal, because $K_1(\tilde{e}_n)$ with $\tilde{e}_n = \sum_{\nu=1}^{n} e_\nu$ represents a properly increasing sequence of faces of K.

T 5.2 For all $y \in \mathscr{B}'$, the spectral theorem

$$y = \sum_{\nu} \lambda_\nu e_\nu \tag{5.1}$$

holds with $e_\nu \in G$, $\sum_\nu e_\nu = \mathbf{1}$ and $\lambda_\nu \neq \lambda_\mu$ for $\nu \neq \mu$; and this representation (5.1) is unique.

Proof. Since each $y \in \mathscr{B}$ can be written $y = \alpha \mathbf{1} - \beta g$ with $g \in L$, we need to prove the theorem only for $g \in L$. With $\lambda_1 = \sup \{\mu(w, g) \mid w \in K\}$, we have $\lambda_1^{-1} g \in [\mathbf{0}, \mathbf{1}] = L$. We define e_1 by $K_1(e_1) = K_1(\lambda_1^{-1} g)$. Since K is compact, $\mu(w, g)$ attains its supremum on K, so that $K_1(\lambda_1^{-1} g) \neq \emptyset$, hence $e_1 \neq \mathbf{0}$. From $K_1(e_1) = K_0(\mathbf{1} - e_1) = K_1(\lambda_1^{-1} g) = K_0(\mathbf{1} - \lambda_1^{-1} g)$ follows $\mathbf{1} - \lambda_1^{-1} g \leqq \mathbf{1} - e_1$ and hence $\lambda_1 e_1 \leqq g$, such that $g_1 = g - \lambda_1 e_1 \in L$. With g_1 just as with g, we can define a λ_2 and an e_2. Let us show $\lambda_2 \leqq \lambda_1$ and $e_2 \perp e_1$. Because $g_1 \leqq g$, we have $\lambda_2 \leqq \lambda$. If we had $\lambda_2 = \lambda_1$, a w_0 would exist with $\lambda_1 = \mu(w_0, g_1) = \mu(w_0, g) - \lambda_1 \mu(w_0, e_1)$. Because $\mu(w_0, g) \leqq \lambda_1$, this gives $\mu(w_0, g) = \lambda_1$ and $\mu(w_0, e_1) = 0$. But $\mu(w_0, g) = \lambda_1$ implies $w_0 \in K_1(e_1)$, which contradicts $\mu(w_0, e_1) = 0$.

For $w \in K_1(\lambda_1^{-1} g) = K_1(e_1)$ follows $\mu(w, g_1) = 0$, i.e. $K_1(e_1) \subset K_0(g_1)$. Because $e_2 \leqq \lambda_2^{-1} g_1$, we have $K_0(e_2) \supset K_0(g_1) \supset K_1(e_1) = K_0(e_1^{\perp})$ and hence $e_2 \leqq e_1^{\perp}$.

Thus we recursively obtain (5.1) since after finitely many steps the process must terminate. The uniqueness also follows easily from this recursive process. □

T 5.3 We have $G = \partial_e L$. All extreme points of L are exposed.

Proof. If we show $\partial_e L \subset G$, then the theorem follows from T 5.1 (v).

For $g \in L$ from (5.1), because $\lambda_\nu > \lambda_{\nu+1}$ follows

$$\sup \{\mu(w, g - (\lambda_1 - \lambda_2) e_1) \mid w \in K\} = \lambda_2 \leqq \lambda_2 + (1 - \lambda_1).$$

Therefore $\lambda_1 - \lambda_2 \neq 1$ yields

$$g' = [1 - (\lambda_1 - \lambda_2)]^{-1} [g - (\lambda_1 - \lambda_2) e_1] \in L.$$

Then $g = (\lambda_1 - \lambda_2) e_1 + [1 - (\lambda_1 - \lambda_2)] g'$ shows that g is not an extreme point. Therefore $g \in \partial_e L$ yields $\lambda_1 - \lambda_2 = 1$, i.e. $\lambda_1 = 1$ and $\lambda_2 = 0$ and hence $\lambda_\nu = 0$ for $\nu \geqq 2$, so that $g \in G$ holds. □

This T 5.3 persists for infinite-dimensional G if also the axioms AVid, AV 3, AV 4s and AV 2f hold, since T 5.3 then follows from the representation theorem T 4.2.5 (see [2] III 6.6).

Presuming AV 3, from § 1 we conclude:

Each $e \in G$ is the sum of finite many atoms; G is an atomic, complete, orthocomplemented and modular lattice. For each $e \in G$, a dimension is defined as the number $d(e)$ of orthogonal atoms p_ν which make $e = \sum_\nu p_\nu$.

Then the well-known representation theorems hold for such lattices (see [9]), but here we shall not state them.

Presuming AV 3 and AV 2f, from T 3.1 we conclude that the set $A(G)$ of the atoms of G is a compact subset of L, from T 3.3 that K_1 is a homeomorphic mapping of $A(G)$ onto $\partial_e K$, and from T 3.4 that $A(G)$ is a connected set. From

T 3.5 and T 3.6 follows that G and all sets G_d are compact, when G_d consists of all e having dimension d.

We now assume only AV 1.1, AV 1.2s and AV 2f. For each $e \in G$ ($e \neq 0$), let us show that the two sets $K_e = K_1(e)$ and $L_e = [\mathbf{0}, e]$ satisfy the same relations as K and L do, i.e. that the following theorems hold.

T 5.4 The set of all $\alpha w_1 - \beta w_2$ with $w_1, w_2 \in K_e$ is a base-normed Banach space $\mathscr{B}_e$ with the basis K_e. The Banach space $\mathscr{B}_e'$ dual to $\mathscr{B}_e$ is just the set of all elements $\alpha e + \beta g$ with $g \in L_e$, where e is the order unit of $\mathscr{B}_e'$ and $L_e = [\mathbf{0}, e]$.

Proof. Since $\mathscr{B}$ is finite-dimensional, the subset of all $\alpha w_1 - \beta w_2$ with $w_1, w_2 \in K_e$ is a closed finite-dimensional vector space $\mathscr{B}_e$, where one can introduce a norm with K_e as basis. For $g \in [\mathbf{0}, e]$ and $w \in K_1(e)$, $\mu(w, g)$ is an affine functional on $K_1(e)$.

Since K is compact, $\mu(w, g)$ attains its supremum $\alpha = \sup_{w \in K} \mu(w, g)$. From $g \leqq e$ (as shown more than once above) also follows $\alpha^{-1} g \leqq e$. With $\mu(w_0, g) = \alpha$ we thus get $\mu(w_0, e) = 1$, i.e. $w_0 \in K_1(e)$. Thus the proof of the spectral theorem T 5.2 shows that the spectral representation of $g \in [\mathbf{0}, e]$ is already determined by the values $\mu(w, g)$ with $w \in K_1(e)$. If $\mu(w, g_1) = \mu(w, g_2)$ with $g_1, g_2 \in [\mathbf{0}, e]$ holds for all $w \in K_1(e)$, we then must have $g_1 = g_2$. Hence the elements $g \in [\mathbf{0}, e]$ can be identified with elements of $\mathscr{B}_e'$.

The space spanned by $[\mathbf{0}, e]$ equals $\mathscr{B}_e'$ if the $g \in [\mathbf{0}, e]$ separate the set $K_e = K_1(e)$, i.e. if $\mu(w_1, g) = \mu(w_2, g)$ for $w_1, w_2 \in K_1(e)$ and for all $g \in [\mathbf{0}, e]$ implies $w_1 = w_2$.

Let $w_1 \neq w_2$ and $\mu(w_1, g) = \mu(w_2, g)$ for all $g \in [\mathbf{0}, e]$; then we also have $\mu(w_1', g) = \mu(w_2', g)$ for all $g \in [\mathbf{0}, e]$ and all w_1', w_2' on the intersection of the straight line $\alpha w_1 + \beta w_2$ with $K_1(e)$. This line segment in $K_1(e)$ has two boundary points $\bar{w}_1$ and $\bar{w}_2$ with $\bar{w}_1 \neq \bar{w}_2$, since we had $w_1 \neq w_2$.

We must have $C(\bar{w}_1) \neq C(\bar{w}_2)$; for otherwise the whole line segment between $\bar{w}_1$ and $\bar{w}_2$ would be in $C(\bar{w}_1)$. By V T 9.2, then $\bar{w}_1$ and $\bar{w}_2$ could not be boundary points of the line segment. From $C(\bar{w}_1) \neq C(\bar{w}_2)$ and AV 2f follows $C(\bar{w}_1) = K_1(e_1)$ and $C(\bar{w}_2) = K_1(e_2)$ with $e_1 \neq e_2$. From $\bar{w}_1 \in K_1(e)$ follows $C(\bar{w}_1) \subset K_1(e)$ and hence $e_1 \leqq e$; likewise we get $e_2 \leqq e$. Therefore $\mu(\bar{w}_2, e_1) = \mu(\bar{w}_1, e_1) = 1$ holds in contradiction to our assumption; thus L_e separates the set K_e.

We must yet show that $[-e, e]$ is the unit ball of $\mathscr{B}_e'$. This follows from the spectral representation of a $y \in \mathscr{B}_e'$, because it has $e_\nu \leqq e$ for $\lambda_\nu \neq 0$. □

T 5.5 $\mathscr{B}_e, \mathscr{B}_e'$ satisfy the axiom AV 1.1.

Proof. By AV 1.1, in $\mathscr{B}, \mathscr{B}'$ we find for $g_1, g_2 \in L_e$, that a $g \in L$ exists with $g \geqq g_1$, $g \geqq g_2$ and $K_0(g) \supseteq K_0(g_1) \cap K_0(g_2)$. Because $g_1, g_2 \in [\mathbf{0}, e]$, we have $K_0(g) \supseteq K_0(e)$ and $K_0(g_2) \supseteq K_0(e)$ and hence $K_0(g) \supseteq K_0(e)$, i.e. $g \leqq e$. □

As T 5.4 has already shown, $L_e = [\mathbf{0}, e]$ holds by definition and hence AV 1.2s is satisfied.

T 5.6 $\mathscr{B}_e, \mathscr{B}_e'$ satisfy AV 2f, i.e. each closed face of K_e is exposed relative to $\mathscr{B}_e'$.

Proof. If F is a closed face of K_e, then F is also a closed face of K; hence there is an $e_1 \in G$ with $F = K_1(e_1)$. Since $K_1(e_1) \subset K_1(e)$ implies $e_1 \leqq e$, we find F exposed relative to $\mathscr{B}'_e$. □

From T 5.3 follows that G_e is the set of extreme points of the order interval $[\mathbf{0}, e]$, i.e. $G_e = \{e' \mid e' \in G, e' \leqq e\}$.

T 5.7 If AV 3 holds for $\mathscr{B}, \mathscr{B}'$, then it also holds for $\mathscr{B}_e, \mathscr{B}'_e$.

Proof. For finite-dimensional faces, $\Delta(e_1, e_3) \neq 0$ is equivalent to $e_1 \neq e_3$. The proof follows immediately from (ix) in VI T 6.2.3. □

Henceforth we assume only AV 1.1, AV 1.2s, AV 2f, i.e. do not use AV 3.

Because $\mathscr{B}_e, \mathscr{B}'_e$ satisfy the same axioms AV 1.1, AV 1.2 s and AV 2f, to each $e \in G$ belongs an operator T_e projecting from $\mathscr{B}'$ onto the subspace $\mathscr{B}'_e \subset \mathscr{B}_e$: To $y \in \mathscr{B}'$, a $y_e \in \mathscr{B}'_e$ is uniquely assigned by $\mu(w, y_e) = \mu(w, y)$ for all $w \in K_1(e)$. Writing $y_e = T_e y$, we obtain $T_e^2 = T_e$; hence T_e projects the space $\mathscr{B}'$ onto $\mathscr{B}'_e$.

One sees immediately $T_e[\mathbf{0}, \mathbf{1}] = [\mathbf{0}, e] = L_e$; hence T_e is positive. (In general, we do not have $T_e e_1 \in G$ for $e_1 \in G$!).

Therefore, the mapping T'_e dual to T_e is an operation $K \to \check{K}$. In particular, for $w \in K_e = K_1(e)$ we find that $\mu(T'_e w, g) = \mu(w, T_e g)$ first yields $T'_e w \in K_1(e)$ and then $T'_e w = w$. For an arbitrary $w \in K$, from $T_e \mathbf{1} = e$ follows $\mu(T_e w, \mathbf{1}) = \mu(w, e)$. Since T'_e is also positive, for $\mu(w, e) \neq 0$ we have $\mu(w, e)^{-1} T'_e w \in K_1(e)$. Therefore, T'_e is a projector from $\mathscr{B}$ onto $\mathscr{B}_e$.

Further theorems about the map $e \to T_e$ can be looked up in [24]. Unfortunately, for infinite-dimensional $\mathscr{B}, \mathscr{B}'$ it has not been possible to prove the existence of such operators T_e without AV 3 and AV 4s. Of course, with AV 3, AV 4s one succeeds when a representation by operators is admitted, because then we get $T_e y = e y e$ and $T'_e w = e w e$.

Appendix

A I Some Theorems for Atoms in the Lattice G

We presume all the axioms AV 1.1, AV 1.2s, AVid, AV 3, AV 4s. Therefore G is an atomic orthomodular lattice. Moreover, e_1, e_2 satisfy the modularity relation $M(e_1, e_2)$, provided at least one of them is finite (VI T 6.3.3 and VIII T 1.3). Furthermore the covering condition (VI T 6.3.4) holds.

T 1 Let p and q be two distinct atoms. If there is an $x \in G$ such that $(p \vee x) \wedge q \neq x \wedge q$, then there is an atom r, distinct from p and q, such that $r \leqq p \vee q$.

Proof. Since $x \wedge q = 0$ or $x \wedge q = q$, from $(p \vee x) \wedge q \neq x \wedge q$ follows $x \wedge q = 0$ and $(p \vee x) \wedge q = q$.

If we had $x \wedge p \neq 0$, i.e. $x \wedge p = p$, it would imply $p \leqq x$ and hence $x \vee p = x$. From this would follow $(p \vee x) \wedge q = x \wedge q = 0$, which contradicts $(p \vee x) \wedge q = q$; therefore $x \wedge p = 0$.

We set $r = (p \vee q) \wedge x$. If we had $r = p \vee q$, we would get $p \vee q \leqq x$. From this would follow $x \wedge q = q$, in contradiction to $x \wedge q = 0$. If we had $r = 0$, the modular relation $M(x, b)$ with $b = p \vee q$ would due to $p \leqq b$ imply $(p \vee x) \wedge b = p \vee (x \wedge b) = p \vee r = p$ and hence $(p \vee x) \wedge b \wedge q = p \wedge q = 0$. With $b \wedge q = q$ follows $(p \vee x) \wedge q = 0$, in contradiction to $(p \vee x) \wedge q = q$.

Thus the covering condition makes r an atom distinct from p and q, with $r < p \vee q$. □

D 1 For an atom q, let S_q be the set of all $a \in G$ for which

$$(a \vee x) \wedge q = x \wedge q$$

holds for all $x \in G$.

T 2 $a \in S_q, c \leqq a \Rightarrow c \in S_q$;
$a \in S_q, b \in S_q \Rightarrow a \vee b \in S_q$.

Proof. Let $(a \vee x) \wedge q = x \wedge q$. For $c \leqq a$ follows

$$(c \vee x) \wedge q \leqq (a \vee x) \wedge q = x \wedge q .$$

That $(c \vee x) \wedge q \geqq x \wedge q$ is trivial.

From $a, b \in S_q$ follows $[(a \vee b) \vee x] \wedge q = [a \vee (b \vee x)] \wedge q = (b \vee x) \wedge q = x \wedge q$. □

T 3 If p is an atom from S_q, then $z = \bigvee_{a \in S_q} a$ is a center element with $0 \neq z \neq 1$ and $p \leqq z, q \leqq z^{\perp}$, so that $p \perp q$.

Proof. From $(a \vee x) \wedge q = x \wedge q$ for all $x \in G$, with $x = a^{\perp}$ follows $q = (a \vee a^{\perp}) \wedge q = a^{\perp} \wedge q$. This gives $q \leqq a^{\perp}$ and hence

$$q \leqq \bigwedge_{a \in S_q} a^{\perp} = \left(\bigvee_{a \in S_q} a\right)^{\perp} = z^{\perp}.$$

Because $p \in S_q$, we have in particular $q \leqq p^{\perp}$. From this also follows $0 \neq z \neq 1$.

Suppose z is not a center element. Then there is a $u \in G$ with $u \neq (u \wedge z) \vee (u \wedge z^{\perp})$. For $y = u - [(u \wedge z) \vee (u \wedge z^{\perp})]$, we have $y \neq 0$ and $y \wedge z = 0 = y \wedge z^{\perp}$. Therefore, there also exists an atom r with $r \wedge z = 0 = r \wedge z^{\perp}$. From $r \wedge z = 0$ follows $r \wedge a = 0$ for all $a \in S_q$. If we had $r \leqq a^{\perp}$ for all $a \in S_q$, also $r \leqq z^{\perp}$ would follow. Hence there is an $a \in S_q$ with $r \wedge a^{\perp} = 0$. We now think of $a \in S_q$ as being so chosen that $r \wedge a = 0 = r \wedge a^{\perp}$. From this follows $a \vee r \neq a$. Hence, there is an atom $t \perp a$ with $a \vee t = a \vee r$. We shall show that $b = (a \vee r) \wedge r^{\perp}$ must be an element of S_q.

Immediately follows $a \vee t = a \vee r = b \vee r$. If we had $b \notin S_q$, there would be a y with $(b \vee y) \wedge q \neq y \wedge q$, i.e. with $(b \vee y) \wedge q = q$ and $y \wedge q = 0$. Because $a \in S_q$, we get $(a \vee t \vee y) \wedge q = (t \vee y) \wedge q$. With $(a \vee t \vee y) \wedge q = (b \vee r \vee y) \wedge q \geqq (b \vee y) \wedge q = q$ therefore follows $(t \vee y) \wedge q = q$, i.e. $q \leqq t \vee y$.

Because $t \leqq a^{\perp}$, the modular relation $a^{\perp} \wedge (y \vee t) = t \vee (a^{\perp} \wedge y)$ holds. With $q \leqq a^{\perp}$ this implies $q \leqq t \vee (a^{\perp} \wedge y)$. Because $a \in S_q$, we further obtain $(a \vee t \vee (a^{\perp} \wedge y)) \wedge q = (t \vee (a^{\perp} \wedge y)) \wedge q = q$ and hence $(a \vee t \vee (a^{\perp} \wedge y)) \wedge q = (a \vee r \vee (a^{\perp} \wedge y)) \wedge q = (r \vee (a^{\perp} \wedge y)) \wedge q = q$. With $y' = a^{\perp} \wedge y$, by $y \wedge q = 0$ we also get $y' \wedge q = 0$. Therefore $q \leqq r \vee y'$, $q \nleq y'$; hence the covering condition gives $q \vee y' = r \vee y'$. Because $q \leqq a^{\perp}$ and $y' \leqq a^{\perp}$, we thus have $r \vee y' \leqq a^{\perp}$ and hence $r \leqq a^{\perp}$. Since this contradicts $r \wedge a^{\perp} = 0$, we find $b \in S_q$.

By T 2, from $b \in S_q$ also follows $a \vee b \in S_q$. Because $b \leqq a \vee r$, we have $a \leqq a \vee b \leqq a \vee r$. Therefore the covering condition yields either $a \vee b = a \vee r$ or $a \vee b = a$. If we had $a \vee b = a$, we would have $b \leqq a$ and hence $t \perp b$. Furthermore, $b \leqq a$ would imply $b \leqq b \vee t \leqq a \vee t = b \vee r$. Because $t \perp b$, this would give $b \vee t = b \vee r$. With $t \perp b$ and $r \perp b$, we would obtain $t = r$ and hence $r \perp a$, in contradiction to $r \wedge a^{\perp} = 0$. Thus $a \vee b = a \vee r$ holds and hence $a \vee r \in S_q$. From this follows $a \vee r \leqq z$ and hence $r \leqq z$, which contradicts $r \wedge z = 0$.

Therefore, z is a center element. □

T 4 If z is a center element and q an atom with $q \leqq z^{\perp}$, they yield $[0, z] \subset S_q$.

Proof. For each $x \in G$ we get $x = (x \wedge z) \vee (x \wedge z^{\perp})$. This implies $z \vee x = z \vee (x \wedge z^{\perp})$; hence $q \leqq z^{\perp}$ gives $(z \vee x) \wedge q = (x \wedge z^{\perp}) \wedge q = x \wedge q$, therefore $z \in S_q$. By T 2 then follows $[0, z] \subset S_q$. □

T 5 If $z = \bigvee_{a \in S_q} a$, then $S_q = [0, z]$ and $z^{\perp}$ is that atom of the center which contains q.

Proof. By T 4 and T 3 we have $[0, z] \subset S_q$. By T 2, from $a \in S_q$ follows $a \vee z \in S_q$ and hence $a \vee z \leqq z$, i.e. $a \leqq z$. If u is that atom of the center which contains q, then

T 4 gives $[0, u^{\perp}] \subset S_q$. But there cannot be any larger center element z than $z = u^{\perp}$, such that $q \leqq z^{\perp}$. □

T 6 If G is irreducible and p, q are two distinct atoms in G, then there is a third atom r distinct from p, q with $r \leqq p \vee q$.

Proof. If G is irreducible, T 3 yields $S_q = \{0\}$. Hence there is an $x \in G$ such that $(p \vee x) \wedge q \neq x \wedge q$. By T 1, the desired atom r then exists. □

T 7 If G is irreducible, the order interval $[0, e]$ is irreducible for each $e \in G$.

Proof. If $[0, e]$ were not irreducible, there would exist a center element z of $[0, e]$ such that $0 \neq z \neq e$. For all $a \leqq e$ we would then have $a = (a \wedge z) \vee [a \wedge (x^{\perp} \wedge e)]$. Therefore, an atom $r \leqq e$ obeys either $r \leqq z$ or $r \leqq z^{\perp} \wedge e$.

Let p be an atom with $p \leqq z$ and q an atom with $q \leqq z^{\perp} \wedge e$. For an atom $r \leqq p \vee q$ follows $r \leqq z$ or $r \leqq z^{\perp}$.

For $r \leqq z$ we would have $r = (p \vee q) \wedge r = p \wedge r$, i.e. $r = p$. For $r \leqq z^{\perp}$ we would likewise get $r = q$. Hence no atom in $p \vee q$ would differ from p, q. Thus by T 6, the lattice G would not be irreducible. □

A II Banach Lattices

D 1 An ordered vector space is called a *vector lattice* if the upper and lower bounds exist for any two elements x, y. Let them be denoted by $x \vee y$ and $x \wedge y$.

T 1 If R is a vector lattice, then

$$x + y = (x \wedge y) + (x \vee y).$$

Proof. Since $z \geqq x$ implies $-z \leqq -x$, we get

$$x \vee y = -[(-x) \wedge (-y)]$$

and

$$x \wedge y = -[(-x) \vee (-y)].$$

Also, $z \geqq x$ implies $z + y \geqq x + y$, hence

$$z + (x \vee y) = (z + x) \vee (z + y).$$

Summarizing, from this follows

$$z - (x \wedge y) = (z - x) \vee (z - y),$$

which for $z = x + y$ becomes

$$x + y - (x \wedge y) = x \vee y. \quad \square$$

D 2 A base-normed Banach space $\mathscr{B}$ has the *Riesz decomposition property* if $[0, x_1] + [0, x_2] = [0, x_3]$. We say briefly: $\mathscr{B}$ is a Riesz space.

T 2 If $\mathscr{B}$ is a Riesz space, $\mathscr{B}'$ is a vector lattice.

Proof. For $l(x) = \langle x, y \rangle$ with $x \in \mathscr{B}_+$ and $y \in \mathscr{B}'$, let us define

$$l_+(x) = \sup \{l(x') \mid x' \in [0, x]\},$$
$$l_-(x) = \sup \{l(x') \mid x' \in [-x, 0]\}.$$

We shall first show that l_+ and l_- are elements of $\mathscr{B}'$. We show this for l_+ since

$$l_-(x) = \sup \{\tilde{l}(x') \mid x' \in [0, x]\},$$

where $\tilde{l}(x) = l(-x) = -l(x)$.

For $\lambda > 0$ and $x \in \mathscr{B}_+$ follows $l_+(\lambda x) = \lambda l_+(x)$. For $x_1, x_2 \in \mathscr{B}_+$, because $[0, x_1] + [0, x_2] = [0, x_1 + x_2]$ follows $l_+(x_1 + x_2) = l_+(x_1) + l_+(x_2)$. Therefore l_+ is an affine functional in $\mathscr{B}_+$. Because $\mathscr{B} = \mathscr{B}_+ - \mathscr{B}_+$, we can extend l_+ as a linear functional to all of $\mathscr{B}$.

With the basis K, we have

$$\sup_{x \in K} |l_+(x)| = \sup_{x' \in \check{K}} |l(x')| = \|y\|,$$

so that $l_+ \in \mathscr{B}'$ and $\|l_+\| \leqq \|y\|$ hold. For this reason we write $l_+ = y_+$ and $l_- = y_-$.

In order to show that $\mathscr{B}'$ is a vector lattice, we need only to show that a supremum exists of 0 and y, since then the supremum of y_1, y_2 equals $y_1 + \sup \{0, y_2 - y_1\}$. Similarly we obtain $\inf(y_1, y_2) = -\sup \{(-y_1), (-y_2)\}$. We now assert $\sup \{0, y\} = y_+$. Immediately from the definition of l_+ follow $y_+ \geqq 0$ and $y_+ \geqq y$. If $y_1 \geqq 0, y$, for $x \in \mathscr{B}_+$ we get

$$\langle x, y_1 \rangle \geqq \langle x, y \rangle = l(x).$$

This gives

$$\langle x', y_1 \rangle \geqq l(x') \quad \text{for} \quad x' \in [0, x],$$

whence $y_1 \geqq y_+$ follows. □

A III The Axiom AVid and the Minimal Decomposition Property

Here let us assume that $\mathscr{B}$ has the *minimal decomposition property*, i.e. each $x \in \mathscr{B}$ is representable as $x = \alpha w_1 - \beta w_2$ with $w_1, w_2 \in K$ and $\|x\| = \alpha + \beta$. Then the following two theorems hold:

T 1 Every exposed point of L is an element of G.

Proof. If g_e is an exposed point of L, then $2 g_e - \mathbf{1}$ is an exposed point of the unit ball $[-\mathbf{1}, \mathbf{1}]$ of $\mathscr{B}'$. Therefore, there is an $x \in \mathscr{B}$ such that $\|x\| = 1$ and $\mu(x, 2 g_e - \mathbf{1}) = 1$ and $\mu(x, 2 g - \mathbf{1}) \neq 1$ for $g \neq g_e$. With the minimal decomposition $x = \alpha w_1 - \beta w_2$ $(\alpha + \beta = 1)$ follows

$$\mu(x, 2 g - \mathbf{1}) = \alpha [2 \mu(w_1, g) - 1] - \beta [2 \mu(w_2, g) - 1].$$

Because $\alpha + \beta = 1$, from $\mu(x, 2 g - \mathbf{1}) = 1$ follows

$$\mu(w_1, g) = 1 \quad \text{and} \quad \mu(w_2, g) = 0.$$

From $\mu(w_2, g) = 0$ follows $g \in L_0(w_2)$ and hence $g \leqq e\,L_0(w_2)$. We briefly write $e_2 = e\,L_0(w_2)$. From $\mu(w_1, g) = 1$ follows $\mathbf{1} - g \in L_0(w_1)$ and thus $\mathbf{1} - g \leqq e\,L_0(w_1) = e_1$. Therefore, $\mathbf{1} - e_1 \leqq g \leqq e_2$.

Conversely, from $\mathbf{1} - e_1 \leqq g \leqq e_2$ also follows $\mu(x, 2g - \mathbf{1}) = 1$. Therefore, g_e is an exposed point if and only if there is a unique element $g \in L$ such that $\mathbf{1} - e_1 \leqq g \leqq e_2$. Hence we have $\mathbf{1} - e_1 = g_e = e_2$, i.e. $g_e \in G$. □

T 2 AVid is equivalent to the fact that G is the set of exposed points of L.

Proof. This follows immediately from T 1 and VI T 3.4.

A IV The Bishop-Phelps Theorem and the Ellis Theorem

In order to spare the reader the trouble of referring to the journals, we shall prove two theorems from [28] and [29].

We proceed from a Banach space $\mathscr{B}$ and consider a convex set K in $\mathscr{B}$. If $x_0 \in K$ and C is a convex cone, we call $x_0 + C$ a *support cone* of K at the point x_0 if $(x_0 + C) \cap K = \{x_0\}$. If C has an interior point, by the separation theorem (see [7] II § 9.1) there is hyperplane H which separates K and $x_0 + C$, i.e. there is a $y \in \mathscr{B}'$ and an α such that $K \subset \{x \mid \langle x, y\rangle \geqq \alpha\}$, $x_0 + C \subset \{x \mid \langle x, y\rangle \leqq \alpha\}$. From this follows $\langle x_0, y\rangle = \alpha$ and $\alpha = \inf_{x \in K} \langle x, y\rangle$. If K is closed, $\{x \mid x \in K, \mu(x, y) = \alpha\}$ is a non-empty, closed face of K. y is a support functional of K. Therefore, in order to find support functionals and closed faces of the closed convex set K it suffices to find a support cone with interior points.

D 1 For $y \in \mathscr{B}'$ with $\|y\| = 1$, and for $k > 0$, let

$$C(y, k) = \{x \mid \|x\| \leqq k\,\langle x, y\rangle\}.$$

This $C(y, k)$ is a closed convex cone. If $k > 1$, then $C(y, k)$ has interior points. In order to show this, we choose an x_1 with $\|x_1\| = 1$ and $\langle x_1, y\rangle > k^{-1} > 1$, which is possible since $\|y\| = 1$. From this follows $\|x_1\| < k\,\langle x_1, y\rangle$. Since the norm $\|x\|$ and the linear functional $\langle x, y\rangle$ are continuous, there is a neighborhood U of x_1 with $\|x\| < k\,\langle x, y\rangle$ for all $x \in U$.

T 1 Let K be a closed, convex subset of $\mathscr{B}$. Let $y \in \mathscr{B}'$ with $\|y\| = 1$ and $\langle x, y\rangle$ bounded on K. Let $k > 0$. For each point $z \in K$, there is an $x_0 \in K$ such that $x_0 \in z + C(y, k)$, and $x_0 + C(y, k)$ is a support cone of K.

Proof. Let $<_C$ be the order on B defined by $C(y, k)$. If I is a totally $<_C$ ordered subset of the closed set $[z + C(y, k)] \cap K$, then $\beta = \sup_{x \in I} \langle x, y\rangle \neq \infty$, since $\langle x, y\rangle$ is bounded on K. Since I is totally ordered, for any two elements $x_1, x_2 \in I$ we have either $x_1 - x_2 \in C(y, k)$ or $x_2 - x_1 \in C(y, k)$. For $x_2 - x_1 \in C(y, k)$ we have $\|x_2 - x_1\| \leqq k\langle x_2 - x_1, y\rangle$. Thus, there is a Cauchy sequence x_ν in I such that $x_\nu \to x' \in [z + C(y, k)] \cap K$ and $\langle x_\nu, y\rangle \to \beta$. For each $x \in I$ we have $x <_C x'$: For, if we had $x_\nu <_C x$ for all x_ν, they would obey $\| x - x_\nu \| \leqq k \langle x - x_\nu, y \rangle = k [\langle x, y \rangle - \langle x_\nu, y \rangle]$. This would yield $\| x - x_\nu \| < \varepsilon$ for sufficiently large ν, i.e. $x = x'$. Since l has a supremum in $[z + C(y, k)] \cap K$, by Zorn's lemma there is a maximal element x_0 in $[z + C(y, k)] \cap K$, i.e. $[x_0 + C(y, k)] \cap K = \{x_0\}$. □

T 2 Let $y_1, y_2 \in \mathscr{B}'$ with $\| y_1 \| = \| y_2 \| = 1$ and $\langle x, y_2 \rangle \leqq \frac{\varepsilon}{2}$ with $\varepsilon > 0$ for all x with $\langle x, y_1 \rangle = 0$ and $\| x \| \leqq 1$. Then either $\| y_1 + y_2 \| \leqq \varepsilon$ or $\| y_1 - y_2 \| \leqq \varepsilon$.

Proof. By the Hahn-Banach theorem, one can find a $z \in \mathscr{B}'$ which on the hyperplane $\{x | \langle x, y_1 \rangle = 0\}$ coincides with y_2 and has the norm $\| z \| = \sup \{ | \langle x, y_2 \rangle | \langle x, y_1 \rangle = 0$ and $\| x \| \leqq 1\}$. Therefore, by assumption, $\| z \| \leqq \frac{\varepsilon}{2}$. Since $\langle x, y_2 - z \rangle = 0$ if $\langle x, y_1 \rangle = 0$, we have $y_2 - z = \alpha y_1$ for a suitable number α. From this follows $\| y_2 - \alpha y_1 \| \leqq \| z \| \leqq \frac{\varepsilon}{2}$.

Let $\alpha \geqq 0$. For $\alpha \geqq 1$, we have $\alpha^{-1} \leqq 1$ and $\| y_2 - y_1 \| \leqq \| y_2 - \alpha^{-1} y_2 \| + \| \alpha^{-1} y_2 - y_1 \| \leqq 1 - \alpha^{-1} + \alpha^{-1} \| y_2 - \alpha y_1 \| \leqq 1 - \alpha^{-1} + \frac{\varepsilon}{2}$. With $\alpha = \| \alpha y_1 \| \leqq \| y_2 \| + \| \alpha y_1 - y_2 \| = 1 + \| y_2 - \alpha y_1 \|$ follows $1 - \alpha^{-1} = 1 - (1 + \| y_2 - \alpha y_1 \|)^{-1} = (1 + \| y_2 - \alpha y_1 \|)^{-1} \| y_2 - \alpha y_1 \| \leqq \| y_2 - \alpha y_1 \| \leqq \frac{\varepsilon}{2}$. We thus finally obtain $\| y_2 - y_1 \| \leqq \varepsilon$.

For $0 \leqq \alpha \leqq 1$ follows

$$\begin{aligned} \| y_2 - y_1 \| &\leqq \| y_2 - \alpha y_1 \| + (1 - \alpha) \| y_1 \| \leqq \tfrac{\varepsilon}{2} + (1 - \alpha) \\ &= \tfrac{\varepsilon}{2} + \| y_2 \| - \| \alpha y_1 \| \leqq \tfrac{\varepsilon}{2} + \| y_2 - \alpha y_1 \| \leqq \varepsilon . \end{aligned}$$

If $\alpha \leqq 0$, one replaces y_1 by $(- y_1)$ and thus obtains $\| y_2 + y_1 \| \leqq \varepsilon$. □

T 3 Let $y_1, y_2 \in \mathscr{B}'$ with $\| y_1 \| = \| y_2 \| = 1$, and $k > 1 + \frac{2}{\varepsilon}$ with $0 < \varepsilon < 1$. If y_2 is not negative on $C(y_1, k)$, then $\| y_1 - y_2 \| \leqq \varepsilon$.

Proof. One can choose an x_1 with $\| x_1 \| = 1$ so that $\langle x_1, y_1 \rangle > k^{-1} (1 + \frac{2}{\varepsilon}) < 1$. Let x_2 be chosen in $\{x | \langle x, y \rangle = 0\}$ so that $\| x_2 \| \leqq \frac{2}{\varepsilon}$. Then follows $\| x_1 \pm x_2 \| \leqq 1 + \frac{2}{\varepsilon} < k \langle x_1, y_1 \rangle = k \langle x_1 + x_2, y_1 \rangle$, therefore $x_1 \pm x_2 \in C(y_1, k)$ and hence by assumption $\langle x_1 \pm x_2, y_2 \rangle \geqq 0$. From this follows $| \langle x_2, y_2 \rangle | \leqq \langle x_1, y_2 \rangle \leqq \| x_1 \| = 1$. For $\langle x, y_1 \rangle = 0$ and $\| x \| = 1$ one can choose $x_2 = \frac{2}{\varepsilon} x$ and obtain $| \langle \frac{2}{\varepsilon} x, y_2 \rangle | \leqq 1$, i.e. $| \langle x, y_2 | \leqq \frac{\varepsilon}{2}$. Therefore T 2 yields $\| y_1 + y_2 \| \leqq \varepsilon$ or $| y_1 - y_2 \| \leqq \varepsilon$. In order to exclude $\| y_1 + y_2 \| \leqq \varepsilon$, we choose an $x_3 \in \mathscr{B}$ with $\| x_3 \| = 1$ and $\langle x_3, y_1 \rangle > \max \{k^{-1}, \varepsilon\}$. From this follows $x_3 \in C(y_1, k)$. By assumption, $\langle x, y_2 \rangle \geqq 0$; so we conclude $\| y_1 + y_2 \| \geqq \langle x_3, y_1 + y_2 \rangle \geqq \langle x_3, y_1 \rangle > \varepsilon$. □

T 4 Let K be a closed convex set in $\mathscr{B}$, let M be a compact, convex set in $\mathscr{B}$. Let $y_1 \in \mathscr{B}'$ with $\| y_1 \| = 1$ and $\sup_{x \in K} \langle x, y_1 \rangle < \inf_{x \in M} \langle x, y_1 \rangle$ (by the separation theorem, there always exists such a y_1 if $K \cap M = \emptyset$; see [7] II § 9.2). Then, for each $\varepsilon > 0$, there exist a $y_2 \in \mathscr{B}'$ and an $x_0 \in K$ with $\| y_1 - y_2 \| \leqq \varepsilon$ and $\langle x_0, y_2 \rangle = \sup_{x \in K} \langle x, y_2 \rangle < \inf_{x \in M} \langle x, y_2 \rangle$.

Proof. For brevity we set $\gamma = \sup_{x \in K} \langle x, y_1 \rangle$, $\delta = \inf_{x \in M} \langle x, y_1 \rangle$. We choose a β with $\gamma < \beta < \delta$. With $\mathscr{B}_{|1|}$ as the unit ball of $\mathscr{B}$, we find that $N = M + (\delta - \beta) \mathscr{B}_{|1|}$ is a neighborhood of M. Since M was assumed compact, N is bounded.

Because $\inf_{x \in \mathscr{B}_{|1|}} \langle x, y_1 \rangle = -1$, we have

$$\inf_{x \in N} \langle x, y_1 \rangle = \inf_{x \in M} \langle x, y_1 \rangle - (\delta - \beta) = \beta.$$

With $\alpha = 1 + \frac{2}{\varepsilon}$, we choose an $x_1 \in K$ with $\gamma - \langle x_1, y_1 \rangle < (2\alpha)^{-1}(\beta - \gamma)$. Let μ be greater than $2^{-1}(\beta - \gamma)$ and $\sup_{x \in N} \| x - x_1 \|$. With $k = 2\alpha\mu(\beta - \gamma)^{-1}$, we have $k > \alpha > 1$. By T 1, there is a point x_0 in K such that $x_0 + C(y_1, k)$ is a support cone of K at the point x_0 and $x_0 - x_1 \in C(y_1, k)$ holds. We have $N \subset x_0 + C(y_1, k)$, because $x \in N$ makes $\| x - x_0 \| \leqq \| x - x_1 \| + \| x_1 - x_0 \| < \mu + \| x_0 - x_1 \| \leqq \mu + k \langle x_0 - x_1, y_1 \rangle \leqq \mu + k(\gamma - \langle x_1, y_1 \rangle) < \mu + k(2\alpha)^{-1}(\beta - \gamma) = 2\mu < 2\alpha\mu = k(\beta - \gamma) \leqq k \langle x - x_0, y \rangle$. By the separation theorem, there is a $y_2 \in B'$ with

$$\sup_{x \in K} \langle x, y_2 \rangle \leqq \inf_{x \in x_0 + C(y_1, k)} \langle x, y_2 \rangle.$$

Since $x_0 + C(y_0, k)$ is a support cone of K at x_0, we have $\sup_{x \in K} \langle x, y_2 \rangle = \langle x_0, y_2 \rangle$. Moreover,

$$\begin{aligned} \langle x_0, y_2 \rangle &\leqq \inf_{x \in x_0 + C(y_1 k)} \langle x, y_2 \rangle \leqq \inf_{x \in N} \langle x, y_2 \rangle \\ &= \inf_{x \in M} \langle x, y_2 \rangle - (\delta - \beta) < \inf_{x \in M} \langle x, y_2 \rangle. \end{aligned}$$

Since

$$\inf_{x \in C(y_1, k)} \langle x, y_2 \rangle = \inf_{x \in x_0 + C(y_1, k)} \langle x, y_2 \rangle - \langle x_0, y_2 \rangle \geqq 0$$

and $k > \alpha = 1 + \frac{2}{\varepsilon}$, from T 2 follows $\| y_1 - y_2 \| \leqq \varepsilon$. □

As a special case of T 4 we immediately obtain:

T 5 If the convex set K is closed and bounded in $\mathscr{B}$, then the set of those $y \in \mathscr{B}'$ for which $\langle x, y \rangle$ attains its supremum on K, is norm-dense in $\mathscr{B}'$.

In a base-normed Banach space $\mathscr{B}$, one can apply the T 4 just derived not only to the basis K. One also can investigate the problem of the minimal decomposition of an $x \in \mathscr{B}$. As defined in IV § 4, by a minimal decomposition of x one understands a decomposition $x = \alpha w_1 - \beta w_2$ with $w_1, w_2 \in K$ and $\| x \| = \alpha + \beta$. One can also express this by: $x = x_1 - x_2$, where $x_1, x_2 \in \mathscr{B}_+$ and $\| x \| = \| x_1 \| + \| x_2 \|$. As a *positive* decomposition of x let us denote a general decomposition $x = x_1 - x_2$ with $x_1, x_2 \in \mathscr{B}_+$. According to IV T 4.6, every x has a positive decomposition. The set of x_1 for which $x = x_1 - x_2 = x_1 - (x_1 - x)$ is a positive decomposition, is $\mathscr{B}_+ \cap [x + \mathscr{B}_+]$.

The cone $\mathscr{B}'_+$ has interior points (in the norm) since $\mathscr{B}'$ is an order unit space. If y is an interior point of $\mathscr{B}'_+$, then $K_y = \{x \mid x \in \mathscr{B}_+ \text{ and } \langle x, y \rangle = 1\}$ is a basis of $\mathscr{B}_+$. Then $x \in \mathscr{B}_+$ and so $\langle x, y \rangle = \lambda > 0$; hence $\lambda^{-1} x \in K_y$. Therefore the initial basis K of $\mathscr{B}$ equals K_1.

One can define a norm $\| x \|_y$ with K_y instead of K, namely as a gauge functional of the set $\mathrm{co}(K_y \cup -K_y)$; for this see IV T 4.8. We also have $\| x \|_y = \sup \{ | \langle x, \tilde{y} \rangle | \mid -y \leqq \tilde{y} \leqq y \}$. From $\| y \|^{-1} y \in [-\mathbf{1}, \mathbf{1}]$ follows $\| x \|_y \leqq \| y \| \, \| x \|$. Since y is an interior point of $\mathscr{B}'_+$, there is an $\varepsilon > 0$ with $y - \varepsilon \mathbf{1} \in \mathscr{B}'_+$, i.e. $\varepsilon^{-1} y \geqq \mathbf{1}$. From this follows $\| x \|_y \geqq \varepsilon \| x \|$. Therefore the norms $\| x \|_y$ and $\| x \|$ are equivalent. Hence $\mathscr{B}$ is also a Banach space relative to the norm $\| x \|_y$.

T 6 An $x \in \mathscr{B}$ has

$$\| x \| = 2 \inf \{\langle z, \mathbf{1} \rangle \mid z \in \mathscr{B}_+ \cap [x + \mathscr{B}_+]\} - \langle x, \mathbf{1} \rangle.$$

A minimal decomposition of x exists if and only if $\mathbf{1}$ is a support functional on $\mathscr{B}_+ \cap [x + \mathscr{B}_+]$. With $x_1 \in \mathscr{B}_+ \cap [x + \mathscr{B}_+]$ and

$$\langle x_1, \mathbf{1} \rangle = \inf \{\langle z, \mathbf{1} \rangle \mid z \in \mathscr{B}_+ \cap [x + \mathscr{B}_+]\},$$

we find the minimal decomposition $x = x_1 - (x_1 - x)$.

Proof. By IV T 4.8 we get

$$\| x \| = \inf \{\| z \| + \| z - x \| \mid z \in \mathscr{B}_+ \cap [x + \mathscr{B}_+]\}.$$

Because $z, z - x \in \mathscr{B}_+$, we have $\| z \| = \langle z, \mathbf{1} \rangle$ and $\| z - x \| = \langle z - x, \mathbf{1} \rangle$. Thus follows

$$\begin{aligned} \| x \| &= \inf \{\langle 2z - x, \mathbf{1} \rangle \mid z \in \mathscr{B}_+ \cap [x + \mathscr{B}_+]\} \\ &= 2 \inf \{\langle z, \mathbf{1} \rangle \mid z \in \mathscr{B}_+ \cap [x + \mathscr{B}_+]\} - \langle x, \mathbf{1} \rangle. \end{aligned}$$

A positive decomposition $x = x_1 - (x_1 - x)$ is minimal if and only if $\| x \| = \| x_1 \| + \| x_1 - x \|$, i.e. if and only if

$$\langle x, \mathbf{1} \rangle = \inf \{\langle z, \mathbf{1} \rangle \mid z \in \mathscr{B}_+ \cap [x + \mathscr{B}_+]\} - \langle x, \mathbf{1} \rangle. \quad \square$$

This T 6 also holds if one replaces K by K_y and $\| x \|$ by $\| x \|_y$, where y is an interior point of $\mathscr{B}_+$. Therefore, x has a minimal decomposition with respect to the basis K_y if and only if y is a support functional of $\mathscr{B}_+ \cap [x + \mathscr{B}_+]$.

Let E_x denote the set of those y in the interior of $\mathscr{B}_+$ which are support functionals on $\mathscr{B}_+ \cap [x + \mathscr{B}_+]$. Then we find

T 7 E_x is norm-dense in $\mathscr{B}'_+$.

Proof. The set $\mathscr{B}_+ \cap [x + \mathscr{B}_+]$ is closed and convex. Let y_1 be an interior point of $\mathscr{B}'_+$. Then

$$\inf \{\langle x', y_1 \rangle \mid x' \in \mathscr{B}_+ \cap [x + \mathscr{B}_+]\}$$

exists. By T 4, for each $\varepsilon > 0$ there is a support functional y_2 of $\mathscr{B}_+ \cap [x + \mathscr{B}_+]$ with $\| y_1 - y_2 \| \leqq \varepsilon$. $\square$

Bibliography

1. G. Ludwig: *Einführung in die Grundlagen der theoretischen Physik*, 4 Vols. (Vieweg, Braunschweig 1974–1979)
2. G. Ludwig: *Foundations of Quantum Mechanics I*, Texts and Monographs in Phys. (Springer, Berlin, Heidelberg, New York, Tokyo 1983); Foundations of Quantum Mechanics II, Texts and Monographs in Phys. (Springer, Berlin, Heidelberg, New York, Tokyo 1985)
3. G. Ludwig: *Grundstrukturen einer physikalischen Theorie* (Springer, Berlin, Heidelberg, New York 1978)
4. G. Ludwig: Makroskopische Systeme und Quantenmechanik. Notes Math. Phys. *5* (Marburg 1972)
5. Bourbaki: *Topologie generale* (Herrmann, Paris 1961)
6. G. Ludwig: "A Theoretical Description of Single Microsystems", in: *The Uncertainty Principle and Foundations of Quantum Mechanics*, ed. by W. C. Price, S. S. Chissik (Wiley, New York 1977)
7. H. H. Schaefer: *Topological Vector Spaces* (Macmillan, New York 1966)
8. N. Dunford, J. T. Schwartz: *Linear Operators* (Interscience, New York 1958)
9. V. S. Varadarajan: *Geometry of Quantum Theory*, Vol. 1 (Von Nostrand, Princeton, NJ 1968)
10. G. Köthe: *Topologische lineare Räume* (Springer, Berlin Göttingen, Heidelberg 1960)
11. A. Hartkämper, H. Neumann: *Foundations of Quantum Mechanics and Ordered Linear Spaces*, Lecture Notes, Vol. 29 (Springer, Berlin, Heidelberg, New York 1974)
12. A. Kolmogoroff: *Grundbegriffe der Wahrscheinlichkeitsrechnung* (Springer, Berlin 1933)
13. N. Zierler: Axioms for non-relativistic quantum mechanics. Pac. J. Math. *11*, 1151 (1961)
14. N. Zierler: On the lattice of closed subspaces of Hilbert-space. Pac. J. Math. *19*, 583 (1966)
15. M. D. Maclaren: Atomic orthocomplemented lattices. Pac. J. Math. *14*, 597 (1964)
16. M. D. MacLaren: Notes on axioms for quantum mechanics. ANL-7065 (1965)
17. E. Weiss, N. Zierler: Locally compact division rings. Pac. J. Math. *8*, 369 (1958)
18. L. S. Pontrjagin: *Topologische Gruppen*, Teil 1 (Teubner, Leipzig 1957)
19. G. Birkhoff, J. v. Neumann: The logic of quantum mechanics. Ann. Math. *37*, 823 (1936)
20. C. Piron: "Axiomatic Quantic", Ph. D. Thesis, Université de Lausanne, Faculté des Sciences (Birkhäuser, 1964)
21. M. Jammer: *The Conceptual Development of Quantum Mechanics* (McGraw-Hill, New York 1966)
    M. Jammer: *The Philosophy of Quantum Mechanics* (Wiley, New York 1977)
    E. Scheibe: *The Logical Analysis of Quantum Mechanics*, (Pergamon, New York 1973)
22. L. Kanthack: in Vorbereitung
23. K. Drühl: A theory of classical limit for quantum theories which are defined by real Lie Algebras. J. Math. Phys. *19*, 1600 (1978)
24. G. Dähn: Attempt of an axiomatic foundation of quantum mechanics IV. Commun. Math. Phys. *9*, 192 (1968)
25. J. v. Neumann: *Mathematische Grundlagen der Quantenmechanik* (Springer, 1932); [English transl.: *Mathematical Foundation of Quantum Mechanics*, translated by R. T. Beyer (Princeton NJ, 1955)]
26. G. Ludwig: Meß- und Präparierprozesse. Notes Math. Phys. *6* (Marburg 1972)

27. H. Neumann: *A Mathematical model for a set of microsystems.* Int. J. Theor. Phys. *17,* 3 (1978)
H. Neumann: „Zur Verdeutlichung der statistischen Interpretation der Quantenmechanik durch ein mathematisches Modell für eine Menge von Mikrosystemen", in: *Grundlagen der Quantentheorie,* ed. by P. Mittelstaedt, J. Pfarr (B. I. Wissenschaftsverlag, Mannheim 1980)
28. E. Bishop, R. R. Phelps: The support functionals of a convex set. Proc. Symp. Pure Math. *7* 27 (1963)
29. A. J. Ellis: Minimal decompositions in partially ordered normed vector spaces. Proc. Cambridge Philos. Soc. *64,* 989 (1968)
30. G. Ludwig: Axiomatische Basis einer physikalischen Theorie und Theoretische Begriffe. Z. allg. Wissenschaftstheorie *12* (1), 55 (1981)
31. F. Riesz, B. Sz. Nagy: *Vorlesungen über Funktionalanalysis* (Deutscher Verlag der Wissenschaften, Berlin 1956)
32. H. Neumann: A new physical characterization of classical systems in quantum mechanics. Int. J. Theor. Phys. *9,* 225 (1974)
33. H. Neumann: Classical systems and observables in quantum mechanics. Commun. Math. Phys. *23,* 100 (1971)
34. P. R. Halmos: *Measure Theory* (Van Nostrand, Princeton, NJ 1950)
35. G. Ludwig: Der Meßprozeß. Z. Phys. *135,* 483 (1953)
G. Ludwig: Zur Deutung der Beobachtung in der Quantenmechanik. Phys. Bl. *11,* 489 (1955)
G. Ludwig: Zum Ergodensatz und zum Begriff der makroskopischen Observablen. Z. f. Naturforsch. *A 12,* 662 (1957)
G. Ludwig: Zum Ergodensatz und zum Begriff der makroskopischen Observablen I. Z. Phys. *150,* 346 (1958)
G. Ludwig: Zum Ergodensatz und zum Begriff der makroskopischen Observablen II, Z. Phys. *152,* 98 (1958)
G. Ludwig: "Axiomatic Quantum Statistics of Macroscopic Systems", in *Ergodic Theories,* ed. by P. Caldirola (Academic, New York 1960), p. 57
G. Ludwig: „Gelöste und ungelöste Probleme des Meßprozesses in der Quantenmechanik", in: *Werner Heisenberg und die Physik unserer Zeit* (Vieweg, Braunschweig 1961), p. 150
G. Ludwig: Zur Begründung der Thermodynamik auf Grund der Quantenmechanik. Z. Phys. *171,* 476 (1963)
G. Ludwig: Zur Begründung der Thermodynamik auf Grund der Quantenmechanik II, Masterequation. Z. Phys. *173,* 232 (1963)
G. Ludwig: Versuch einer axiomatischen Grundlegung der Quantenmechanik und allgemeinerer physikalischer Theorien. Z. Phys. *181,* 233 (1964)
G. Ludwig: "An Axiomatic Foundation of Quantum Mechanics on a Nonsubjective Basis", *Quantum Theory and Reality,* ed. by M. Bunge (Springer, Berlin, Heidelberg, New York 1967), p. 98
G. Ludwig: Attempt of an axiomatic foundation of quantum mechanics and more general theories II. Commun. Math. Phys. *4,* 331 (1967)
G. Ludwig: Hauptsätze über das Messen als Grundlage der Hilbert-Raum-Struktur der Quantenmechanik, Z. Naturforsch. *A 22,* 1303 (1967)
G. Ludwig: Ein weiterer Hauptsatz über das Messen als Grundlage der Hilbert-Raum-Struktur der Quantenmechanik. Z. Naturforsch. *A 22,* 1324 (1967)
G. Ludwig: Attempt of an axiomatic foundation of quantum mechanics and more generel theories III. Commun. Math. Phys. *9,* (1968)
G. Dähn: [24]
P. Stolz: Attempt of an axiomatic foundation of quantum mechanics and more general theories V. Commun. Math. Phys. *11,* 303 (1969)
G. Ludwig: Deutung des Begriffs „physikalische Theorie" und axiomatische Grundlegung der Hilbert-Raum-Struktur der Quantenmechanik durch Hauptsätze des Messens. Lecture Notes Phys., Vol. 4 (Springer, Berlin, Heidelberg, New York 1970)
P. Stolz: Attempt of an axiomatic foundation of quantum mechanics and more general theories VI. Commun. Math. Phys. *23,* 117 (1971)

G. Ludwig: The Measuring Process and an Axiomatic Foundation of Quantum Mechanics, in *Foundations of Quantum Mechanics*, ed. by B. d'Espagnat (Academic, New York 1971), p. 287
G. Ludwig: A physical interpretation of an axiom within an axiomatic approach to quantum mechanics and a new formulation of this axiom as a general covering condition. Notes in Math. Phys. 1 (Marburg 1971)
G. Ludwig: Transformationen von Gesamtheiten und Effekten. Notes Math. Phys. *4* (Marburg 1971)
G. Ludwig: [4]
G. Ludwig: [26]
G. Ludwig: An Improved formulation of some theorems and axioms in the axiomatic foundation of the Hilbert space structure of quantum mechanics. Commun. Math. Phys. *26*, 78 (1972)
G. Ludwig: "Why a New Approach to Found Quantum Theory?" in *The Physicist's Conception of Nature*, ed. by J. Mehra (Reidel, Dordrecht, 1973), p. 702
G. Ludwig: "Measuring and Preparing Processes", in *Foundation of Quantum Mechanics and Ordered Linear Spaces*, ed. by A. Hartkämper, H. Neumann. Lecture Notes Phys. Vol. 29 (Springer, Berlin, Heidelberg, New York 1974), p. 122
G. Ludwig: Measurement as a process of interaction between macroscopic systems. Notes Math. Phys. *14* (Marburg 1974)
G. Ludwig: [6]
G. Ludwig: [47]
G. Ludwig: "An Axiomatic Basis of Quantum Mechanics", in: *Interpretations and Foundations of Quantum Theory*, ed. by H. Neumann (B. I. Wissenschaftsverlag, Mannheim 1981), p. 49
G. Ludwig: Quantum theory as a theory of interactions between macroscopic systems which can be described objectively. Erkenntnis *16*, 359 (1981)
G. Ludwig: "The Connection Between the Objective Description of Macrosystems and Quantum Mechanics of 'Many Particles'", in: *Old and New Questions in Physics, Cosmology, Philosophy and Theoretical Biology*, ed. by A. van der Merwe (Plenum, New York 1983), p. 243

36. B. Mielnik: Geometry of quantum states. Commun. Math. Phys. *9*, 55 (1968)
B. Mielnik: Theory of filters. Commun. Math. Phys. *15*, 1 (1969)
B. Mielnik: Generalized quantum mechanics. Commun. Math. Phys. *31*, 221 (1974)
B. Mielnik: "Quantum Logic: Is It Necessarily Orthocomplemented?" in *Quantum Mechanics, Determinism, Causality and Particles*, ed. by M. Flato, Z. Maric, A. Milojevic, D. Sternheimer, J. P. Vigier (Reidel, Dordrecht 1976)
37. A. Lande: *Foundation of Quantum Theory* (Yale University Press, New Haven 1955)
A. Lande: *New Foundations of Quantum Mechanics* (Cambridge University Press, Cambridge 1965)
38. J. S. Bell: "Introduction to the Hidden-Variable Question", in *Foundations of Quantum Mechanics*, ed. by B. d'Espagnat (Academic, New York 1971), p. 171
39. O. M. Nikodym: Sur l'existance d'une mesure parfaitement additive et non separable Mem. Acac. Roy. Belg. *17* (1939)
O. M. Nikodym: *The Mathematical Apparatures for Quantum Theories* (Springer, Berlin, Heidelberg, New York 1966), Chapt. A.
40. G. Ludwig: "Imprecision in Physics", in *Structure and Approximation in Physical Theories*, ed. by A. Hartkämper, H.-J. Schmidt (Plenum, New York 1980)
41. H. Neumann: The description of preparation and registration of physical systems and conventional probability theory. Found. Phys. *13*, 761 (1983)
42. P. Janich: *Die Protophysik der Zeit* (Suhrkamp, Frankfurt/Main 1980)
43. R. Werner: "Quantum Harmonic-Analysis on Phase Space", Preprint, Universität Osnabrück (1983)
A. Barchielli, L. Lanz, G. M. Prosperi: Statistics of continous trajectories in quantum mechanics: operation-valued stochastic-processes. Found. Phys. *13*, 779 (1983)
44. R. Werner: "The Concept of Embeddings in Statistical Mechanics", Ph. D. Thesis, Marburg (1982)

45. R. Werner: Physical uniformities and the state space of nonrelativistic quantum mechanics. Found. Phys. *13,* 859 (1983)
46. H. Gerstberger, H. Neumann, R. Werner: „Makroskopische Kausalität und relativistische Quantenmechanik", in *Grundprobleme der modernen Physik,* ed. by J. Nitsch, J. Pfarr, E. W. Stachow (B. I. Wissenschaftsverlag, Mannheim 1981), p. 205
H. Neumann, R. Werner: Causality between preparation and registration processes in relativistic quantum theory. Int. J. Theor. Phys. *22,* 781 (1983)
47. G. Ludwig: Axiomatische Basis der Quantenmechanik, Notes Math. Phys. *16, 17, 18,* Marburg (1980)
48. G. Ludwig: "Restriction and Embedding", in *Reduction in Science, Structure, Examples, Philosophical Problems,* ed. by W. Balzer, D. Pearce, H.-J. Schmidt (Reidel, Dordrecht 1984), p. 17

List of Frequently Used Symbols

List of Axioms

Index

G. Ludwig

Foundations of Quantum Mechanics I

Translated from the German by C. A. Hein

1983. XII, 426 pages
(Texts and Monographs in Physics)
ISBN 3-540-11683-4

Contents:

Springer-Verlag
Berlin
Heidelberg
New York
Tokyo

G. Ludwig

Foundations of Quantum Mechanics II

Translated from the German by C. A. Hein

1985. 54 figures. Approx 430 pages
(Texts and Monographs in Physics)
ISBN 3-540-13009-8

Contents: Representations of Hilbert Spaces by Function Spaces. - Equations of Motion. - The Spectrum of One-Electron Systems. - Spectrum of Two Electron Systems. - Selection Rules and the Intensity of Spectral Lines. - Spectra of Many-Electron Systems. - Molecular Spectra and the Chemical Bond. - Scattering Theory. - The Measurement Process and the Preparation Process. - Quantum Mechanics, Macrophysics and Physical World-Views. - Appendix V: Groups and Their Representations. - References. - Index.

Springer-Verlag
Berlin
Heidelberg
New York
Tokyo

A. Böhm

Quantum Mechanics

2nd edition. 1985. 1 figure.
Approx. 550 pages
(Texts and Monographs in Physics)
ISBN 3-540-13985-0